T0335663

MATHEMATICAL ASPECTS OF QUANTUM FIELD THEORY

Over the last century quantum field theory has made a significant impact on the formulation and solution of mathematical problems and has inspired powerful advances in pure mathematics. However, most accounts are written by physicists, and mathematicians struggle to find clear definitions and statements of the concepts involved. This graduate-level introduction presents the basic ideas and tools from quantum field theory to a mathematical audience. Topics include classical and quantum mechanics, classical field theory, quantization of classical fields, perturbative quantum field theory, renormalization, and the standard model.

The material is also accessible to physicists seeking a better understanding of the mathematical background, providing the necessary tools from differential geometry on such topics as connections and gauge fields, vector and spinor bundles, symmetries, and group representations.

Edson de Faria is a Professor in the Instituto de Matemática e Estatística at the Universidade de São Paulo.

Welington de Melo is a Professor in the Instituto de Matemática Pura e Aplicada in Rio de Janeiro.

CAMBRIDGE STUDIES IN ADVANCED MATHEMATICS

All the titles listed below can be obtained from good booksellers or from Cambridge University Press. For a complete series listing visit: http://www.cambridge.org/series/sSeries.asp?code=CSAM

Already published

78 V. Paulsen *Completely bounded maps and operator algebras*
79 F. Gesztesy & H. Holden *Soliton equations and their algebro-geometric solutions, I*
81 S. Mukai *An introduction to invariants and moduli*
82 G. Tourlakis *Lectures in logic and set theory, I*
83 G. Tourlakis *Lectures in logic and set theory, II*
84 R. A. Bailey *Association schemes*
85 J. Carlson, S. Müller-Stach & C. Peters *Period mappings and period domains*
86 J. J. Duistermaat & J. A. C. Kolk *Multidimensional real analysis, I*
87 J. J. Duistermaat & J. A. C. Kolk *Multidimensional real analysis, II*
89 M. C. Golumbic & A. N. Trenk *Tolerance graphs*
90 L. H. Harper *Global methods for combinatorial isoperimetric problems*
91 I. Moerdijk & J. Mrčun *Introduction to foliations and Lie groupoids*
92 J. Kollár, K. E. Smith & A. Corti *Rational and nearly rational varieties*
93 D. Applebaum *Lévy processes and stochastic calculus (1st Edition)*
94 B. Conrad *Modular forms and the Ramanujan conjecture*
95 M. Schechter *An introduction to nonlinear analysis*
96 R. Carter *Lie algebras of finite and affine type*
97 H. L. Montgomery & R. C. Vaughan *Multiplicative number theory, I*
98 I. Chavel *Riemannian geometry (2nd Edition)*
99 D. Goldfeld *Automorphic forms and L-functions for the group GL(n,R)*
100 M. B. Marcus & J. Rosen *Markov processes, Gaussian processes, and local times*
101 P. Gille & T. Szamuely *Central simple algebras and Galois cohomology*
102 J. Bertoin *Random fragmentation and coagulation processes*
103 E. Frenkel *Langlands correspondence for loop groups*
104 A. Ambrosetti & A. Malchiodi *Nonlinear analysis and semilinear elliptic problems*
105 T. Tao & V. H. Vu *Additive combinatorics*
106 E. B. Davies *Linear operators and their spectra*
107 K. Kodaira *Complex analysis*
108 T. Ceccherini-Silberstein, F. Scarabotti & F. Tolli *Harmonic analysis on finite groups*
109 H. Geiges *An introduction to contact topology*
110 J. Faraut *Analysis on Lie groups: An Introduction*
111 E. Park *Complex topological K-theory*
112 D. W. Stroock *Partial differential equations for probabilists*
113 A. Kirillov, Jr *An introduction to Lie groups and Lie algebras*
114 F. Gesztesy *et al. Soliton equations and their algebro-geometric solutions, II*
115 E. de Faria & W. de Melo *Mathematical tools for one-dimensional dynamics*
116 D. Applebaum *Lévy processes and stochastic calculus (2nd Edition)*
117 T. Szamuely *Galois groups and fundamental groups*
118 G. W. Anderson, A. Guionnet & O. Zeitouni *An introduction to random matrices*
119 C. Perez-Garcia & W. H. Schikhof *Locally convex spaces over non-Archimedean valued fields*
120 P. K. Friz & N. B. Victoir *Multidimensional stochastic processes as rough paths*
121 T. Ceccherini-Silberstein, F. Scarabotti & F. Tolli *Representation theory of the symmetric groups*
122 S. Kalikow & R. McCutcheon *An outline of ergodic theory*
123 G. F. Lawler & V. Limic *Random walk: A modern introduction*
124 K. Lux & H. Pahlings *Representations of groups*
125 K. S. Kedlaya p-*adic differential equations*
126 R. Beals & R. Wong *Special functions*

Mathematical Aspects of Quantum Field Theory

EDSON DE FARIA
Universidade de São Paulo

WELINGTON DE MELO
IMPA, Rio de Janeiro

CAMBRIDGE
UNIVERSITY PRESS

University Printing House, Cambridge CB2 8BS, United Kingdom

Cambridge University Press is part of the University of Cambridge.

It furthers the University's mission by disseminating knowledge in the pursuit of education, learning and research at the highest international levels of excellence.

www.cambridge.org
Information on this title: www.cambridge.org/9780521115773

First published 2010

A catalogue record for this publication is available from the British Library

Library of Congress Cataloguing in Publication data
Faria, Edson de.
Mathematical aspects of quantum field theory / Edson de Faria, Welington de Melo.
p. cm. – (Cambridge studies in advanced mathematics ; 127)
Includes bibliographical references and index.
ISBN 978-0-521-11577-3 (hardback)
1. Quantum field theory – Mathematics. I. Melo, Welington de. II. Title. III. Series.
QC174.45.F37 2010
530.14'30151 – dc22 2010012779

ISBN 978-0-521-11577-3 Hardback

Contents

Foreword

Mathematicians really understand what mathematics is. Theoretical physicists really understand what physics is. No matter how fruitful the interplay between the two subjects, the deep intersection of these two understandings seems to me to be quite modest. Of course, many theoretical physicists know a lot of mathematics. And many mathematicians know a fair amount of theoretical physics. This is very different from a deep understanding of the other subject. There is great advantage in the prospect of each camp increasing its appreciation of the other's goals, desires, methodology, and profound insights. I do not know how to really go about this in either case. However, the book in hand is a good first step for the mathematicians.

The method of the text is to explain the meaning of a large number of ideas in theoretical physics via the splendid medium of mathematical communication. This means that there are descriptions of objects in terms of the precise definitions of mathematics. There are clearly defined statements about these objects, expressed as mathematical theorems. Finally, there are logical step-by-step proofs of these statements based on earlier results or precise references. The mathematically sympathetic reader at the graduate level can study this work with pleasure and come away with comprehensible information about many concepts from theoretical physics . . . quantization, particle, path integral . . . After closing the book, one has not arrived at the kind of understanding of physics referred to above; but then, maybe, armed with the information provided so elegantly by the authors, the process of infusion, assimilation, and deeper insight based on further rumination and study can begin.

Dennis Sullivan
East Setauket, New York

Preface

In this book, we attempt to present some of the main ideas of quantum field theory (QFT) for a mathematical audience. As mathematicians, we feel deeply impressed – and at times quite overwhelmed – by the enormous breadth and scope of this beautiful and most successful of physical theories.

For centuries, mathematics has provided physics with a variety of tools, oftentimes on demand, for the solution of fundamental physical problems. But the past century has witnessed a new trend in the opposite direction: the strong impact of physical ideas not only on the formulation, but on the very solution of mathematical problems. Some of the best-known examples of such impact are (1) the use of renormalization ideas by Feigenbaum, Coullet, and Tresser in the study of universality phenomena in one-dimensional dynamics; (2) the use of classical Yang–Mills theory by Donaldson to devise invariants for four-dimensional manifolds; (3) the use of quantum Yang–Mills by Seiberg and Witten in the construction of new invariants for 4-manifolds; and (4) the use of quantum theory in three dimensions leading to the Jones–Witten and Vassiliev invariants. There are several other examples.

Despite the great importance of these physical ideas, mostly coming from quantum theory, they remain utterly unfamiliar to most mathematicians. This we find quite sad. As mathematicians, while researching for this book, we found it very difficult to absorb physical ideas, not only because of eventual lack of rigor – this is rarely a priority for physicists – but primarily because of the absence of clear definitions and statements of the concepts involved. This book aims at patching some of these gaps of communication.

The subject of QFT is obviously incredibly vast, and choices had to be made. We follow a more or less chronological path from classical mechanics in the opening chapter to the Standard Model in Chapter 9. The basic mathematical principles of quantum mechanics (QM) are presented in Chapter 2, which also contains an exposition of Feynman's path integral approach to QM. We use several non-trivial facts about the spectral theory of self-adjoint operators and C^* algebras, but everything we use is presented with complete proofs in Appendix A. Rudiments of special relativity are given in Chapter 3, where Dirac's fundamental insight leading to relativistic field theory makes its entrance.

Classical field theory is touched upon in Chapter 5, after a mathematical interlude in Chapter 4 where the necessary geometric language of bundles and connections is introduced. The quantization of classical *free* fields, which can be done in a mathematically rigorous and constructive way, is the subject of Chapter 6. As soon as non-trivial *interactions* between fields are present, however, rigorous quantization becomes a very difficult and elusive task. It can be done in spacetimes of dimensions 2 and 3, but we do not touch this subject (which may come as a disappointment to some). Instead, we present the basics of perturbative quantum field theory in Chapter 7, and then briefly discuss the subject of renormalization in Chapter 8. This approach to quantization of fields shows the Feynman path integral in all its glory on center stage.

Chapter 9 serves as an introduction to the Standard Model, which can be regarded as the crowning achievement of physics in the twentieth century, given the incredible accuracy of its predictions. We only present the semi-classical model (i.e. before quantization), as no one really knows how to quantize it in a mathematically rigorous way.

The book closes with two appendices, one on Hilbert spaces and operators, the other on C^* algebras. Taken together, they present a complete proof of the spectral theorem for self-adjoint operators and other non-trivial theorems (e.g. Stone, Kato–Rellich) that are essential for the proper foundations of QM and QFT. The last section of Appendix B presents an extremely brief introduction to *algebraic* QFT, a very active field of study that is deeply intertwined with the theory of von Neumann algebras.

We admit to being perhaps a bit uneven about the prerequisites. For instance, although we do *not* assume that the reader knows any functional analysis on Hilbert spaces (hence the appendices), we *do* assume familiarity with the basic concepts of differentiable manifolds, differential forms and tensors on manifolds, etc. A previous knowledge of the differential-geometric concepts of principal bundles, connections, and curvature would be desirable, but in any case these notions are presented briefly in Chapter 4. Other mathematical subjects such as representation theory or Grassmann algebras are introduced on the fly.

The first version of this book was written as a set of lecture notes for a short course presented by the authors at the 26th Brazilian Math Colloquium in 2007.

For this Cambridge edition, the book was completely revised, and a lot of new material was added.

We wish to thank Frank Michael Forger for several useful discussions on the Standard Model, and also Charles Tresser for his reading of our manuscript and his several remarks and suggestions. We have greatly benefited from discussions with several other friends and colleagues, among them Dennis Sullivan, Marco Martens, Jorge Zanelli, Nathan Berkovits, and Marcelo Disconzi. To all, and especially to Dennis Sullivan for his beautiful foreword, our most sincere thanks.

1
Classical mechanics

This chapter presents a short summary of the classical mechanics of particle systems. There are three distinct formulations of classical mechanics: Newtonian, Lagrangian, and Hamiltonian. As we shall see in later chapters, the paradigms offered by the Lagrangian and Hamiltonian formulations are both extremely important in modern quantum physics and beyond.

1.1 Newtonian mechanics

1.1.1 Newtonian spacetime

From a mathematical standpoint, Newtonian spacetime S is a four-dimensional Euclidean (affine) space where the translation group $\mathbb{R}^4$ acts transitively, and in which a surjective linear functional $\tau : S \to \mathbb{R}$ is defined (intuitively corresponding to time). The points of S are called *events*. Two events $p, q \in S$ are said to be *simultaneous* if $\tau(p) = \tau(q)$. The difference $\Delta\tau(p, q) = \tau(q) - \tau(p)$ is called the *time interval* between the events p and q. If two events p, q are simultaneous, their *spatial displacement* $\Delta s(p, q)$ is by definition the Euclidean distance between p and q in S. The structure on Newtonian spacetime provided by the time interval and spatial displacement functions is called a *Galilean structure*. A Galilean transformation of S is an affine transformation that preserves the Galilean structure.

Alternatively, a Galilean transformation can be viewed as an affine change of coordinates between two inertial reference frames. An *inertial frame* on Newtonian spacetime S is an affine map $\alpha : S \to \mathbb{R}^4$ providing an identification of S with $\mathbb{R}^4 = \mathbb{R}^3 \times \mathbb{R}$ that preserves time intervals and spatial displacements. More precisely, if $p, q \in S$ and $\alpha(p) = (x^1, x^2, x^3, t)$ and

$\alpha(q) = (y^1, y^2, y^3, s)$, then we have

$$\Delta\tau(p,q) = s - t$$
$$\Delta s^2(p,q) = (x^1 - y^1)^2 + (x^2 - y^2)^2 + (x^3 - y^3)^2 .$$

Given two inertial frames α, β, the map $\beta \circ \alpha : \mathbb{R}^4 \to \mathbb{R}^4$ is a Galilean transformation. With this definition, it is clear that the Galilean transformations form a group. We leave it as an exercise for the reader to verify that the most general Galilean transformation $(x, t) \to (x', t')$ is of the form

$$t' = t + t_0$$
$$x' = Ax + a + tv ,$$

where $t_0 \in \mathbb{R}$, $a, v \in \mathbb{R}^3$, and A is a rotation in $\mathbb{R}^3$ (i.e., an orthogonal matrix). From this fact, it follows that the Galilean group is a 10-dimensional Lie group.

In a given inertial frame, a *uniform linear motion* is a path of the form $t \to (x(t), t + t_0) \in \mathbb{R}^3 \times \mathbb{R}$, where $x(t) = x_0 + tv$ (with $t_0 \in \mathbb{R}$ and $x_0, v \in \mathbb{R}^3$ constants). It is an easy exercise to verify the following fact.

Proposition 1.1 (Galileo's principle of inertia) *Galilean transformations map uniform motions to uniform motions.*

From a physical point of view, uniform motions are precisely the motions of *free particles*. So another way of stating the above principle is this: The time evolution of free particles is invariant under the Galilean group.

1.1.2 Newtonian configuration space

In Newtonian mechanics, the configuration space for a system of N unconstrained particles is

$$M = \{(x_1, x_2, \ldots, x_N) : \; x_i \in \mathbb{R}^3 \text{ for all } i\} \equiv \mathbb{R}^{3N} .$$

Newtonian *determinism* asserts that the time evolution of such a system is completely known once the initial positions and velocities of all particles are given. Mathematically, the time evolution is governed by a system of second-order ordinary differential equations (ODE's) coming from Newton's second law of motion,

$$m_i \ddot{x}_i = F_i(x_1, x_2, \ldots, x_N; \dot{x}_1, \dot{x}_2, \ldots, \dot{x}_N) \quad (i = 1, 2, \ldots, N) . \tag{1.1}$$

Here, m_i is the mass of the ith particle, and F_i is the (resultant) force acting on the ith particle. We regard $F_i : M \to \mathbb{R}^3$ as a smooth vector function.

A Newtonian system is said to be *conservative* if there exists a (smooth) function $V: M \to \mathbb{R}$, called the *potential*, such that

$$F_i = -\nabla_i V, \quad \text{where } \nabla_i = \left(\frac{\partial}{\partial x_i^1}, \frac{\partial}{\partial x_i^2}, \frac{\partial}{\partial x_i^3} \right).$$

The name "conservative" stems from the fact that, if we define the *kinetic energy* of such a system by

$$T = \frac{1}{2} \sum_{i=1}^{N} m_i \left[\left(\dot{x}_i^1\right)^2 + \left(\dot{x}_i^2\right)^2 + \left(\dot{x}_i^3\right)^2 \right],$$

and its total energy by $E = T + V$, then we have conservation of energy, namely,

$$\frac{dE}{dt} = \sum_{i=1}^{N} m_i \langle \dot{x}_i, \ddot{x}_i \rangle + \sum_{i=1}^{N} \langle \nabla_i V(x_1, x_2, \ldots, x_N), \dot{x}_i \rangle = 0 .$$

Remark 1.2 We would like to emphasize that the forces appearing in (1.1) may indeed depend quite explicitly on the velocities $v_i = \dot{x}_i$. The best example of a physically meaningful situation where this happens is the classical (non-relativistic) electrodynamics of a single electrically charged particle. If such a charged particle moves about in space in the presence of a magnetic field $\boldsymbol{B}$ and an electric field $\boldsymbol{E}$ in $\mathbb{R}^3$, it is acted upon by the so-called *Lorentz force*, given by

$$\boldsymbol{F} = q\,(\boldsymbol{E} + \boldsymbol{v} \wedge \boldsymbol{B}) ,$$

where q is the particle's charge and $\boldsymbol{v} = \dot{\boldsymbol{x}}$ is the particle's velocity. Here $\wedge$ denotes the standard cross product of vectors, which in cartesian coordinates is given by

$$\boldsymbol{v} \wedge \boldsymbol{B} = \begin{vmatrix} \boldsymbol{e}_1 & \boldsymbol{e}_2 & \boldsymbol{e}_3 \\ v_1 & v_2 & v_3 \\ B_1 & B_2 & B_3 \end{vmatrix},$$

where $\{\boldsymbol{e}_1, \boldsymbol{e}_2, \boldsymbol{e}_3\}$ is the canonical basis of $\mathbb{R}^3$. Assuming that the particle has mass m, Newton's second law gives us the second-order ODE

$$m\ddot{\boldsymbol{x}} = q\,(\boldsymbol{E} + \dot{\boldsymbol{x}} \wedge \boldsymbol{B}) ,$$

or, in coordinates, the system

$$\begin{aligned} m\ddot{x}_1 &= q\,(E_1 + \dot{x}_2 B_3 - \dot{x}_3 B_2) \\ m\ddot{x}_2 &= q\,(E_2 + \dot{x}_3 B_1 - \dot{x}_1 B_3) \\ m\ddot{x}_3 &= q\,(E_3 + \dot{x}_1 B_2 - \dot{x}_2 B_1) . \end{aligned}$$

This system can be recast in Lagrangian form (in the sense of Section 1.2). The reader can work it out as an exercise.

1.2 Lagrangian mechanics

Lagrangian mechanics was born out of the necessity to deal with *constrained* systems of particles. Such systems are called *Lagrangian*.

1.2.1 Lagrangian systems

In a Lagrangian system, the configuration space is an embedded submanifold M of dimension $n \leq 3N$ of $\mathbb{R}^{3N}$ (where N is the number of particles). The number n is the number of degrees of freedom of the system. Local coordinates for M are usually denoted by $q = (q^1, q^2, \ldots, q^n)$ and are referred to as generalized coordinates. The Lagrangian function (or simply the Lagrangian) of such a system is a smooth function $L : TM \to \mathbb{R}$. Here, TM is the tangent bundle of M, and it is called the (Lagrangian) phase space. Associated to the Lagrangian function, we have an *action functional* S defined on the space of paths on M as follows. Given a differentiable path $\gamma : I \to M$, where $I = [a, b]$ is a time interval, consider its lift to TM given by $(\gamma, \dot{\gamma})$, and let

$$S(\gamma) = \int_a^b L(\gamma(t), \dot{\gamma}(t))\, dt \;.$$

1.2.2 The least action principle

The underlying principle behind Lagrangian mechanics was first discovered by Maupertuis (who was inspired by earlier work of Fermat in optics) and later mathematically established by Euler in 1755. In the nineteenth century, it was formulated in more general terms by Hamilton, and it is known even today as the *least action principle*. In a nutshell, this principle states that the physical trajectories of a Lagrangian system are extrema of the action functional. Thus, the Newtonian problem of solving a system of second-order, typically non-linear ODE's is tantamount in Lagrangian mechanics to the *variational problem* of finding the critical points of an action functional. This variational problem gives rise to the so-called Euler–Lagrange equations.

1.2.3 The Euler–Lagrange equations

Let us describe the standard procedure in the calculus of variations leading to the Euler–Lagrange equations. We define a *variation* of a smooth curve $\gamma : I \to M$ to be a smooth map $\tilde{\gamma} : (-\epsilon, \epsilon) \times I \to M$ such that $\tilde{\gamma}(0, t) = \gamma(t)$

for all t and $\tilde{\gamma}(s,a) = \gamma(a)$ and $\tilde{\gamma}(s,b) = \gamma(b)$ for all $s \in (-\epsilon, \epsilon)$. In other words, a variation of γ is a family of curves $\gamma_s = \tilde{\gamma}(s,\cdot) : I \to M$ having the same endpoints as γ and such that $\gamma_0 = \gamma$. The corresponding (first) variation of the action functional S at γ is by definition

$$\delta S(\gamma) = \frac{\partial}{\partial s}\bigg|_{s=0} S(\gamma_s) .$$

A curve γ is said to be a critical point for the action functional S if $\delta S(\gamma) = 0$ for all possible variations of γ. We leave it as an exercise for the reader to check that

$$\delta S(\gamma) = \int_a^b \delta L(\gamma, \dot{\gamma})\, dt , \tag{1.2}$$

where

$$\delta L(\gamma, \dot{\gamma}) = DL(\gamma, \dot{\gamma}) \cdot (\delta\gamma, \delta\dot{\gamma}) ,$$

and where, in turn,

$$\delta\gamma = \frac{\partial}{\partial s}\bigg|_{s=0} \gamma_s ,$$

as well as

$$\delta\dot{\gamma} = \frac{\partial}{\partial s}\bigg|_{s=0} \dot{\gamma}_s .$$

Using generalized coordinates $q, \dot{q}$, and writing $\delta\gamma = (\delta q^1, \delta q^2, \ldots, \delta q^n)$ as well as $\delta\dot{\gamma} = (\delta\dot{q}^1, \delta\dot{q}^2, \ldots, \delta\dot{q}^n)$, we see that

$$DL(\gamma, \dot{\gamma})(\delta\gamma, \delta\dot{\gamma}) = \sum_{i=1}^{n} \left(\frac{\partial L}{\partial q^i}(q, \dot{q})\delta q^i + \frac{\partial L}{\partial \dot{q}^i}(q, \dot{q})\delta\dot{q}^i \right) .$$

Putting this expression back into (1.2), we have

$$\delta S(\gamma) = \int_a^b \sum_{i=1}^{n} \left(\frac{\partial L}{\partial q^i}(q, \dot{q})\delta q^i + \frac{\partial L}{\partial \dot{q}^i}(q, \dot{q})\delta\dot{q}^i \right) dt .$$

But because $\delta\dot{q}^i = d(\delta q^i)/dt$ for all i, a simple integration by parts on the right-hand side yields

$$\delta S(\gamma) = \sum_{i=1}^{n} \int_a^b \left[\frac{\partial L}{\partial q^i}(q, \dot{q}) - \frac{d}{dt}\left(\frac{\partial L}{\partial \dot{q}^i}(q, \dot{q}) \right) \right] \delta q^i\, dt .$$

Because $\delta S(\gamma)$ must be equal to zero at a critical point γ for all possible variations, it follows that each expression in brackets on the right-hand side of the above equality must vanish. In other words, we arrive at the Euler–Lagrange equations,

$$\frac{\partial L}{\partial q^i}(q, \dot{q}) - \frac{d}{dt}\left(\frac{\partial L}{\partial \dot{q}^i}(q, \dot{q})\right) = 0\,.$$

1.2.4 Conservative Lagrangian systems

In classical mechanics, the Lagrangian of a conservative system is not an arbitrary function, but rather takes the form

$$L(q, v) = T_q(v) - V(q)\,, \tag{1.3}$$

where $V : M \to \mathbb{R}$ is the potential and $T_q : TM_q \to \mathbb{R}$ is a quadratic form on the vector space TM_q. More precisely, a conservative Lagrangian system with n degrees of freedom consists of the following data:

1. A Riemannian manifold M of dimension n.
2. A smooth function $V : M \to \mathbb{R}$ called the potential.
3. For each $q \in M$, a quadratic form $T_q : TM_q \to \mathbb{R}$ called the kinetic energy, given by

$$T_q(v) = \frac{1}{2}\langle v, v\rangle_q = \frac{1}{2}\|v\|_q^2\,.$$

Here, $\langle\cdot,\cdot\rangle : TM_q \times TM_q \to \mathbb{R}$ is the Riemannian inner product at $q \in M$.
4. A Lagrangian function $L : TM \to \mathbb{R}$ given by (1.3).

A basic example is provided by the unconstrained conservative Newtonian system of N particles given in Section 1.1. In that case we have $M = \mathbb{R}^{3N}$, $TM = \mathbb{R}^{3N} \times \mathbb{R}^{3N}$. The Riemannian structure on M is given by the inner product

$$\langle v, w\rangle_q = \sum_{i=1}^{N} m_i(v_i^1 w_i^1 + v_i^2 w_i^2 + v_i^3 w_i^3)\,.$$

Note that this inner product does not depend on the point $q = (x_1^1, x_1^2, x_1^3, \ldots, x_N^1, x_N^2, x_N^3) \in \mathbb{R}^{3N}$. The Lagrangian is therefore

$$L(q, v) = \frac{1}{2}\sum_{i=1}^{N} m_i\left[(v_i^1)^2 + (v_i^2)^2 + (v_i^3)^2\right] - V(q)\,.$$

If we write down the Euler–Lagrange equations for this Lagrangian, we recover, not surprisingly, Newton's second law of motion.

1.3 Hamiltonian mechanics

As we saw, in Lagrangian mechanics the Euler–Lagrange equations of motion are deduced from a variational principle, but they are still second-order (and usually non-linear) ODE's. In Hamiltonian mechanics, the equations of motion become first-order ODE's, and the resulting Hamiltonian flow is in principle more amenable to dynamical analysis. This reduction of order is accomplished by passing from the tangent bundle of the configuration space (Lagrangian phase space) to the cotangent bundle (Hamiltonian phase space).

1.3.1 The Legendre transformation

Suppose we have a Lagrangian system with n degrees of freedom with configuration space given by a Riemannian manifold M and Lagrangian $L : TM \to \mathbb{R}$. The Lagrangian L gives rise to a map $P : TM \to T^*M$ between tangent and cotangent bundles, the so-called *Legendre transformation*. The map P is defined by

$$P(q, v) = \left(q, \frac{\partial L}{\partial \dot{q}}(q, v)\right) ,$$

for all $q \in M$ and all $v \in TM_q$. Its derivative at each point has the form

$$DP(q, v) = \begin{bmatrix} I & * \\ 0 & \left(\frac{\partial^2 L}{\partial \dot{q}^i \partial \dot{q}^j}(q, v)\right) \end{bmatrix} .$$

Hence, P will be a local diffeomorphism provided that

$$\det \left(\frac{\partial^2 L}{\partial \dot{q}^i \partial \dot{q}^j}(q, v)\right)_{n \times n} \neq 0 .$$

We leave it as an exercise for the reader to check that if this condition holds at all points $(q, v) \in TM$ then P is in fact a global diffeomorphism (a bundle isomorphism, in fact).

In local coordinates, P transforms each generalized velocity $\dot{q}^i$ into a generalized *momentum*

$$p_i = \frac{\partial L}{\partial \dot{q}^i} .$$

What do we gain by passing to the cotangent bundle? The answer lies in the fact that the cotangent bundle of every differentiable manifold carries a natural *symplectic structure*, as we shall see.

1.3.2 Symplectic manifolds

A symplectic structure on a differentiable manifold N is a closed 2-form $\omega \in \wedge^2(N)$ that is non-degenerate in the sense that if $X \in \mathcal{X}(N)$ is a vector field on N such that $\omega(X, Y) = 0$ for every vector field $Y \in \mathcal{X}(N)$, then $X = 0$. A manifold N together with a symplectic 2-form ω is called a *symplectic manifold*. Symplectic manifolds are necessarily even-dimensional (exercise).

In a symplectic manifold (N, ω), there is a natural isomorphism between 1-forms and vector fields,

$$I_\omega : \wedge^1(N) \longrightarrow \mathcal{X}(N)$$
$$\alpha \mapsto X_\alpha ,$$

where X_α is defined by

$$\omega(X_\alpha, Y) = \alpha(Y) \quad \text{for all } Y \in \mathcal{X}(N) .$$

The non-degeneracy of ω guarantees that I_ω is indeed well-defined and an isomorphism.

Given any function $f : N \to \mathbb{R}$, we can use I_ω to associate to f a vector field on N. We simply define $X_f = I_\omega(df)$, where df is the differential 1-form of f. The vector field X_f is called the Hamiltonian vector field associated to f. The flow ϕ_t on N generated by this vector field (which we assume to be globally defined) is called the *Hamiltonian flow* of f (or X_f). This flow preserves the symplectic form ω, in the sense that $\phi_t^*\omega = \omega$ for all t (exercise).

Taking the exterior product of ω with itself n times, we get a $2n$-form on N, i.e. a volume form $\lambda = \omega \wedge \omega \wedge \cdots \wedge \omega$. This volume form is called the Liouville form, and it is also preserved by Hamiltonian flows. This makes the subject very rich from the point of view of ergodic theory.

1.3.3 Hamiltonian systems

Going back to mechanics, we claim that the cotangent bundle T^*M of the configuration space M has a canonical symplectic structure. Indeed, on T^*M, there is a canonical 1-form $\theta \in \wedge^1(T^*M)$, the so-called Poincaré 1-form, defined as follows. Let $\pi : T^*M \to M$ be the projection onto the base. Let $(q, p) \in T^*M$, and let $(\xi, \eta) \in T(T^*M)$ be a tangent vector at (q, p). Note that $p \in T^*M_q$ is a linear functional on TM_q. Hence, we can take

$$\theta_{q,p}(\xi, \eta) = p\,(D\pi(q, p)(\xi, \eta)) .$$

In local coordinates $(q^1, \ldots, q^n; p_1, \ldots, p_n)$ on T^*M, this 1-form is given by

$$\theta = p_i \, dq^i .$$

Taking the exterior derivative of θ yields our desired symplectic form,

$$\omega = d\theta = -\, dp_i \wedge dq^i .$$

Now, given a function $f : T^*M \to \mathbb{R}$, let X_f be the Hamiltonian vector field associated to f (via the symplectic structure in T^*M just introduced). It is easy to write down explicitly the components X^i_f, $i = 1, 2, \ldots, 2n$, of X_f in local coordinates, in terms of f. Indeed, if Y is any vector field on T^*M with components Y^i, then from

$$\omega(X_f, Y) = df(Y)$$

we deduce that

$$dp_i(X_f)dq^i(Y) - dp_i(Y)dq^i(X_f) = -df(Y)\,.$$

In terms of the components of both vector fields, this means that

$$\sum_{i=1}^{n}\left(X_f^{i+n}Y^i - Y^{i+n}X_f^i\right) = -\sum_{i=1}^{n}\left(\frac{\partial f}{\partial q^i}Y^i + \frac{\partial f}{\partial p_i}Y^{i+n}\right)\,.$$

Because this holds true for every Y, comparing terms on both sides yields, for all $i = 1, 2, \ldots, n$,

$$X_f^{i+n} = -\frac{\partial f}{\partial q^i}\ ;\ X_f^i = \frac{\partial f}{\partial p_i}\,.$$

In other words, the Hamiltonian vector field associated to f is given by

$$X_f = \left(\frac{\partial f}{\partial p}, -\frac{\partial f}{\partial q}\right)\,.$$

The corresponding flow on T^*M is therefore given by the solutions to the following system of first-order ODE's:

$$\begin{aligned} \dot{q}^i &= \frac{\partial f}{\partial p_i} \\ \dot{p}_i &= -\frac{\partial f}{\partial q^i}\,. \end{aligned} \tag{1.4}$$

Next, we ask the following question: is there a choice of f, let us call it H, for which the above flow in the cotangent bundle is the image under the Legendre map P of the Lagrangian time evolution in the tangent bundle? The answer is yes, and the function H is called *the* Hamiltonian of our particle system. To see what H looks like, let us write its total differential using (1.4) for $f = H$:

$$\begin{aligned} dH &= \frac{\partial H}{\partial q^i}dq^i + \frac{\partial H}{\partial p_i}dp_i \\ &= -\dot{p}_i dq^i + \dot{q}^i dp_i\,. \end{aligned}$$

Using the fact, coming from the Euler–Lagrange equations, that

$$\dot{p}_i = \frac{\partial L}{\partial q^i}$$

and also the fact that

$$\dot{q}^i dp_i = d(p_i \dot{q}^i) - \frac{\partial L}{\partial \dot{q}^i} d\dot{q}^i ,$$

we see that

$$dH = -\frac{\partial L}{\partial q^i} dq^i + d(p_i q^i) - \frac{\partial L}{\partial \dot{q}^i} d\dot{q}^i = d(p_i \dot{q}^i - L) .$$

Therefore, we can take $H(q, p) = p_i \dot{q}^i - L(q, \dot{q})$. In this expression, of course, we implicitly assume that the generalized velocities $\dot{q}^i$ are expressed as functions of the generalized momenta p_j, which is certainly possible, because we have assumed from the beginning that the Legendre map P is invertible.

Summarizing, we have shown how to pass from a Lagrangian formulation of a mechanical system, which lives in the tangent bundle to the configuration space, to a Hamiltonian formulation, which lives in the cotangent bundle. The Hamiltonian dynamical system is given by a system of first-order ODE's, namely

$$\dot{q}^i = \frac{\partial H}{\partial p_i}$$

$$\dot{p}_i = -\frac{\partial H}{\partial q^i} .$$

1.3.4 Hamiltonian of a conservative system

When our original (Lagrangian) system is conservative, with Lagrangian given by $L(q, v) = T_q(v) - V(q)$ as before, the above passage to the corresponding Hamiltonian system yields the Hamiltonian $H : T^*M \to \mathbb{R}$ given by $H = T + V$, or more precisely, $H(q, p) = T_q(p) + V(q)$. Such a system is called a conservative Hamiltonian system.

1.4 Poisson brackets and Lie algebra structure of observables

A classical observable is simply a differentiable function $f : T^*M \to \mathbb{R}$ in the phase space of our Hamiltonian system. Given a pair of observables $f, g : T^*M \to \mathbb{R}$ in phase space, we define their *Poisson bracket* by

$$\{f , g\} = \sum_{i=1}^{n} \left(\frac{\partial f}{\partial q_i} \frac{\partial g}{\partial p_i} - \frac{\partial f}{\partial p_i} \frac{\partial g}{\partial q_i} \right) .$$

This notion of Poisson bracket endows the space of all observables in phase space with the structure of a *Lie algebra*. Because the Hamiltonian H is itself an observable, it is easy to recast Hamilton's equations with the help of the Poisson bracket as

$$\dot{q} = \{q\,,\,H\}\,,\ \dot{p} = \{p\,,\,H\}\,.$$

Now, every observable has a time evolution dictated by the Hamiltonian flow. Using the above form of Hamilton's equations and the definition of Poisson bracket, the reader can check as an exercise that the time evolution of an observable f satisfies the first-order equation

$$\dot{f} = \{f\,,\,H\}\,.$$

The reader will not fail to notice that there are *two* distinct algebraic structures in the space of all observables of a Hamiltonian system: the Lie algebra structure given by the Poisson bracket, and the commutative algebra (actually a commutative C^*-algebra; see Chapter 2) structure given by ordinary pointwise multiplication of functions. These structures are quite independent. From an algebraic point of view, one can think of the process of *quantization* of a classical system, to be described in Chapter 2, as a way to transform the space of observables into a non-commutative algebra in such a way that the Lie algebra structure and this non-commutative algebra structure become compatible.

1.5 Symmetry and conservation laws: Noether's theorem

Suppose we have a Lagrangian system in $\mathbb{R}^n$ whose Lagrangian function $L : \mathbb{R}^n \times \mathbb{R}^n \to \mathbb{R}$ does not depend on a particular generalized coordinate q^j (for some $1 \leq j \leq n$). In other words, L is invariant under translations along the q^j coordinate axis; i.e.

$$L(q^1, \ldots, q^j + \tau, \ldots, q^n; \dot{q}^1, \ldots, \dot{q}^n) = L(q^1, \ldots, q^j, \ldots, q^n; \dot{q}^1, \ldots, \dot{q}^n)$$

for all $\tau \in \mathbb{R}$. Then, from the Euler–Lagrange equations, we deduce that

$$\dot{p}_j = \frac{d}{dt}\left(\frac{\partial L}{\partial \dot{q}^j}\right) = 0\,.$$

This means that the jth generalized momentum is constant along the physical trajectories in phase space. In other words, p_j is an example of what physicists call an *integral of motion*.

The underlying principle here was strongly generalized by Noether. Her theorem states that to each one-parameter group of symmetries in configuration space there corresponds an integral of motion in phase space.

Theorem 1.3 (Noether) *Let $L : TM \to \mathbb{R}$ be the Lagrangian of a Lagrangian system, and let*

$$G = \{\phi_s : M \to M : s \in \mathbb{R}\}$$

be a one-parameter group of diffeomorphisms such that

$$L(\phi_s(q), D\phi_s(q)v) = L(q, v) \tag{1.5}$$

for all $q \in M$ and all $v \in TM_q$. Then there exists a function $I_G : TM \to \mathbb{R}$ that is constant on trajectories; i.e.

$$\frac{d}{dt} I_G(q(t), \dot{q}(t)) = 0 \, . \tag{1.6}$$

Proof. Define I_G as follows:

$$I_G(q, v) = \frac{\partial L}{\partial \dot{q}}(q, v) \, \frac{\partial}{\partial s}\bigg|_{s=0} \phi_s(q) \, .$$

Then

$$\frac{d}{dt} I_G(q, \dot{q}) = \frac{d}{dt}\left(\frac{\partial L}{\partial \dot{q}}(q, \dot{q})\right) \frac{\partial}{\partial s}\bigg|_{s=0} \phi_s(q) + \frac{\partial L}{\partial \dot{q}}(q, \dot{q}) \, \frac{\partial}{\partial s}\bigg|_{s=0} \frac{d}{dt} \phi_s(q) \, .$$

We may use the Euler–Lagrange equations in the first term on the right-hand side. Note also that

$$\frac{d}{dt} \phi_s(q) = D\phi_s(q) \, \dot{q} \, .$$

Therefore we have, using (1.5),

$$\begin{aligned} \frac{d}{dt} I_G(q, \dot{q}) &= \frac{\partial L}{\partial q}(q, \dot{q}) \, \frac{\partial}{\partial s}\bigg|_{s=0} \phi_s(q) + \frac{\partial L}{\partial \dot{q}}(q, \dot{q}) \, \frac{\partial}{\partial s}\bigg|_{s=0} D\phi_s(q)\dot{q} \\ &= \frac{\partial}{\partial s}\bigg|_{s=0} L(\phi_s(q), D\phi_s(q)\dot{q}) \\ &= \frac{\partial}{\partial s}\bigg|_{s=0} L(q, \dot{q}) = 0 \, . \end{aligned}$$

This establishes (1.6) and finishes the proof. □

Besides conservation of linear momentum, another simple application of Noether's theorem occurs when the Lagrangian is rotationally invariant (i.e., invariant under the orthogonal group). In this case we deduce the law of conservation of angular momentum; see Exercise 1.2 below and the discussion in the context of quantum mechanics in Section 2.7. Noether's theorem survives as an important principle in modern physics, namely, that the infinitesimal symmetries of a system yield conservation laws. Its importance is felt especially in field theory, through the concept of gauge invariance (see Chapter 5).

Exercises

1.1 The electromagnetic field in a region Ω of 3-space can be represented by a 4-vector potential $(\phi, \boldsymbol{A})$, where $\phi : \Omega \to \mathbb{R}$ is the scalar potential and $\boldsymbol{A} : \Omega \to \mathbb{R}^3$ is the vector potential, and

$$\boldsymbol{B} = \nabla \wedge \boldsymbol{A}\,, \quad \boldsymbol{E} = -\nabla\phi - \frac{\partial \boldsymbol{A}}{\partial t}$$

are the magnetic and electric fields, respectively (see Chapter 5). Consider a particle with charge q and mass m subject to this field.

(a) Show that the Euler–Lagrange equations applied to

$$L(\boldsymbol{x}, \dot{\boldsymbol{x}}) = \frac{1}{2} m \langle \dot{\boldsymbol{x}}, \dot{\boldsymbol{x}} \rangle + q \langle \dot{\boldsymbol{x}}, \boldsymbol{A} \rangle - q\phi$$

yield the non-relativistic equation of motion

$$m\ddot{\boldsymbol{x}} = q\left(\boldsymbol{E} + \dot{\boldsymbol{x}} \wedge \boldsymbol{B}\right)\,.$$

(b) Letting $\boldsymbol{p} = m\dot{\boldsymbol{x}} + q\boldsymbol{A}$, show that the Hamiltonian of this system is

$$H(\boldsymbol{p}, \boldsymbol{x}) = \frac{1}{2m} \|\boldsymbol{p} - q\boldsymbol{A}\|^2 + q\phi\,.$$

1.2 Suppose we have a Lagrangian system of particles whose Lagrangian is rotationally invariant. Using Noether's theorem, show that such a system satisfies the law of conservation of angular momentum.

2
Quantum mechanics

In this chapter, we present the basic principles of the quantum mechanics of particle systems from a modern mathematical perspective. The proper study of quantum mechanics brings together several interesting mathematical ideas, ranging from group representations to the spectral theory of unbounded self-adjoint operators in Hilbert spaces and the theory of C^* algebras. The facts from analysis of operators in Hilbert spaces and C^* algebras needed in this chapter are presented in detail in the appendices at the end of the book. More on the mathematical foundations of quantum mechanics can be found in the books by L. Fadeev and O. Yakubovskii [FY], G. Mackey [M], and F. Strocchi [Str] listed in the bibliography.

2.1 The birth of quantum theory

The origins of quantum theory can be traced back to M. Planck's study, published in 1900, of so-called black-body radiation. It is well known that matter glows when heated, going from red-hot to white-hot as the temperature rises. The color (or frequency of radiation) of the radiating body is independent of its surface, and for a black body it is found to be a function of temperature alone. The behavior of the radiation energy as a function of temperature seemed quite different at high temperatures and low temperatures. In a desperate attempt to fit these different behaviors into a single law, Planck introduced the hypothesis that energy was not continuous, but came in discrete packets or *quanta*.

At the beginning of the twentieth century, a number of experiments pointed to a breakdown of the laws of classical mechanics at very small (subatomic) scales. Among them were the following:

(i) *Rutherford's scattering of α-particles.* This experiment at first seemed to support the classical picture of an atom as a microscopic planetary system, held through an analogy between Newton's and Coulomb's laws. However, it lead to some unexpected conclusions:
   (a) Unlike planetary systems, all atoms have about the same size ($\sim 10^{-8}$ cm).
   (b) According to the Maxwellian theory of electromagnetic radiation, an electron orbiting around a nucleus should emit radiation, thereby loosing energy and ultimately colliding with the nucleus. This is in flagrant contradiction to the obvious stability of matter that is observed in nature.
   (c) The energy absorbed or emitted by an atom was seen to vary in *discrete* quantities, and not continuously, as expected by the classical theory. From a classical standpoint, the energy of radiation of an electron in orbit varies continuously with the size (radius) of the orbit. This leads to the surprising conclusion that only a discrete set of orbit sizes are allowed.

(ii) *The photoelectric effect.* When light waves of sufficiently high frequency ν hit the surface of a metal, electrons are emitted. For this to happen the light frequency has to be above a frequency threshold ν_0 that depends on the metal. The experiments show that the kinetic energy of each emitted electron is directly proportional to the difference $\nu - \nu_0$, but is *independent* of the intensity of the radiation. This is in sharp contrast to the classical theory of radiation. In 1905, Einstein proposed an explanation of this phenomenon. He introduced the idea that, microscopically, a light wave with frequency ν has energy that is not continuous, but comes in discrete *quanta*, or *photons*, each with energy $E = h\nu$, where h is the same constant found by Planck in his work on black-body radiation.[1] Thus, although behaving like a wave, light seemed to have a corpuscular nature at very small scales.

(iii) *Matter waves.* The radical hypothesis that material particles could behave like waves was first put forth in 1923 by de Broglie, who even suggested that a particle with momentum p should be assigned a *wavelength* equal to h/p, where h is Planck's constant. The first experimental confirmation of this *wave–particle duality of matter* came soon afterwards. In 1927, Davisson and Germer conducted an experiment in which a well-focused beam of electrons, all with approximately the same energy E, was scattered off the plane surface of a crystal whose

[1] In CGS units, Planck's constant is $h \simeq 6.6 \times 10^{-27}\,\mathrm{cm} \cdot \mathrm{g/s}$.

planar layers are separated by a distance d. They observed that an interference pattern – indicating wave behavior – emerged when the beam's incidence angle θ was such that $2d \sin\theta = nh/p$, where n is a positive integer and $p = \sqrt{2mE}$ is the electron's momentum, consistent with de Broglie's hypothesis.

These experimental facts and a number of others (such as those on polarization properties of light, or the Stern–Gerlach experiment on *spin*) called for a revision of the laws of mechanics. The first proposed theory, the so-called old quantum theory of Bohr, Sommerfeld, and others, was a mixture of *ad hoc* quantum rules with Newtonian laws – with which the quantum rules were frequently in conflict – and therefore conceptually not satisfactory. Then, in 1926, Heisenberg proposed his *uncertainty principle*, according to which the position and momentum of a microscopic particle cannot be measured simultaneously. The classical idea of particles describing well-defined paths in phase space should be abandoned altogether. At the same time, Schrödinger proposed that electrons and other particles should be described by *wave functions* satisfying a partial differential equation (PDE) that now bears his name (presumably guided by the idea that, where there is a wave, there must be a wave equation). The new quantum theory that emerged – quantum mechanics – was developed by Born, Jordan, Pauli, Dirac, and others.

In this chapter, we present a short account of the mathematical ideas and structures underlying quantum mechanics, from a modern mathematical viewpoint. The foundations of quantum mechanics were set on firm mathematical ground in the hands, above all, of von Neumann – who realized that the appropriate setting for quantum mechanics should be a theory of self-adjoint operators in a Hilbert space (and who single-handedly developed most of that theory) – and also Wigner, whose study of quantum symmetries plays a decisive role today.

2.2 The basic principles of quantum mechanics

2.2.1 States and observables

In quantum mechanics, one postulates the existence of a separable complex Hilbert space $\mathcal{H}$ that helps represent the possible states of a given quantum system. This Hilbert space can be finite-dimensional (as in the case of a system consisting of a single photon, or some fermion), but more often it is an infinite-dimensional space. The *pure states* of the system are the complex one-dimensional subspaces, or *rays*, of the Hilbert space $\mathcal{H}$. Because each such ray is completely determined once we give a unit vector belonging to it, pure states are also represented by unit vectors in $\mathcal{H}$. This representation is not unique, however: any two unit vectors that differ by a *phase* (i.e. by a complex scalar of modulus one) represent the same state. Thus, the pure state space of quantum mechanics is the projective space $\mathcal{S} = P^1(\mathcal{H})$.

The *observables* of a quantum system are represented by self-adjoint operators on $\mathcal{H}$. It turns out that in many situations we must allow these operators to be unbounded (defined over a dense subspace of $\mathcal{H}$). As we shall see, this is unavoidable, for instance, when we perform the so-called canonical quantization of a classical system of particles. The class of all observables of a quantum system will be denoted by $\mathcal{O}$.

Let us recall the spectral theorem for self-adjoint operators.

Theorem 2.1 *Let A be a self-adjoint operator on a Hilbert space $\mathcal{H}$, with domain $D_A \subseteq \mathcal{H}$. Then there exist a measure space (Ω, μ), a unitary operator $T : \mathcal{H} \to L^2(\Omega, \mu)$, and a measurable function $\alpha : \Omega \to \mathbb{R}$ such that the diagram*

$$\begin{array}{ccc} \mathcal{H} \supseteq D_A & \xrightarrow{A} & \mathcal{H} \\ {\scriptstyle T}\big\downarrow & & \big\downarrow{\scriptstyle T} \\ L^2(\Omega, \mu) \supset D_{M_\alpha} & \xrightarrow[M_\alpha]{} & L^2(\Omega, \mu) \end{array}$$

commutes. Here M_α denotes the multiplication operator given by $M_\alpha(f) = \alpha \cdot f$, and $D_{M_\alpha} = \{f \in L^2(\Omega, \mu) : \alpha \cdot f \in L^2(\Omega, \mu)\}$ is its natural domain of definition.

For a complete proof of this theorem, see Appendix A (Theorem A.22 in Section A.5). The spectral theorem allows us to associate to each observable A and each state $\psi \in \mathcal{H}$ a Borel probability measure on the real line, in the following way. Let $E \subseteq \mathbb{R}$ be a Borel set, and let χ_E be its characteristic function. Then the operator $M_{\chi_E \circ \alpha} : L^2(\Omega, \mu) \to L^2(\Omega, \mu)$ is an orthogonal projection (check!). Hence the operator $P_E = T^{-1} \circ M_{\chi_E \circ \alpha} \circ T$ is an orthogonal projection on $\mathcal{H}$. Using this orthogonal projection, we define

$$\mathbb{P}_{A,\psi}(E) = \langle \psi, P_E \psi \rangle \in \mathbb{R}^+ .$$

This gives us a probability measure on the real line. The physical interpretation of the non-negative number $\mathbb{P}_{A,\psi}(E)$ is that it represents the probability that a measurement of the observable A when the system is in the state ψ will be a number belonging to E. We therefore have a map $\mathcal{O} \times \mathcal{S} \to \mathcal{P}(\mathbb{R})$ (the space of Borel probability measures on the real line) given by $(A, \psi) \mapsto \mathbb{P}_{A,\psi}$.

Besides pure states, there are also composite or *mixed* states of a quantum system. Roughly speaking, these correspond to convex combinations of pure states. The mathematical definition is as follows. First, we define a *density operator* (also called a density *matrix*) to be a self-adjoint operator M on our Hilbert space $\mathcal{H}$ that is *positive* and *trace class*, with trace equal to one. Positivity means that for all $\xi \in \mathcal{H}$ we have $\langle M\xi, \xi\rangle \geq 0$. We say that M is trace class if, given any orthonormal basis $\{\psi_i\}$ for $\mathcal{H}$, we have $\sum_i \langle M\psi_i, \psi_i\rangle < \infty$. One can easily show that this sum, when finite, is independent of the choice of orthonormal basis, and therefore it is a property of M alone; it is called the *trace* of M, and denoted $\mathrm{Tr}(M)$. A density operator is normalized so that $\mathrm{Tr}(M) = 1$. It turns out that if M is trace class and A is any self-adjoint operator, then MA is trace class. A *mixed state* is a linear functional ω on the space of observables such that $\omega(A) = \mathrm{Tr}(MA)$ for all observables A, where M is a fixed density operator. It should be apparent that mixed states are more general than pure states. Indeed, if a unit vector $\psi \in \mathcal{H}$ represents a pure state, we can take M to be the orthogonal projection of $\mathcal{H}$ onto the subspace generated by ψ. This M is a density operator, and its corresponding mixed state ω satisfies $\omega(A) = \langle \psi, A\psi\rangle$ for all observables A (see Exercise 2.1).

2.2.2 Time evolution

Another important postulate of quantum mechanics is that the time evolution of a quantum mechanical system with Hilbert space $\mathcal{H}$ is given by a one-parameter group of unitary operators $U_t \in \mathcal{U}(\mathcal{H})$. Thus, if the system is initially in state ψ_0, then its state at time t will be $\psi_t = U_t(\psi_0)$.

Theorem 2.2 (Stone) *Every one-parameter group of unitary operators $U_t : \mathcal{H} \to \mathcal{H}$ of a separable complex Hilbert space $\mathcal{H}$ possesses an infinitesimal generator. More precisely, there exists a self-adjoint operator H on $\mathcal{H}$ such that $U_t = e^{-itH}$ for all $t \in \mathbb{R}$.*

Again, the proof of this theorem can be found in Appendix A (see Theorem A.33 in Section A.9). In the case of our quantum system, this special observable H, normalized by a constant $\hbar$ called Planck's constant, is called the *Hamiltonian* of the system. From Stone's theorem, it follows at once that the time evolution of states ψ_t satisfies the following first-order differential equation:

$$\frac{d\psi}{dt} = -\frac{i}{\hbar} H(\psi) .$$

This is known as Schrödinger's equation.

2.2.3 Symmetries

According to Wigner, a symmetry of a quantum system is a map in pure state space that preserves the transition probabilities between states. More precisely, given two rays $r, r' \in \mathcal{S} = P^1(\mathcal{H})$, we let

$$P(r, r') = \frac{|\langle z, z' \rangle|}{\|z\| \|z'\|} ,$$

where $z, z' \in \mathcal{H}$ are any two unit vectors belonging to r and r' respectively. Then we have the following definition.

Definition 2.3 *A* symmetry *of the quantum system with Hilbert space $\mathcal{H}$ is a continuous bijection $S : \mathcal{S} \to \mathcal{S}$ such that $P(S(r), S(r')) = P(r, r')$ for all pairs of rays $r, r' \in \mathcal{S}$.*

Symmetries of quantum systems correspond to unitary or anti-unitary operators in the Hilbert space of the system. This is the content of the following theorem, due to Wigner.

Theorem 2.4 *If S is a symmetry, then there exists a unitary or anti-unitary operator $\hat{S} : \mathcal{H} \to \mathcal{H}$ such that the following diagram commutes:*

$$\begin{array}{ccc} \mathcal{H} & \xrightarrow{\hat{S}} & \mathcal{H} \\ {\scriptstyle \pi}\big\downarrow & & \big\downarrow{\scriptstyle \pi} \\ \mathcal{S} & \xrightarrow[S]{} & \mathcal{S} \end{array}$$

where $\pi : \mathcal{H} \to \mathcal{S}$ is the obvious projection map.

Proof. A proof can be found in [We, I, pp. 91–6]. □

2.2.4 Heisenberg's uncertainty principle

The uncertainty principle discovered by Heisenberg reveals a major difference between classical and quantum mechanics. Whereas in classical mechanics the observables are functions on phase space and therefore constitute a commutative algebra, in quantum mechanics the observables are operators, and their algebra is *non-commutative*. Heisenberg's principle, stated as a mathematical theorem below, asserts that two observables of a quantum system cannot be measured simultaneously with absolute accuracy unless they commute as operators. There is an intrinsic uncertainty in their simultaneous measurement, one that is not due simply to experimental errors.

Let us suppose that A is an observable of a given quantum system. Thus, A is a (densely defined) self-adjoint operator on a Hilbert space $\mathcal{H}$. Let $\psi \in \mathcal{H}$ with $\|\psi\| = 1$ represent a state of the system. Then

$$\langle A \rangle_\psi = \langle \psi, A\psi \rangle$$

is the *expected value* of the observable A in the state ψ. The *dispersion* of A in the state ψ is given by the square root of its *variance*; that is

$$\Delta_\psi A = \langle (A - \langle A \rangle_\psi I)^2 \rangle_\psi^{1/2} = \| A\psi - \langle A \rangle_\psi \psi \| \,.$$

Note that if ψ is an eigenvector of A with eigenvalue λ, then $\langle A \rangle_\psi = \lambda$ and $\Delta_\psi A = 0$.

Theorem 2.5 (Heisenberg's uncertainty principle) *If A and B are observables of a quantum system, then for every state ψ common to both operators we have*

$$\Delta_\psi A \, \Delta_\psi B \geq \frac{1}{2} \left| \langle [A, B] \rangle_\psi \right| \,. \tag{2.1}$$

Proof. Subtracting a multiple of the identity from A and another such multiple from B does not change their commutator, and it does not affect their variances. Hence we may suppose that $\langle A \rangle_\psi = \langle B \rangle_\psi = 0$. Now, we have, using the self-adjointness of both operators,

$$\begin{aligned} \left| \langle [A, B] \rangle_\psi \right| &= \left| \langle \psi, (AB - BA)\psi \rangle \right| \\ &= \left| \langle A\psi, B\psi \rangle - \langle B\psi, A\psi \rangle \right| \\ &= 2 \left| \operatorname{Im} \langle A\psi, B\psi \rangle \right| \,. \end{aligned} \tag{2.2}$$

But the Cauchy–Schwarz inequality tells us that

$$\left| \operatorname{Im} \langle A\psi, B\psi \rangle \right| \leq \| A\psi \| \, \| B\psi \| \,. \tag{2.3}$$

Combining (2.2) with (2.3) yields (2.1), as desired. ☐

A simple consequence of this result is the fact that in a quantum particle system, the position and momentum of a particle cannot be measured simultaneously with absolute certainty.

2.2.5 The von Neumann axioms

Let us summarize the above discussion by stating, in an informal manner, the basic axioms of von Neumann describing a quantum mechanical system.

(i) The pure states of a quantum mechanical system are given by rays (equivalently, by unit vectors up to a phase) in a complex, separable Hilbert space $\mathcal{H}$.

(ii) The observables of the system are (densely defined) self-adjoint operators on $\mathcal{H}$.

(iii) Given a Borel set $E \subseteq \mathbb{R}$, the probability that an observable A has a value in E when the system is in the state $\psi \in \mathcal{H}$ is $\langle \psi, P_E \psi \rangle$, where

P_E is the spectral projection of A associated to E (via the spectral theorem).

(iv) If the system is in a state ψ at time 0, then its state at time t is given by $\psi(t) = \exp(-iHt/\hbar)\psi$, where H is the Hamiltonian (or energy) observable of the system.

2.3 Canonical quantization

There is an informal, heuristic principle in quantum mechanics called the *correspondence principle*. According to this principle, to each classical Hamiltonian system there should correspond a quantum system whose "classical limit" is the given classical system. The process of constructing such a quantum version of a classical system is called *quantization*. The construction involves finding a suitable Hilbert space from which the quantum states of the system are made, and suitable self-adjoint operators on this Hilbert space corresponding to the classical observables of the system.

We shall describe here the so-called *canonical quantization* of a Hamiltonian system consisting of N particles in Euclidean space.

The classical system here has states given by points in $\mathbb{R}^n$, where $n = 3N$, and *position* observables $q_1, q_2, \ldots, q_n$. These are functions on phase space, and as such they satisfy the canonical commutation relations $\{q_i, q_j\} = 0$ (Poisson brackets here). After quantization, we should have a Hilbert space $\mathcal{H}$ and position observables $Q_1, Q_2, \ldots, Q_n$ which are now self-adjoint operators on $\mathcal{H}$. These should still satisfy the commutation relations $[Q_i, Q_j] = 0$.

These commutation relations combined with the spectral theorem imply that there exist a common measure space (Ω, μ), a unitary equivalence $T : \mathcal{H} \to L^2(\Omega, \mu)$, and functions $q_1, q_2, \ldots, q_n$ on Ω such that each of the diagrams

$$\begin{array}{ccc} \mathcal{H} \supseteq D_{Q_i} & \xrightarrow{\;Q_i\;} & \mathcal{H} \\ {\scriptstyle T}\downarrow & & \downarrow{\scriptstyle T} \\ L^2(\Omega, \mu) \supseteq D_{M_{q_i}} & \xrightarrow[M_{q_i}]{} & L^2(\Omega, \mu) \end{array}$$

commutes. What space should (Ω, μ) be? No uniqueness is expected here, unless we make an extra hypothesis: we assume that the observables Q_j form a *complete system* in the sense that if A is an observable that commutes with all Q_j then $A = g(Q_1, Q_2, \ldots, Q_n)$ for some Borel map g. (This is not always physically reasonable: systems of particles that exhibit spin after quantization, for instance – a purely quantum property – can never be complete in this sense.)

If the hypothesis is satisfied, however, then it is possible to take the measure space to be the classical configuration space itself, namely $\Omega = \mathbb{R}^n$, with μ equal to Lebesgue measure. Hence the Hilbert space of the system is identified with $L^2(\Omega, \mu)$. The position operators become multiplication operators Q_j : $\psi \mapsto q_j \psi$.

Now, we have for each j a representation $T_j : \mathbb{R} \to \mathcal{U}(L^2(\Omega, \mu))$ of the translation group of the real line in our Hilbert space, given by

$$T_j^t \psi(q_1, q_2, \ldots, q_j, \ldots, q_n) = \psi(q_1, \ldots, q_j - t, \ldots, q_n) .$$

Such a representation is unitary (exercise). Hence by Stone's Theorem 2.2, there exists an infinitesimal generator P_j such that

$$T_j^t \psi = e^{itP_j} \psi .$$

In fact, we have

$$P_j = -i \frac{\partial}{\partial q_j} .$$

The operators P_j defined in this way are the *momentum* operators.

Proposition 2.6 *The position and momentum operators defined above satisfy the Heisenberg commutation relations*

$$[Q_j, Q_k] = 0 \; ; \; [P_j, P_k] = 0 \; ; \; [P_j, Q_k] = -\delta_{jk} i I .$$

Proof. Exercise. □

We note *en passant* that the last equality in the above proposition can be combined with Theorem 2.5 to yield a more familiar statement of Heisenberg's uncertainty principle.

Proposition 2.7 *The dispersions of the position operator P_j and the momentum operator Q_j satisfy the inequality*

$$\Delta_\psi P_j \Delta_\psi Q_j \geq \frac{1}{2}$$

for every state ψ belonging to the domains of both operators.

Proof. Immediate from Theorem 2.5 and the fact that $[P_j, Q_j] = -iI$. □

Next we ask ourselves the following question: what should be the Hamiltonian operator of this system? To answer this question, we have to go back to the general setting of the previous section and understand the time evolution of an observable. If $U_t = e^{-itH}$ denotes the one-parameter group generated by the Hamiltonian H and A is an observable, then heuristically we have

$$\dot{A} = \frac{d}{dt}\Big|_{t=0} \left(e^{itH} A e^{-itH}\right) = i(HA - AH) ;$$

in other words,

$$\dot{A} = i[H, A] .$$

We emphasize that this is purely formal; in particular, the right-hand side does not have a meaning yet. But we proceed heuristically: if $A = Q_j$, then we expect to have

$$\dot{Q}_j = i[H, Q_j] = \frac{1}{m_j} P_j ,$$

where $m_j > 0$ is a constant corresponding to the mass of the particle with classical position coordinate q_j. This yields the formal equality

$$HQ_j - Q_j H = -\frac{1}{m_j} \frac{\partial}{\partial q_j} . \tag{2.4}$$

The Hamiltonian operator H that we are looking for should satisfy these relations. Now we have a simple lemma.

Lemma 2.8 *For each $j = 1, 2, \ldots, n$ we have*

$$\frac{\partial^2}{\partial q_j^2} Q_j - Q_j \frac{\partial^2}{\partial q_j^2} = 2 \frac{\partial}{\partial q_j} \tag{2.5}$$

as operators (densely defined on $L^2(\mathbb{R}^n)$).

Proof. If $\psi \in L^2(\mathbb{R}^n)$ is twice differentiable, then we have on one hand

$$\left(\frac{\partial^2}{\partial q_j^2} Q_j \right) \psi = \frac{\partial}{\partial q_j} \left(\psi + q_j \frac{\partial \psi}{\partial q_j} \right) = 2 \frac{\partial \psi}{\partial q_j} + q_j \frac{\partial^2 \psi}{\partial q_j^2} ,$$

and on the other hand

$$\left(Q_j \frac{\partial^2}{\partial q_j^2} \right) \psi = q_j \frac{\partial^2 \psi}{\partial q_j^2} .$$

Subtracting these equalities yields (2.5). □

This motivates us to define an operator H_T as follows:

$$H_T = -\sum_{j=1}^{n} \frac{1}{2m_j} \frac{\partial^2}{\partial q_j^2} .$$

By Lemma 2.8 and (2.4), we have $H_T Q_j - Q_j H_T = H Q_j - Q_j H$ for all j. This shows that the operator $\hat{V} = H - H_T$ commutes with each Q_j; i.e. $[\hat{V}, Q_j] = 0$ for all j. Due to our assumption that the position operators are complete, this means that $\hat{V} = \hat{V}(Q_1, Q_2, \ldots, Q_n)$. Because each Q_j is a multiplication operator in $L^2(\mathbb{R}^n)$ (by the function q_j) and they commute, the spectral theorem implies that there exists a (measurable) function V on $\mathbb{R}^n$ such that $\hat{V}\psi = V \cdot \psi$.

We arrive at the following expression for the Hamiltonian operator:

$$H\psi = -\frac{1}{2}\sum_{j=1}^{n} \frac{1}{m_j} \frac{\partial^2 \psi}{\partial q_j^2} + V\psi .$$

2.4 From classical to quantum mechanics: the C^* algebra approach

We would like to say a few words about the more modern mathematical approach to quantum mechanics via C^* algebras, even though this viewpoint will not really be used in the remainder of this book, except in a brief discussion of algebraic quantum field theory at the end of Appendix B. The reader is advised to consult that appendix for all the basic definitions and results about C^* algebras that are relevant in the discussion to follow.

In classical (Hamiltonian) mechanics, the states of a system are described by points in a phase space Γ. The observables – physical quantities that are measurable in experiments – are described by real- (or complex-) valued functions on phase space. As it turns out, the observables form a commutative C^* algebra. As became clear at the end of the 19th century, this description of a mechanical system is quite inadequate if the system has too many particles. In the new approach proposed by Gibbs and Boltzmann, one only talks about the probability that a system is in a given state. More precisely, in statistical mechanics the states are taken to be probability distributions on phase space Γ, whereas the observables become random variables on Γ (note, however, that they still form an abelian C^* algebra). The states in statistical mechanics are, therefore, more general than classical states: the latter, also called pure states, correspond to (Dirac) point-mass distributions concentrated at one given point in phase space. In the statistical-mechanical description, when a system is in a given state, all one really measures about an observable is its expected value with respect to the probability distribution of that state. Thus, a state can be viewed as a positive linear functional on the C^* algebra of observables. This establishes a certain duality between states and observables. In the early part of the twentieth century, it became apparent that even this more general model of the world given by statistical mechanics was insufficient to describe phenomena at the microscopic (sub-atomic) scale. On this scale, Heisenberg's uncertainty principle rules: in many situations, two given observables cannot be accurately measured simultaneously in a given state – not just experimentally, but in principle. In addition, the measured values of observables (say, energy) are oftentimes discrete quantities, not continuous, as one might expect.

The new description of the world that emerges from this picture – quantum mechanics – is a radical departure from either classical or statistical mechanics. Whereas in classical or statistical mechanics the observables form a commutative C^* algebra, in quantum mechanics this C^* algebra is non-commutative. The states of a quantum system are defined by duality – in analogy to what happens in classical or statistical mechanics – as positive linear functionals on the C^* algebra of observables.

The abstract algebraic structures described above, however nice, are not entirely satisfactory for the description of concrete physical systems. To actually measure and predict, we need a concrete realization of such abstract structures. In the classical case, the algebra of observables is commutative: it is $C(\Gamma)$ for some (compact) space Γ. Conversely, if one is given only a commutative C^* algebra, one can reconstruct the phase space Γ: this is the abelian version of the Gelfand–Naimark theorem, according to which every commutative C^* algebra (with unity) is isometrically isomorphic to $C(X)$ for some compact Hausdorff space X. In the quantum case, the C^* algebra of observables is non-commutative. Here a concrete representation of such a C^* algebra is offered by the full Gelfand–Naimark theorem.

Theorem 2.9 (Gelfand–Naimark) *Every C^* algebra is isometrically isomorphic to an algebra of (bounded) linear operators in some Hilbert space.*

This theorem is presented with a complete proof in Appendix B, Theorem B.44. Thus, the observables of a quantum system are represented by linear operators on a certain Hilbert space. By the above duality between states and observables, it can be shown (the so-called Gelfand–Naimark–Segal construction – see Appendix B, Theorem B.42) that in this representation the pure states correspond to rays in the Hilbert space. Once this Hilbert space picture is in place, one can study the dynamical evolution of the system. This can be done through the analysis of either the time evolution of states – the Schrödinger evolution equation – or the time evolution of observables – the Heisenberg evolution equation. These are dual to each other.

Remark 2.10 We warn the reader that this rather simplistic outline omits several important points. For example, when trying to quantize a classical system, say a system with a single particle, we run into the difficulty that the position (q) and momentum (p) observables are not (and can never be made into) bounded operators, and we seemingly fall outside the scope of the above discussion. This difficulty was resolved by Weyl a long time ago. The idea is to replace q and p with their complex exponentials e^{iq} and e^{ip} (these will be bounded as long as q and p are essentially self-adjoint) and look at the C^* algebra generated by the algebra of polynomials on these. The resulting C^* algebra is called the Weyl algebra. Von Neumann has shown that all regular, irreducible representations of the Weyl algebra are unitarily equivalent. See Section 2.5 below for more on these facts. A good reference for the C^* algebra approach to quantum mechanics is [Str]. But see also our Appendix B.

2.5 The Weyl C^* algebra

As we saw in Section 2.3, the quantization of a classical mechanical system consisting, say, of a single particle yields position and momentum observables Q and P respectively, satisfying Heisenberg's commutator relations

$$[P, P] = 0 = [Q, Q]\,; \;\; [P, Q] = -i\hbar I\,.$$

We know that such observables are represented by pairs of self-adjoint operators defined on *some* Hilbert space $\mathcal{H}$. Such a Hilbert space cannot be finite-dimensional; otherwise P and Q would be represented by matrices satisfying $PQ - QP = -i\hbar I$, which is impossible – to see why, simply take the traces of both sides of this equality. Thus, $\mathcal{H}$ is infinite-dimensional. Could P and Q both be bounded operators on $\mathcal{H}$? The following result shows that the answer is no.

Lemma 2.11 *Let P, Q be self-adjoint elements of a C^* algebra such that $[P, Q] = \alpha I$ for some $\alpha \in \mathbb{C}$. Then $\alpha = 0$.*

Proof. Note that for all $n \geq 1$ we have

$$\begin{aligned}[P, Q^n] &= [P, Q]Q^{n-1} + Q[P, Q^{n-1}] \\ &= \alpha Q^{n-1} + Q[P, Q^{n-1}]\,.\end{aligned}$$

From this it follows by induction that

$$[P, Q^n] = n\alpha Q^{n-1}\,.$$

Taking norms on both sides yields

$$n|\alpha|\, \|Q^{n-1}\| = \|PQ^n - Q^nP\| \leq 2\|P\|\,\|Q\|\,\|Q^{n-1}\|\,. \tag{2.6}$$

But now, because Q is self-adjoint, we have $\|Q^{n-1}\| = \|Q\|^{n-1}$. Moreover, the commutator relation $[P, Q] = \alpha I$ tells us that either $\alpha = 0$, in which case we are done, or else $\|Q\| \neq 0$. In the latter case we can divide both the left and right sides of (2.6) by $\|Q^{n-1}\|$, getting the inequality $2\|P\| \cdot \|Q\| \geq n|\alpha|$. Because this holds for all $n \geq 1$ and $\|P\|$, $\|Q\|$ are bounded, we deduce that $\alpha = 0$. □

This lemma shows that, when one quantizes a classical system, the appearance of unbounded operators as observables is unavoidable. This fact introduces certain technical difficulties into the study of quantum mechanical systems.

One way out of such difficulties was devised by Weyl. Instead of the self-adjoint operators P, Q, Weyl proposed to consider the one-parameter groups

$$U(\alpha) = e^{i\alpha P} \quad \text{and} \quad V(\alpha) = e^{i\alpha Q}\,, \quad \text{where } \alpha \in \mathbb{R}\,.$$

These are *unitary* operators, and therefore bounded. They are called *Weyl operators*. The physical motivation behind Weyl's idea is the fact that what one really wants to understand is the time evolution of an observable, not so much the observable *per se*.

The Heisenberg commutator relations for the operators P and Q translate into new commutation relations for the corresponding Weyl operators. This will be stated precisely below. First we need an identity, the so-called Baker–Hausdorff formula. The statement below depends on the following concept. If A is a self-adjoint operator on the Hilbert space $\mathcal{H}$, we say that a vector $\psi \in \mathcal{H}$ is *analytic* if $A^n\psi$ is well-defined for all n, and also the exponential $e^A\psi$.

Lemma 2.12 (Baker–Hausdorff) *Let A, B be self-adjoint operators on a Hilbert space. If the domains of A, B, and $A + B$ have a common dense subspace D of analytic vectors, then in D we have*

$$e^A e^B = e^{A+B+\frac{1}{2}[A,B]}\,, \tag{2.7}$$

provided $[A, B]$ commutes with both A and B.

Proof. Note that the right-hand side of (2.7) is equal to

$$e^{A+B} e^{\frac{1}{2}[A,B]}\,,$$

because $[A, B]$ commutes with $A + B$. Let us consider the function of a real variable α given by

$$F(\alpha) = e^{\alpha A} e^{\alpha B} e^{-\alpha(A+B)} e^{-\frac{\alpha^2}{2}[A,B]}\,. \tag{2.8}$$

To prove (2.7), it suffices to show that $F(1) = I$ (identity operator). We will show in fact that $F(\alpha) = I$ for all α. Because we clearly have $F(0) = I$, all we have to do is to show that the derivative $F'(\alpha) = 0$ for all α. Calculating the derivative explicitly, we get

$$F'(\alpha) = e^{\alpha A} e^{\alpha B} \left(e^{-\alpha B} A e^{\alpha B} - A - \alpha[A, B] \right) e^{-\alpha(A+B)} e^{-\frac{\alpha^2}{2}[A,B]} . \quad (2.9)$$

We claim that the expression between parentheses in (2.9) vanishes. To see why, let $\Phi(\alpha) = e^{-\alpha B} A e^{\alpha B}$. Then

$$\Phi'(\alpha) = e^{-\alpha B}[A, B]e^{\alpha B} = [A, B] ,$$

because $[A, B]$ commutes with B, and hence with $e^{\alpha B}$. This shows that $\Phi(\alpha) = \alpha[A, B] + C$ for some C. But $\Phi(0) = A$, so $C = A$, and therefore $\Phi(\alpha) = \alpha[A, B] + A$. In other words, we have

$$e^{-\alpha B} A e^{\alpha B} - \alpha[A, B] - A = 0 .$$

This proves our claim; the right-hand side of (2.9) vanishes identically, and we are done. □

With this lemma at hand, we are now in a position to establish the Weyl commutation relations.

Theorem 2.13 *The Weyl one-parameter groups of unitary operators satisfy the following relations, for all* $\alpha, \beta \in \mathbb{R}$:

(i) $U(\alpha)U(\beta) = U(\beta)U(\alpha) = U(\alpha + \beta)$;
(ii) $V(\alpha)V(\beta) = V(\beta)V(\alpha) = V(\alpha + \beta)$;
(iii) $U(\alpha)V(\beta) = e^{-i\hbar\alpha\beta} V(\beta)U(\alpha)$.

Proof. To prove (i), apply Lemma 2.12 with $A = i\alpha P$ and $B = i\beta P$ (note that in this case $[A, B] = 0$). Similarly, to prove (ii), apply the same lemma with $A = i\alpha Q$ and $B = i\beta Q$ (again, $[A, B] = 0$). Finally, to prove (iii), take $A = i\alpha P$ and $B = i\beta Q$ and once again apply Lemma 2.12. This time $[A, B] = -\alpha\beta[P, Q] = -i\hbar\alpha\beta$. Therefore, on the one hand, we have

$$U(\alpha)V(\beta) = e^{-\frac{i\hbar}{2}\alpha\beta} e^{i(\alpha P + \beta Q)} ,$$

and on the other hand,

$$V(\beta)U(\alpha) = e^{\frac{i\hbar}{2}\alpha\beta} e^{i(\alpha P + \beta Q)} .$$

Comparing these last two equalities immediately yields (iii). □

The algebra over the complex numbers generated by abstract elements $U(\alpha)$ and $V(\beta)$ ($\alpha, \beta \in \mathbb{R}$) satisfying the Weyl commutation relations given in Theorem 2.13 is called the *Weyl algebra*, and is denoted by $\mathcal{A}_W$. One can define an involution $* : \mathcal{A}_W \to \mathcal{A}_W$ by letting

$$U(\alpha)^* = U(-\alpha)\,, \qquad V(\beta)^* = V(-\beta)$$

and extending it to the whole algebra in the obvious way. The elements $U(\alpha)$, $V(\beta)$ are unitary, in the sense that

$$U(\alpha)^*U(\alpha) = U(\alpha)U(\alpha)^* = I$$

by the Weyl relations, and similarly for $V(\beta)$, $V(\beta)^*$. Finally, one can define a (unique) norm over $\mathcal{A}_W$ in such a way that $U(\alpha)$, $V(\beta)$, and $U(\alpha)V(\beta)$ all have norm equal to one (by the Weyl relations, every monomial in the generators can be reduced, up to multiplication by a complex number of unit modulus, to the form $U(\alpha)V(\beta)$, and therefore also has norm one). It is possible to prove that the completion of $\mathcal{A}_W$ with respect to this norm is a C^* algebra, still denoted $\mathcal{A}_W$. It is called, not surprisingly, the *Weyl C^* algebra.*

Thus, a quantum system consisting of a single particle is described by the Weyl C^* algebra. The possible quantum states of such a one-particle system are given by the *representations* of the C^* algebra $\mathcal{A}_W$ in a suitable Hilbert space. The task of finding such representations is greatly facilitated by a theorem of Von Neumann, stated below.

Definition 2.14 *A unitary representation ρ of the Weyl C^* algebra into a separable Hilbert space $\mathcal{H}$ is said to be* regular *if $\alpha \mapsto \rho(U(\alpha))$ and $\beta \mapsto \rho(V(\beta))$ are strongly continuous maps.*

Theorem 2.15 (Von Neumann) *All regular irreducible representations of the Weyl C^* algebra $\mathcal{A}_W$ are unitarily equivalent.*

We will not prove this theorem here. The interested reader can consult, for instance, [Str, pp. 61–2].

2.6 The quantum harmonic oscillator

Let us now see how the quantization scheme described in the previous section can be applied to the harmonic oscillator, the simplest and most important example of a Hamiltonian system. The classical Hamiltonian in generalized coordinates is

$$H(q, p) = \frac{p^2}{2m} + \frac{1}{2}m\omega^2q^2\,, \tag{2.10}$$

where m is the *mass* and ω is a positive constant measuring the *frequency* of oscillation. Let us assume henceforth that $m = 1$, and let us take Planck's constant $\hbar$ to be equal to 1 also. Our quantization scheme dictates that p and

q be promoted to self-adjoint operators P and Q acting on *some* Hilbert space $\mathcal{H}$. Accordingly, the quantum Hamiltonian

$$H = \frac{1}{2}\left(P^2 + \omega^2 Q^2\right) \tag{2.11}$$

becomes a self-adjoint operator as well.

2.6.1 Computing the spectrum

We want to find the *spectrum* of H in (2.11). We will in fact find all eigenvalues and corresponding eigenvectors. In order to do so, we make the following assumptions.

(i) The operators P and Q act irreducibly on the Hilbert space $\mathcal{H}$. In other words, $\mathcal{H}$ cannot be decomposed into a non-trivial direct sum of subspaces that are invariant under both P and Q.

(ii) The operator H given in (2.11) has an eigenvalue λ.

We shall deal with these assumptions in due course. For now, the interesting thing is that, as soon as we have these facts at hand, we can determine the full spectrum of H in an essentially algebraic fashion, with the help of Heisenberg's commutator relation

$$[Q, P] = i\ . \tag{2.12}$$

To do this, we introduce the following operators:

$$\begin{aligned} a &= \frac{1}{\sqrt{2\omega}}(\omega Q + iP)\ , \\ a^* &= \frac{1}{\sqrt{2\omega}}(\omega Q - iP)\ . \end{aligned}$$

These are called the *annihilation* and *creation* operators, respectively. Note that, because P and Q are self-adjoint, a^* is equal precisely to the adjoint of a (i.e. $a^\dagger = a^*$). An easy computation using (2.12) yields

$$\begin{aligned} aa^* &= \frac{1}{2\omega}(\omega Q + iP)(\omega Q - iP) \\ &= \frac{1}{2\omega}\left\{(P^2 + \omega^2 Q^2) + i\omega[P, Q]\right\} \\ &= \frac{1}{\omega}\left(H + \frac{\omega}{2}\right)\ . \end{aligned} \tag{2.13}$$

Similarly, we have

$$a^* a = \frac{1}{\omega}\left(H - \frac{\omega}{2}\right)\ . \tag{2.14}$$

Combining (2.13) with (2.14), we deduce that

$$[a^*, a] = 1 \,. \tag{2.15}$$

Here are some purely algebraic consequences of this last identity. We have

$$\begin{aligned}\left[a, (a^*)^2\right] &= aa^*a^* - a^*a^*a \\ &= aa^*a^* - a^*aa^* + a^*aa^* - a^*a^*a \\ &= [a, a^*]a^* + a^*[a, a^*] = -2a^* \,.\end{aligned}$$

Thus, $[(a^*)^2, a] = 2a^*$. By induction we get, for all $n \geq 1$,

$$\left[(a^*)^n, a\right] = n(a^*)^{n-1} \,. \tag{2.16}$$

Two other identities are immediate from (2.15), namely

$$[H, a] = -\omega a \quad \text{and} \quad [H, a^*] = \omega a^* \,. \tag{2.17}$$

Now, using the second of our assumptions above, let $\psi \in \mathcal{H}$ be an eigenvector of H belonging to the eigenvalue λ, i.e. $H\psi = \lambda\psi$. Then, using (2.14), we have

$$\omega\left(a^*a + \frac{1}{2}\right)\psi = \lambda\psi \,.$$

Taking the inner product on the left by ψ, we get

$$\omega\langle\psi, a^*a\psi\rangle + \frac{\omega}{2}\|\psi\|^2 = \lambda\|\psi\|^2 \,.$$

Because a^* is the adjoint of a, this last equality becomes

$$\omega\|a\psi\|^2 + \frac{\omega}{2}\|\psi\|^2 = \lambda\|\psi\|^2 \,,$$

from which it follows that

$$\left(\lambda - \frac{\omega}{2}\right)\|\psi\|^2 = \omega\|a\psi\|^2 \geq 0 \,.$$

This shows that $\lambda \geq \omega/2$, with equality if and only if $a\psi = 0$. This proves that the spectrum of H, which we know is real because H is self-adjoint, is bounded from below by $\omega/2 > 0$. Now, from the fact that $H\psi = \lambda\psi$ and using (2.17), we see that

$$aH\psi = \lambda(a\psi)$$

implies, that

$$(Ha + \omega a)\psi = \lambda(a\psi) \,,$$

and therefore

$$H(a\psi) = (\lambda - \omega)a\psi \ .$$

We deduce that either $a\psi = 0$, or else $a\psi$ is an eigenvector of H with eigenvalue $\lambda - \omega$, and thus $\lambda - \omega \geq \omega/2$. Proceeding inductively, we get a sequence of eigenvectors

$$\psi, a\psi, a^2\psi, \ldots, a^N\psi, \ldots$$

with corresponding eigenvalues

$$\lambda, \lambda - \omega, \lambda - 2\omega, \ldots, \lambda - N\omega, \ldots .$$

This cannot keep going down forever, because the eigenvalues of H are all positive. Therefore there exists $N \geq 0$ such that $\psi_0 = a^N\psi$ satisfies $H\psi_0 = \lambda_0\psi_0$ for some $\lambda_0 > 0$ but $a\psi_0 = 0$. It is an easy exercise to see that $\lambda_0 = \omega/2$.

The above *ladder reasoning* shows that the assumption that H has an eigenvalue is equivalent to the existence of a non-zero vector $\psi_0 \in \mathcal{H}$ such that $a\psi_0 = 0$. This vector, which we normalize to be a unit vector, is called the *ground state* of H.

In the above argument we have reached the ground state by stepping down the ladder. Now we can climb back up it to get *all* the eigenstates of H and their corresponding eigenvalues. Starting with the ground state ψ_0, we define $\phi_n = (a^*)^n\psi_0$ for each $n \geq 0$. We claim that each ϕ_n is an eigenvector of H. Indeed, using the second equality in (2.17), we see that

$$H\phi_n = Ha^*(a^*)^{n-1}\psi_0 = (a^*H + \omega a^*)\phi_{n-1} \ ,$$

where

$$H\phi_n = a^*H\phi_{n-1} + \omega\phi_n \ . \tag{2.18}$$

It follows easily by induction from (2.18) that ϕ_n is an eigenvector of H for each $n \geq 0$ (note that $\phi_0 = \psi_0$). In fact, denoting the corresponding eigenvalue by λ_n, we see that (2.18) also implies that $\lambda_n = \lambda_{n-1} + \omega$ for each $n \geq 1$. Because $\lambda_0 = \omega/2$, we deduce that $\lambda_n = (n + \frac{1}{2})\omega$ for all $n \geq 0$.

It is an easy exercise to check that the subspace of $\mathcal{H}$ spanned by $\{\phi_n : n \geq 0\}$ is invariant under the operators P and Q (see Exercise 2.4). But now the irreducibility assumption (i) above implies that $\mathcal{H}$ must be equal to this subspace. In particular, we have proved that the spectrum of H is

$$\sigma(H) = \left\{ \left(n + \frac{1}{2} \right) \omega \ : \ n = 0, 1, 2, \ldots \right\} \ . \tag{2.19}$$

The vectors ϕ_n are pairwise orthogonal, because they are eigenvectors of a self-adjoint operator belonging to distinct eigenvalues. Note, however, from (2.16) that

$$\begin{aligned}\|\phi_n\|^2 &= \left\langle (a^*)^n \psi_0, (a^*)^n \psi_0 \right\rangle \\ &= \left\langle a^n (a^*)^n \psi_0, \psi_0 \right\rangle \\ &= \left\langle a^{n-1}(n(a^*)^{n-1} + (a^*)^{n-1} a)\psi_0, \psi_0 \right\rangle \\ &= n \left\langle a^{n-1}(a^*)^{n-1} \psi_0, \psi_0 \right\rangle .\end{aligned}$$

This shows that

$$\|\phi_n\|^2 = n \left\langle (a^*)^{n-1} \psi_0, (a^*)^{n-1} \psi_0 \right\rangle = n \|\phi_{n-1}\|^2 ,$$

and, therefore, by induction,

$$\|\phi_n\|^2 = n! \|\psi_0\|^2 = n! .$$

Thus, an orthonormal basis of eigenvectors for H is given by

$$\psi_n = \frac{1}{\sqrt{n!}} (a^*)^n \psi_0 , \quad n = 0, 1, 2, \ldots . \tag{2.20}$$

2.6.2 Concrete coordinate representation

We have determined the spectrum of H under the assumptions (i) and (ii) above. Let us now exhibit a concrete representation of H, on a concrete Hilbert space $\mathcal{H}$, that satisfies both assumptions. Everything boils down to finding $\psi_0 \in \mathcal{H}$ such that $a\psi_0 = 0$. Let $\mathcal{H} = L^2(\mathbb{R})$, and consider the Schwartz space $\mathcal{S}(\mathbb{R}) \subset L^2(\mathbb{R})$, which is dense in $L^2(\mathbb{R})$. Let $Q, P : \mathcal{S}(\mathbb{R}) \to L^2(\mathbb{R})$ be given by

$$\begin{aligned}(Q\psi)(x) &= x\psi(x) , \\ (P\psi)(x) &= -i \frac{\partial \psi}{\partial x}(x) .\end{aligned}$$

These operators are essentially self-adjoint (see Appendix A). Accordingly, the operator a acts on $\mathcal{S}(\mathbb{R})$ by the formula

$$(a\psi)(x) = \frac{1}{\sqrt{2\omega}} \left(\omega x \psi(x) + \frac{d\psi}{dx}(x) \right) .$$

A similar formula holds true for $a^*\psi$. In order to find the desired eigenvector for H, we need to solve $a\psi_0 = 0$. This is an ordinary differential equation, namely

$$\psi_0'(x) = -\omega x \psi(x) .$$

Its general solution is $\psi_0(x) = Ce^{-(1/2)\omega x^2}$, for some (real) constant C. This constant is chosen so that ψ_0 has unit L^2-norm. Thus,

$$1 = \|\psi_0\|^2 = C^2 \int_{\mathbb{R}} \left(e^{-\frac{1}{2}\omega x^2}\right)^2 dx ,$$

and therefore

$$C^2 = \left(\int_{-\infty}^{\infty} e^{-\omega x^2}\, dx\right)^{-1} = \sqrt{\frac{\omega}{\pi}} .$$

We deduce that

$$\psi_0(x) = \left(\frac{\omega}{\pi}\right)^{\frac{1}{4}} e^{-\frac{1}{2}\omega x^2} .$$

Note that ψ_0 is indeed an element of $\mathcal{S}(\mathbb{R})$. With ψ_0 at hand, we can now write down explicitly each ψ_n of the orthonormal basis of eigenvectors that completely diagonalizes H. We get

$$\psi_n(x) = \left(\frac{\omega}{\pi}\right)^{\frac{1}{4}} \frac{(2\omega)^{-n/2}}{\sqrt{n!}} \left(\omega x + \frac{d}{dx}\right)^n e^{-\frac{1}{2}\omega x^2} .$$

When the action of the differential operator appearing in this expression is explicitly worked out, the final result will involve, for each n, a polynomial $H_n(x)$ of degree n times the Gaussian function $e^{-(1/2)\omega x^2}$. The polynomials $H_n(x)$ are the classical *Hermite polynomials*.

2.6.3 Holomorphic representation

Here is an alternate, concrete representation of the Hamiltonian operator H (which must obviously be isomorphic to the one given in the previous subsection). It turns out that the assumptions (i) and (ii) we made in Section 2.6.1 are equivalent to assuming the existence of a separable Hilbert space $\mathcal{H}$ and a complete pair of operators P and Q on this space satisfying the Heisenberg commutator relation (2.12). Let us now prove once again that they exist.

As Hilbert space $\mathcal{H}$, consider the complex vector space of all holomorphic functions $f : \mathbb{C} \to \mathbb{C}$ such that

$$\frac{1}{2i} \int_{\mathbb{C}} |f(z)|^2 e^{-z\bar{z}}\, dz d\bar{z} < \infty ,$$

endowed with the inner product

$$\langle f, g\rangle = \frac{1}{2i} \int_{\mathbb{C}} f(z)\overline{g(z)} e^{-z\bar{z}}\, dz d\bar{z} .$$

As the reader can check, the polynomials $\varphi_n \in \mathcal{H}$ given by $\varphi_n(z) = z^n$ ($n = 0, 1, \ldots$) are mutually orthogonal in $\mathcal{H}$ with respect to this inner product.

Moreover, if $f = \sum f_n z^n$ and $g = \sum g_n z^n$ are holomorphic functions in $\mathcal{H}$, then

$$\langle f, g \rangle = \sum_{n=0}^{\infty} \alpha_n f_n g_n ,$$

where

$$\alpha_n = 2\pi \int_0^{\infty} r^{2n+1} e^{-r^2} \, dr .$$

Lemma 2.16 *With the inner product defined above, $\mathcal{H}$ is a complex Hilbert space.*

Proof. An exercise for the reader. □

Next, define the (unbounded) linear operators P and Q on this Hilbert space $\mathcal{H}$ as follows:

$$Qf(z) = zf(z) ,$$

$$Pf(z) = -i \frac{\partial f}{\partial z} .$$

Lemma 2.17 *The pair of operators P, Q is complete, and it satisfies the Heisenberg commutator relation.*

Proof. Let T be a linear operator on $\mathcal{H}$ that commutes with both P and Q. Since the kernel of P is the subspace of constant functions, from $PT = TP$ we see that T maps constants to constants. Now let $\xi_n = T(\varphi_n)$. From $TQ\varphi_n = QT\varphi_n$ we deduce that $\xi_{n+1}(z) = z\xi_n(z)$. Therefore

$$T\varphi_n = \xi_n = z^n \xi_0(z) = c\varphi_n(z) ,$$

where $c = T(1) \in \mathbb{C}$. This shows that $T = cI$, so the pair P, Q is complete. We leave it as a very easy exercise for the reader to check that $[P, Q] = -iI$. □

2.7 Angular momentum quantization and spin

As a second important example of quantization, let us consider the angular momentum of a particle system. To simplify the discussion, we shall only treat here the case of a single particle in 3-space. Recall that the *classical* angular momentum is given by $\Omega = \boldsymbol{x} \wedge \boldsymbol{p}$, where $\boldsymbol{x} = (x_1, x_2, x_3)$ is the particle's position and $\boldsymbol{p} = (p_1, p_2, p_3)$ is its momentum. The components $\Omega_1, \Omega_2, \Omega_3$

of the angular momentum are given by

$$\Omega_j = \varepsilon_{jkl} x_k p_l \ , \tag{2.21}$$

where ε_{jkl} is the totally symmetric symbol satisfying $\varepsilon_{jkl} = 1$ if jkl is an even permutation of 123, $\varepsilon_{jkl} = -1$ if that permutation is odd, and $\varepsilon_{jkl} = 0$ otherwise. Calculating the Poisson bracket $\{\Omega_1, \Omega_2\}$, we get

$$\begin{aligned} \{\Omega_1, \Omega_2\} &= \sum_{j=1}^{3} \left(\frac{\partial \Omega_1}{\partial x_j} \frac{\partial \Omega_2}{\partial p_j} - \frac{\partial \Omega_1}{\partial p_j} \frac{\partial \Omega_2}{\partial x_j} \right) \\ &= x_1 p_2 - x_2 p_1 = \Omega_3 \ . \end{aligned}$$

Thus, $\{\Omega_1, \Omega_2\} = \Omega_3$, and similarly $\{\Omega_2, \Omega_3\} = \Omega_1$ and $\{\Omega_3, \Omega_1\} = \Omega_2$. One recognizes here the relations defining the Lie algebra of $SO(3)$.

Now the idea is that, when we quantize this system, we should obtain operators L_1, L_2, L_3 corresponding to $\Omega_1, \Omega_2, \Omega_3$ respectively (and an operator L corresponding to Ω). The Poisson bracket relations should translate into similar relations for the *commutators* of these operators. Due to (2.21), we realize that

$$L_j = \varepsilon_{jkl} \hat{x}_k \hat{p}_l \ , \tag{2.22}$$

where $\hat{x}_k$ are the position operators and $\hat{p}_l$ are the momentum operators. Using the Heisenberg commutation relations $[\hat{x}_k, \hat{p}_l] = i\hbar\delta_{kl} I$, we deduce (exercise) that

$$[L_1, L_2] = i\hbar L_3 \ ; \quad [L_2, L_3] = i\hbar L_1 \ ; \quad [L_3, L_1] = i\hbar L_2 \ . \tag{2.23}$$

Thus, we can think of the quantization procedure as giving rise to a representation of the Lie algebra of $SO(3)$ into the self-adjoint operators of some Hilbert space, with the generators being mapped to operators L_j satisfying (2.23). We shall assume that such representation is irreducible (this is consistent with Wigner's definition of *particle*; see Chapter 4).

We are interested in the spectral properties of the operators L_j. At this point, we could consider the one-parameter groups of unitary operators generated by the L_j's and invoke the Peter–Weyl theorem, according to which the irreducible representations of a compact Lie group (such as $SO(3)$) are all *finite-dimensional*. We prefer instead to proceed in elementary fashion. For this purpose, it is convenient to consider the operator

$$L^2 = L_1^2 + L_2^2 + L_3^2 \ .$$

Note that, because each L_j is self-adjoint, L_j^2 is a positive operator. Therefore L^2, being a sum of positive operators, is positive also.

Lemma 2.18 *The operator L^2 commutes with each L_j.*

Proof. To prove this lemma, it is convenient to use the identity

$$[AB, C] = A[B, C] + [A, C]B \; . \tag{2.24}$$

Applying (2.24) with $A = B = L_j$ and $C = L_k$, we get

$$[L_j^2, L_k] = L_j[L_j, L_k] + [L_j, L_k]L_j \; .$$

Hence we have, using (2.23),

$$\begin{aligned} [L_1^2, L_1] = 0 \; ; \quad [L_1^2, L_2] &= i\hbar(L_3L_1 + L_1L_3) \\ [L_1^2, L_3] &= -i\hbar(L_1L_2 + L_2L_1) \; . \end{aligned} \tag{2.25}$$

Similarly, we have

$$\begin{aligned} [L_2^2, L_1] = -i\hbar(L_2L_3 + L_3L_2) \; ; \quad [L_2^2, L_2] = 0 \\ [L_2^2, L_3] = i\hbar(L_1L_2 + L_2L_1) \; , \end{aligned} \tag{2.26}$$

as well as

$$\begin{aligned} [L_3^2, L_1] = i\hbar(L_2L_3 + L_3L_2) \; ; \quad 1[L_3^2, L_2] = -i\hbar(L_3L_1 + L_1L_3) \\ [L_2^2, L_3] = 0 \; . \end{aligned} \tag{2.27}$$

Adding up the first relations in (2.25), (2.26), and (2.27), we get

$$\begin{aligned} [L^2, L_1] &= [L_1^2, L_1] + [L_2^2, L_1] + [L_3^2, L_1] \\ &= 0 - i\hbar(L_2L_3 + L_3L_2) + i\hbar(L_2L_3 + L_3L_2) = 0 \; . \end{aligned}$$

Similarly, we get $[L^2, L_2] = 0$, and also $[L^2, L_3] = 0$. □

This shows that L^2 commutes with every element of the Lie algebra generated by the L_j's. Because we are assuming that this representation is irreducible, it follows that L^2 *belongs* to this Lie algebra. Therefore L^2 must be a multiple of the identity: $L^2 = \lambda I$ for some $\lambda \geq 0$ (recall that L^2 is a positive operator). It is easy to exclude the possibility that $\lambda = 0$: the representation would be trivial in this case. Hence we can assume that $\lambda > 0$.

Lemma 2.19 *The operators L_j are bounded.*

Proof. We claim that L_j^2 is a bounded operator. This follows easily from the fact that L_j^2 is a positive operator and from the fact that $L^2 \geq L_j^2$. This is turn shows that the spectrum of L_j is contained in the interval $[-\sqrt{\lambda}, \sqrt{\lambda}]$, and therefore L_j is a bounded operator also. □

We can actually describe the spectrum of each L_j quite explicitly. Let us do it for $j = 3$. For this purpose, we introduce the following (bounded, but not self-adjoint) operators $L_+ = L_1 + iL_2$ and $L_- = L_1 - iL_2$. In what follows, these operators will play a role akin to that of the creation and annihilation operators introduced in the analysis of the harmonic oscillator.

It is an easy exercise to check that $[L_3, L_\pm] = \pm\hbar L_\pm$.

Lemma 2.20 *Let A, B be elements of a Lie algebra such that $[A, B] = B$. Then for all $\alpha \in \mathbb{C}$ we have*

$$e^{\alpha A}B = e^{\alpha}Be^{\alpha A} .$$

Proof. Because $AB = BA + B = B(I + A)$, we have by induction

$$A^n B = B(I + A)^n .$$

Multiplying both sides of this equality by $\alpha^n/n!$ and adding up the resulting series, we get $e^{\alpha A}B = Be^{\alpha(I+A)} = e^{\alpha}Be^{\alpha A}$. □

From applying this lemma to $A = \pm\hbar^{-1}L_3$, $B = L_\pm$, and $\alpha = is\hbar$, it follows that

$$e^{isL_3}L_\pm e^{-isL_3} = e^{is\hbar}L_\pm .$$

Taking $s = 2\pi/\hbar$, we see from this relation that the unitary operator $e^{2\pi iL_3/\hbar}$ commutes with L_+, L_-, and L_3. As before, we deduce that it is a (phase) multiple of the identity; in other words,

$$e^{2\pi iL_3/\hbar} = e^{i\theta}I .$$

This equality allows us to prove that the spectrum of the operator L_3 is *discrete*. We need the following more general result.

Lemma 2.21 *Let A be a self-adjoint operator such that $e^{2\pi iA} = e^{i\theta}I$, for some θ. Then A has a discrete spectrum.*

Proof. Let us consider the one-parameter group of unitary operators given by

$$U(s) = e^{isA}e^{-is\theta/2\pi} = e^{is\Phi} ,$$

where

$$\Phi = A - \frac{\theta}{2\pi}I .$$

Applying the spectral theorem, we have the integral representation

$$U(s) = \int_{\sigma(\Phi)} e^{is\lambda}\, dE(\lambda) .$$

From this and the functional calculus, we can write

$$|U(2\pi) - I| = \int_{\sigma(\Phi)} |e^{2\pi i\lambda} - 1|\, dE(\lambda)\,.$$

But the left-hand side of this last equality is zero. Therefore $\sigma(\Phi) \subseteq \mathbb{Z}$. This shows that

$$\sigma(A) \subseteq \left\{ n + \frac{\theta}{2\pi}\ :\ n \in \mathbb{Z} \right\}.$$

This shows that the spectrum of A is discrete as claimed. □

Lemma 2.22 *The operator L_3 has a discrete spectrum.*

Proof. By applying Lemma 2.21 to $A = L_3/\hbar$, the result follows. We deduce in fact that

$$\sigma(L_3) \subseteq \left\{ \left(n + \frac{\theta}{2\pi} \right) \hbar\ :\ n \in \mathbb{Z} \right\}.$$

The inclusion is actually strict, because we know already that L_3 is bounded. □

Obviously, the same result holds for the other two angular momentum operators L_1, L_2. Summarizing, we know so far that these operators are bounded with finite spectrum. Each element of the spectrum must therefore be an eigenvalue. Now, if $\lambda \in \sigma(L_3)$ and ψ is an eigenvector with eigenvalue λ, then we have

$$L_3 L_\pm \psi = (\lambda \pm \hbar)\, L_\pm \psi\ .$$

Thus, the operators L_+ and L_- have the effect of raising and lowering the eigenvalues. Using the fact that the spectrum of L_3 is finite, we deduce that there exist integers k and l with $k < l$ such that the eigenvalues of L_3 are $\lambda + k\hbar, \lambda + (k+1)\hbar, \ldots, \lambda + (l-1)\hbar, \lambda + l\hbar$. We leave it as an exercise to deduce from these facts that there exists a positive integer ℓ such that *either*

$$\sigma(L_3) = \left\{ n\hbar\ :\ n = -\ell, \cdots, -1, 0, 1, \cdots, \ell \right\}.$$

or

$$\sigma(L_3) = \left\{ \left(n + \frac{1}{2} \right) \hbar\ :\ n = -\ell, \cdots, -1, 0, 1, \cdots, \ell \right\}.$$

Note that the entire analysis so far *has not used* (2.22) *at all!*. One can use (2.22) to rule out the second possibility for the spectral properties of the orbital angular momentum operators.

This still leaves open the possibility that there are irreducible representations for which the generators of the associated Lie algebra have spectra given by the second of the two options above. And indeed there are. The quantum observables, however, do *not* correspond to any classical observable. They are the so-called *spin operators* S_1, S_2, S_3. They arise from irreducible unitary representations of the Lie group $SU(2)$, which is a double covering of $SO(3)$ and whose Lie algebra is the same as that of $SO(3)$.

2.8 Path integral quantization

In this section, we present the method of quantization of particle systems via path integrals, first devised by Feynman. The basic idea, however, had already appeared in the work of Dirac. We shall present the mathematically rigorous results first, postponing the physical motivation.

2.8.1 The Trotter product formula

For simplicity, we deal with the case of one particle in Euclidian space $\mathbb{R}^n$, subject to a potential V. We use coordinates x for position and p for momentum. The classical Hamiltonian is given by

$$H_{\text{cl}} = \frac{p^2}{2m} + V(x) .$$

The canonical quantization procedure described in Section 2.3 yields the Hilbert space $L^2(\mathbb{R}^n)$ as the space of quantum states, and the quantum Hamiltonian becomes the operator given by

$$H = -\frac{\hbar^2}{2m}\Delta + V .$$

If V is reasonable (at least square integrable), the domain of this operator is a (dense) subspace $D_H \subseteq L^2(\mathbb{R}^n)$ containing the Schwartz space $\mathcal{S}(\mathbb{R}^n)$, and for $\psi \in D_H$ we have of course

$$H\psi = -\frac{\hbar^2}{2m}\Delta\psi + V\psi .$$

For more reasonable potentials, the operator H will be self-adjoint. We will have a precise sufficient condition later.

Let us write $H = H_0 + V$, where H_0 is the free particle Hamiltonian

$$H_0 = -\frac{\hbar^2}{2m}\Delta .$$

From this point on, we shall use units for which $\hbar = 1$. When $V = 0$, the time evolution of the free particle is given by

$$(e^{-itH_0}\psi)(x) = (4\pi i t)^{-n/2} \int_{\mathbb{R}^n} e^{i|x-y|^2/4t}\psi(y)\,dy .$$

Here we see the *free propagator* $K_0(x, y; t) = (4\pi i t)^{-n/2} e^{i|x-y|^2/4t}$.

We also know the time evolution of the (multiplication) operator V. It is a one-parameter group of multiplication operators, given simply by

$$(e^{-itV}\psi)(x) = e^{-itV(x)}\psi(x) .$$

The real problem is how to obtain the time evolution of the combined operator $H = H_0 + V$. This is where the Trotter product formula comes to the rescue. Let us first state a version for bounded operators. We need the following lemma.

Lemma 2.23 *Let C and D be bounded operators on a Banach space. Then we have*

$$e^{C+D} - e^C e^D = -\frac{1}{2}[C, D] + R ,$$

where R is an absolutely convergent series of monomials in C and D of degrees greater than or equal to three.

Proof. Compute the left-hand side by writing out the power series expansions of the exponentials. The details are left as an exercise. □

In fact, a much more explicit statement is given by the Baker–Campbell–Hausdorff formula; see [Ros].

Theorem 2.24 (Trotter's formula I) *If A and B are bounded operators in a Banach space, then*

$$e^{A+B} = \lim_{N\to\infty} \left(e^{A/N}e^{B/N}\right)^N ,$$

where the limit is taken in the operator norm topology.

Proof. Let us consider, for each n, the operators $S_n = e^{(A+B)/n}$ and $T_n = e^{A/n}e^{B/n}$. To prove Trotter's formula, we need to show that $\|S_n^n - T_n^n\| \to 0$ as $n \to \infty$ (where $\|\cdot\|$ denotes the operator norm).

First note that if S and T are operators, we have the identity

$$S^n - T^n = \sum_{j=0}^{n-1} S^j(S-T)T^{n-j-1} .$$

This is easily proved by induction on n. This implies that

$$\|S^n - T^n\| \leq \sum_{j=0}^{n-1} \|S\|^j \, \|S-T\| \, \|T\|^{n-j-1} .$$

Applying this inequality to $S = S_n$ and $T = T_n$ and taking into account that $\|e^C\| \leq e^{\|C\|}$ for every bounded operator C, we get

$$\|S_n^n - T_n^n\| \leq \|S_n - T_n\| \sum_{j=0}^{n-1} e^{\frac{j}{n}\|A+B\|} e^{\frac{n-j-1}{n}(\|A\|+\|B\|)}$$

$$\leq n e^{\|A\|+\|B\|} \|S_n - T_n\| .$$

This reduces our task to proving that $\|S_n - T_n\|$ converges to zero faster than n^{-1}. But this is now a consequence of Lemma 2.23. Applying that lemma to $C = A/n$ and $D = B/n$, we see that

$$S_n - T_n = -\frac{1}{2n^2}[A, B] + \frac{1}{n^3} R_n ,$$

where R_n is a bounded operator whose norm is uniformly bounded in n. This shows that $\|S_n - T_n\| = O(n^{-2})$ and finishes the proof. □

For unbounded operators (such as the ones we have here), the situation is not quite as nice, but still sufficiently nice.

Theorem 2.25 (Trotter's formula II) *Let A and B be self-adjoint operators on a separable Hilbert space.*

(a) If $A + B$ is essentially self-adjoint in $D_A \cap D_B$, then

$$\text{s-lim}_{N\to\infty} \left(e^{iA/N} e^{iB/N}\right)^N = e^{i(A+B)} .$$

(b) If A and B are bounded from below, then

$$\text{s-lim}_{N\to\infty} \left(e^{-A/N} e^{-B/N}\right)^N = e^{-(A+B)} .$$

Proof. See [RS1, p. 297]. □

Using the above version of the Trotter product formula, one can write down the time evolution of the Hamiltonian $H = H_0 + V$ as a limit, as follows.

Theorem 2.26 *If the potential can be written as a sum $V = V_1 + V_2$, where $V_1 \in L^2(\mathbb{R}^3)$ and $V_2 \in L^\infty(\mathbb{R}^3)$, then $H = H_0 + V$ is essentially self-adjoint. Moreover, for all $\psi \in D_H$ and every $x_0 \in \mathbb{R}^3$ we have*

$$e^{-it(H_0+V)}\psi(x_0) = \lim_{N\to\infty} \left(\frac{4\pi it}{N}\right) \int_{\mathbb{R}^3}\int_{\mathbb{R}^3} \cdots \int_{\mathbb{R}^3} e^{iS_N(x_0,x_1,\cdots,x_N;t)}$$

$$\times \psi(x_N)\, dx_1 dx_2 \ldots dx_N ,$$

where

$$S_N(x_0, x_1, \cdots, x_N; t) = \sum_{k=1}^{N} \frac{t}{N} \left[\frac{1}{4} \left(\frac{x_k - x_{k-1}}{t/N} \right)^2 - V(x_k) \right] .$$

Proof. The first assertion follows immediately from a theorem of Kato and Rellich, providing a criterion for self-adjointness. The Kato–Rellich theorem is presented with proof in Appendix A (see Theorem A.35 in Section A.10). The second assertion is a consequence of Theorem 2.25 (a) above. □

2.8.2 The heuristic path integral

The Trotter product formula and the resulting formula for the time evolution of the Hamiltonian operator were presented in a mathematically rigorous way in the previous section. But what can one say about their physical meaning? What follows is a discussion of the motivation behind these formulas, namely the notion of *path integral*. We emphasize that the discussion below is predominantly heuristic.

We assume that $\hat{H}$ is the quantum Hamiltonian operator of a one-dimensional system, corresponding to a classical Hamiltonian H (via a canonical quantization procedure, say). We let $\hat{q}$ and $\hat{p}$ be the position and momentum operators. We shall use Dirac's bra and ket notations throughout. Thus, $|q\rangle$ and $|p\rangle$ denote the eigenstates of $\hat{q}$ and $\hat{p}$ with eigenvalues q and p, respectively, whereas $\langle q|$ and $\langle p|$ denote their "dual eigenstates" (linear functionals). The reader has the right to be puzzled that we speak of eigenvectors for the position operator. By way of clarification, in the holomorphic representation given in Section 2.6.3, the plane wave function $|p\rangle = \psi_p(z) = e^{ipz}$ is an eigenfunction of $\hat{p}$ with eigenvalue p, and the Fourier transform of such eigenfunctions give us delta distributions $\delta(z - q) = |q\rangle$ as eigenfunctions of the position operator $\hat{q}$. These of course live *outside* the underlying Hilbert space. The way to ascribe precise mathematical meaning to these generalized eingenfunctions is to introduce the concept of *rigged* Hilbert space. To avoid a lengthy digression, we refrain from doing so, but see [Ti]. Instead, we ask the reader to believe that a rigorous treatment can be done, and to accept the following facts regarding these generalized eigenfunctions:

(i) $\langle q|q'\rangle = \delta(q - q')$, and $\langle p|p'\rangle = \delta(q - q')$.

(ii) $\langle q|p\rangle = e^{ipq}$.

(iii) $1 = \int_{\mathbb{R}} dq\, |q\rangle\langle q|$.

(iv) $1 = \frac{1}{2\pi} \int_{\mathbb{R}} dp\, |p\rangle\langle p|$.

We remark that the identities in (iii) and (iv) are operator identities, and they come from the fact that $\{|q\rangle : q \in \mathbb{R}\}$ and $\{|p\rangle : p \in \mathbb{R}\}$ are orthogonal bases of the (rigged) Hilbert space. Their meaning is that, if $|\psi\rangle$ is a state in Hilbert space, then we have the orthogonal decomposition

$$|\psi\rangle = \int_{\mathbb{R}} \langle q|\psi\rangle \, |q\rangle \, dq \, .$$

This can be made precise with the help of the spectral theorem (applied to $\hat{q}$).

Now, in the Schrödinger representation, we know that the time evolution of a state $|\psi\rangle$ is given by

$$|\psi\rangle(t) = e^{-i\hat{H}t}|\psi\rangle \, .$$

Here we are assuming that $\hbar = 1$. In the Heisenberg representation, the states are time-independent, and we look instead at the time evolution of observables. Hence, if A is a self-adjoint operator, then $A(t) = e^{i\hat{H}t}Ae^{-i\hat{H}t}$. We are interested in the case when $A = \hat{q}$ (or $\hat{p}$). Let $|q,t\rangle$ be the eigenstate of $\hat{q}(t)$ with eigenvalue q, so that

$$\hat{q}(t)|q,t\rangle = q \, |q,t\rangle \, .$$

These eigenstates remain mutually orthogonal for all t, and we have a generalized version of (iii), namely

$$1 = \int_{\mathbb{R}} |q,t\rangle\langle q,t| \, dq \, . \tag{2.28}$$

Let us look at the time evolution of the position operator of our system between an initial time t_i and a final time t_f. The transition probability amplitude between an initial state $|q_i, t_i\rangle$ and a final state $|q_f, t_f\rangle$ is given by the inner product $\langle q_f, t_f|q_i, t_i\rangle$. Our goal is to compute this amplitude. Note that

$$\langle q_f, t_f|q_i, t_i\rangle = \langle q_f|e^{-i\hat{H}(t_f - t_i)}|q_i\rangle \, .$$

The idea behind the computation is to partition the interval $[t_i, t_f]$ into N subintervals of equal length $\epsilon_N = (t_f - t_i)/N$ through the points $t_j = t_i + j\epsilon_N$, $j = 0, \ldots, N$. Using the identity (2.28) with $t = t_1$ and $q = q_1$, we can write

$$\langle q_f, t_f|q_i, t_i\rangle = \int_{\mathbb{R}} \langle q_f, t_f|q_1, t_1\rangle \, \langle q_1, t_1|q_i, t_i\rangle \, dq_1 \, .$$

This process can be repeated inductively for $t_2, \ldots, t_N$. We get the representation

$$\langle q_f, t_f | q_i, t_i \rangle = \int_{\mathbb{R}^N} \prod_{j=0}^{N-1} \langle q_{j+1}, t_{j+1} | q_j, t_j \rangle \, dq_1 \cdots dq_N \, .$$

Taking into account that

$$\langle q_{j+1}, t_{j+1} | q_j, t_j \rangle = \langle q_{j+1} | e^{-i\hat{H}(t_{j+1}-t_j)} | q_j \rangle \, ,$$

we have

$$\mathcal{A} = \langle q_f, t_f | q_i, t_i \rangle = \int_{\mathbb{R}^N} \prod_{j=0}^{N-1} \langle q_{j+1} | e^{-i\hat{H}\epsilon_N} | q_j \rangle \, dq_1 \cdots dq_N \, . \tag{2.29}$$

Now, the point is that, when ϵ is small, we have an operator expansion of the form

$$e^{-i\hat{H}\epsilon} = 1 - i\epsilon\hat{H} + O(\epsilon^2) \, .$$

When this expansion is used in each term in the product integrand in (2.29), we are faced with the problem of evaluating $\langle q_{j+1} | \hat{H} | q_j \rangle$ for each j. Here we invoke the identity (iv), using the momentum variable $p = p_j$, so that

$$\langle q_{j+1} | \hat{H} | q_j \rangle = \frac{1}{2\pi} \int \langle q_{j+1} | p_j \rangle \, \langle p_j | \hat{H} | q_j \rangle \, dp_j \, . \tag{2.30}$$

To compute $\langle p_j | \hat{H} | q_j \rangle$, we have to remember that $\hat{H} = \hat{H}(\hat{q}, \hat{p})$ is an operator involving the position and momentum operators $\hat{q}$, $\hat{p}$, which do not commute. They can be made to act on either side of the inner product, but it is necessary to specify an order in which this is to be done. To circumvent potential ambiguities, we suppose that, whenever we have products of the non-commuting operators $\hat{q}$, $\hat{p}$ in the expression defining $\hat{H}$, the $\hat{p}$ factors always appear to the left of the $\hat{q}$ factors. Under this assumption, and taking into account that $\langle p_j |$ and $| q_j \rangle$ are eigenstates (for $\hat{p}$ and $\hat{q}$ respectively), we see that

$$\langle p_j | \hat{H} | q_j \rangle = H(q_j, p_j) \langle p_j | q_j \rangle \, ,$$

where $H(q, p)$ is the classical Hamiltonian evaluated at (q, p). Using the identities in (i), we have

$$\langle p_j | q_j \rangle = \overline{\langle q_j | p_j \rangle} = e^{-p_j q_j} \, , \quad \text{as well as } \langle q_{j+1} | p_j \rangle = e^{i p_j q_{j+1}} \, .$$

Putting these data back into (2.30) for each j, we get

$$\langle q_{j+1} | \hat{H} | q_j \rangle = \frac{1}{2\pi} \int H(q_j, p_j) e^{i p_j (q_{j+1} - q_j)} dp_j \, .$$

Taking these expressions to (2.29), we see that the amplitude $\mathcal{A} = \langle q_f, t_f | q_i, t_i \rangle$ can be written as

$$\mathcal{A} = \frac{1}{(2\pi)^N} \int_{\mathbb{R}^N} \int_{\mathbb{R}^N} \exp\left\{ i \sum_{j=0}^{N-1} p_j (q_{j+1} - q_j) \right\}$$
$$\times \prod_{j=0}^{N-1} \left(1 - i\epsilon_N H(q_j, p_j) + O(\epsilon_N^2)\right) dq_1 \cdots dq_N \, dp_1 \cdots dp_N \, .$$

We leave to the reader, as an exercise, the task of verifying that

$$\prod_{j=0}^{N-1} \left(1 - i\epsilon_N H(q_j, p_j) + O(\epsilon_N^2)\right)$$
$$= (1 + O(\epsilon_N)) \exp\left\{ -i\epsilon_N \sum_{j=0}^{N-1} H(q_j, p_j) \right\} .$$

Therefore the amplitude $\mathcal{A}$ can be rewritten as

$$\mathcal{A} = \frac{1}{(2\pi)^N} \int_{\mathbb{R}^N} \int_{\mathbb{R}^N} \exp\left\{ i\epsilon_N \sum_{j=0}^{N-1} \left[p_j \frac{q_{j+1} - q_j}{\epsilon_N} - H(q_j, p_j) \right] \right\}$$
$$\times (1 + O(\epsilon_N)) \, dq_1 \cdots dq_N \, dp_1 \cdots dp_N \, .$$

Because this must hold for all N, we can take the limit on the right-hand side as $N \to \infty$ to get

$$\mathcal{A} = \lim_{N \to \infty} \frac{1}{(2\pi)^N} \int_{\mathbb{R}^N} \int_{\mathbb{R}^N} \exp\left\{ i\epsilon_N \sum_{j=0}^{N-1} \left[p_j \frac{q_{j+1} - q_j}{\epsilon_N} - H(q_j, p_j) \right] \right\} \tag{2.31}$$
$$\times \, dq_1 \cdots dq_N \, dp_1 \cdots dp_N \, .$$

A heuristic interpretation of this limit can be given as follows. We can think of the points (t_j, q_j) as determining a continuous, piecewise linear path joining (t_i, q_i) to (t_f, q_f), the slope in the jth linear piece being $(q_{j+1} - q_j)/\epsilon_N$ (because $t_{j+1} - t_j = \epsilon_N$). We also have a piecewise linear path in momentum space, interpolating the points (t_j, p_j). The integrand in (2.31) is a function of these two paths, and the integration process happens over *all* such pairs of paths. Let us imagine, rather naively, that as $N \to \infty$ the piecewise linear paths in coordinate space converge to differentiable paths $q(t)$ $(t_i \leq t \leq t_f)$ and that their linear slopes converge to the time derivative $\dot{q}(t)$. Let us also imagine that the product Lebesgue measures $dq_1 \cdots dq_N$ and $dp_1 \cdots dp_N/(2\pi)^N$ both converge, in some sense, to certain heuristic measures:

$$\prod_{j=0}^{N-1} dq_j \to \mathcal{D}q \ ; \quad \prod_{j=0}^{N-1} dp_j/(2\pi)^N \to \mathcal{D}p \, .$$

Then, because

$$i\epsilon_N \sum_{j=0}^{N-1} \left(p_j \frac{q_{j+1} - q_j}{\epsilon_N} - H(q_j, p_j) \right) \to i \int_{t_i}^{t_f} (p\dot{q} - H(q, p)) \, dt \, ,$$

as $N \to \infty$, it follows from (2.31) that

$$\mathcal{A} = \int\!\!\int \exp\left\{ i \int_{t_i}^{t_f} (p\dot{q} - H(q, p)) \, dt \right\} \mathcal{D}q\mathcal{D}p \, .$$

This heuristic expression is called the *path integral in Hamiltonian form.* Of course, it lacks a rigorous mathematical meaning because

(i) The above heuristic infinite-dimensional measures do not exist.

(ii) Even if we could define such measures in the space of all paths, we would have to ascribe meaning to the integrand. This seems quite problematic, because the typical (continuous) path $q(t)$ is nowhere differentiable.

Physicists do not stop at such mathematical difficulties. Let us then press on, and recast the above formula in another form. For this purpose, let us consider a Hamiltonian of the form

$$H(q, p) = \frac{p^2}{2m} + V(q) \, ,$$

where V is a suitable potential. In this case, the dependence on the momentum variable p is quadratic. This allows us, even before taking the limit in (2.31), to perform the integrations in p_j, because they are simply Gaussian integrals. Let us write, for simplicity, $\dot{q}_j = (q_{j+1} - q_j)/\epsilon_N$ for each j. The Gaussian integrals that appear in (2.31) are

$$I_j = \frac{1}{2\pi} \int_{\mathbb{R}} e^{-i\epsilon_N (p_j^2/2m - p_j \dot{q}_j)} dp_j \, .$$

The lemma on Gaussian integration that is needed here, which involves analytic continuation, is presented with proof in Chapter 7. Using that lemma and the standard technique of completing the square, we find that

$$I_j = \left(\frac{2\pi i \epsilon_N}{m} \right)^{-\frac{1}{2}} e^{i\epsilon_N m \dot{q}_j^2/2} \, .$$

Putting this information back into (2.31), we deduce that

$$\mathcal{A} = \lim_{N\to\infty} \left(\frac{2\pi i \epsilon_N}{m}\right)^{-\frac{N}{2}} \int_{\mathbb{R}^N} \exp\left\{ i\epsilon_N \sum_{j=0}^{N-1} \left(\frac{m\dot{q}_j^2}{2} - V(q_j) \right) \right\} dq_1 \cdots dq_N \,.$$

But now, as before, we note that for a differentiable limiting path $q(t)$, we have

$$\lim_{N\to\infty} \epsilon_N \sum_{j=0}^{N-1} \left(\frac{m\dot{q}_j^2}{2} - V(q_j) \right) = \int_{t_i}^{t_f} \left(\frac{m\dot{q}^2}{2} - V(q) \right) dt \,.$$

This last expression is precisely the classical action

$$S(q,\dot{q}) = \int_{t_i}^{t_f} L(q,\dot{q})\, dt \,,$$

where $L(q,\dot{q})$ is the Lagrangian. We deduce at last that, at a purely heuristic level,

$$\mathcal{A} = \langle q_f, t_f | q_i, t_i \rangle = \mathcal{N} \int e^{iS(q,\dot{q})} \mathcal{D}q \,.$$

This is the *path integral in Lagrangian form*. Here $\mathcal{N}$ is a normalizing (infinite!) constant. It is not as dangerous as it might seem, because when computing the actual correlation between the initial and final states, we have to divide the inner product $\langle q_f, t_f | q_i, t_i \rangle$ by the product of the norms of the vectors $|q_i, t_i\rangle$ and $|q_f, t_f\rangle$, and in the process the constant goes away. The real difficulty, of course, lies in the path integral itself. This "sum over histories," as physicists since Feynman like to call it, is an oscillatory integral (due to the purely imaginary exponent in the integrand), and a great deal of cancelation is expected, if a finite result is to be obtained. There are only a few simple situations where the path integral can be explicitly evaluated. One is the case of a free particle ($V = 0$); another is the case of the harmonic oscillator. The reader is invited to try these cases as (perhaps challenging) exercises.

Despite its mathematical difficulties (some of which were dealt with in the previous subsection) the path integral, in its Lagrangian formulation, provides us in principle with a way to perform the quantization of a particle system without any reference to operators or Hilbert space. Only the classical action intervenes, and in principle all quantum correlations could be computed. The Lagrangian path integral is especially useful when we have to study systems with *constraints*. This point of view is extremely fruitful, and can be used in the quantization of fields, as we shall see in Chapter 7.

2.9 Deformation quantization

We close this chapter with a brief discussion of *deformation quantization*. We recall that the classical observables, which are real-valued functions on the phase space, have the structure of a commutative associative algebra under ordinary multiplication, and also the structure of a Lie algebra given by the Poisson bracket. On the other hand, the quantum observables are operators on a Hilbert space and therefore do not commute in general. The idea developed in [BFLS] is that we can look at the non-commutative algebra of quantum observables as a deformation of the commutative algebra of classical observables. In this deformation of algebras we lose commutativity, but we gain a closer connection between the associative product and the Lie product, which is just the commutator. Under this viewpoint, a quantum observable is obtained from a classical observable by a sequence of quantum corrections; i.e. it is a formal power series in Planck's constant whose coefficients are real functions on the classical phase space, and the product, called the $*$-product, is such that the commutator reduces to the Poisson bracket as Planck's constant goes to zero (the classical limit). The reader can find in [BFLS] a detailed discussion of these ideas and in [Kon] and [WL] deep results on the existence and uniqueness of this deformation theory. A possible interpretation of these results is that classical mechanics is an unstable theory that can be deformed into quantum mechanics, which is stable.

Exercises

2.1 Let the unit vector $\psi \in \mathcal{H}$ represent a pure state of a quantum system with Hilbert space $\mathcal{H}$. Let $P_\psi : \mathcal{H} \to \mathcal{H}$ denote the orthogonal projection onto the one-dimensional subspace of $\mathcal{H}$ generated by ψ.

(a) Show that P_ψ is a self-adjoint, positive, and trace-class operator whose trace is equal to one.

(b) Show that the mixed state associated to the density operator $M = P_\psi$ equals the linear functional on observables $A \mapsto \langle \psi, A\psi \rangle$ determined by the pure state ψ; i.e. it can be identified with ψ itself.

2.2 Show that the application of the Gram–Schmidt orthogonalization method to the sequence of complex polynomials $\{1, z, z^2, \ldots\}$, with respect to the inner product given in 2.6.3 yields the (complex) Hermite polynomials H_n.

2.3 Prove that the coordinate representation and the holomorphic representation of the quantum harmonic oscillator are isomorphic. [*Hint:* Use the previous exercise.]

2.4 Recall the operators a and a^* associated to the Hamiltonian H of the quantum harmonic oscillator, and the eigenvectors ψ_n of H. Using the fact that

$$Q = \frac{1}{\sqrt{2}}(a + a^*) \quad \text{and} \quad P = -i\sqrt{\frac{\omega}{2}}(a - a^*)\,,$$

show that both P and Q leave invariant the linear subspace of Hilbert space spanned by $\{\psi_0, \psi_1, \ldots, \psi_n, \ldots\}$.

2.5 Prove the commutator relations (2.23) involving the angular momentum operators.

3
Relativity, the Lorentz group, and Dirac's equation

In this chapter, our goal is to show how quantum mechanics had to be modified to make it into a *relativistic* theory. The theory developed in Chapter 2 is not compatible with Einstein's special relativity: Schrödinger's equation is not relativistically invariant. The attempt by Dirac to make both theories compatible – Dirac's equation – showed that one had to abandon the idea of a physical system having a fixed number of particles. Dirac's theory allows creation and destruction of particles, forcing one to take up instead, as fundamental, the idea of *quantum fields*, of which particles become physical manifestations (eigenstates). This was the birth of quantum field theory.

3.1 Relativity and the Lorentz group

At the end of the nineteenth century, the Michelson–Morley experiments showing that light travels at a speed that is independent of the motion of the observer relative to its source, plus the discovery by Lorentz that the Maxwell equations are invariant under a large group of transformations, exposed a contradiction between Newtonian mechanics and Maxwellian electromagnetism. This led Einstein to reformulate the laws of mechanics (keeping the notion of inertial reference frame and Newton's first law; cf. Chapter 1).

3.1.1 Postulates

In essence, the basic postulates of Einstein's special relativity theory are the following.

(i) *Principle of relativity*: The laws of physics are the same in all inertial frames.

(ii) *Invariance of uniform motion*: If a particle has constant velocity in a given inertial frame, then its velocity is constant in every inertial frame.

(iii) *Invariance of the speed of light*: The speed of light is invariant across all inertial frames.

Just as in Newtonian mechanics, the space of events $\mathcal{E}$ here is four-dimensional. An inertial frame provides an identification of this event space with the standard *Minkowski spacetime* $M = \mathbb{R}^{1,3} \equiv \mathbb{R}^4$. This vector space is endowed with an inner product given by

$$\langle x, y \rangle_M = x_0 y_0 - x_1 y_1 - x_2 y_2 - x_3 y_3 .$$

Here, we write $x = (x^0, x^1, x^2, x^3)$, with (x^1, x^2, x^3) denoting the *spatial* coordinates of x and $x^0 = ct$ denoting its *temporal* coordinate (c denotes the speed of light). Writing vectors in M as column vectors and denoting by x^t the transpose of x, we see that the Minkowski inner product can be written as

$$\langle x, y \rangle_M = x^t G y ,$$

where $G = (g_{ij})$ is the matrix

$$G = \begin{pmatrix} 1 & 0 & 0 & 0 \\ 0 & -1 & 0 & 0 \\ 0 & 0 & -1 & 0 \\ 0 & 0 & 0 & -1 \end{pmatrix} .$$

This matrix is called *Minkowski's metric tensor*.

The physical reason for Minkowski's inner product structure lies in the third postulate. Suppose we are given an inertial frame O. If a light source is placed at $(0, 0, 0)$ at $t = 0$ then it sends out a spherical wavefront that – because the speed of light is c – will reach a given point (x^1, x^2, x^3) in space at time $t = x^0/c$ such that

$$(x^0)^2 - (x^1)^2 - (x^2)^2 - (x^3)^2 = 0 . \tag{3.1}$$

This is the equation of a cone in $\mathbb{R}^4$, called the *light cone*. Thus, the wave with source at the origin traces out the light cone. Let us now consider another inertial frame $\overline{O}$ with a common spacetime origin with O, and denote the coordinates in $\overline{O}$ by $\overline{x} = (\overline{x}^0, \overline{x}^1, \overline{x}^2, \overline{x}^3)$. Then, *because the speed of light in* $\overline{O}$ *is also* c, the equation of the light cone in this new inertial frame is still the same,

$$(\overline{x}^0)^2 - (\overline{x}^1)^2 - (\overline{x}^2)^2 - (\overline{x}^3)^2 = 0 .$$

In other words, the light cone in event space has the same equation across all inertial frames (with a common origin).

3.1.2 Lorentz transformations

The observation we have just made can be recast in the following way. Let $\mathcal{T} : \mathcal{E} \to M$ and $\overline{\mathcal{T}} : \mathcal{E} \to M$ denote the two frame mappings defining the two given inertial frames $O, \overline{O}$. Then the bijection $\Lambda : M \to M$ given by $\Lambda = \overline{\mathcal{T}} \circ \mathcal{T}^{-1}$ represents the change of reference frame. Because we have a common event in $\mathcal{E}$ that was assumed to be the origin in both frames, this map satisfies $\Lambda(0) = 0$. Moreover, we have established above that Λ must preserve the light cone C given by (3.1), i.e. $\Lambda(C) \subseteq C$. Now, the second postulate implies that straight lines representing uniform motions in one reference frame must correspond to straight lines in the other frame. Hence Λ maps lines to lines.[1] Under some mild continuity assumptions, it follows (exercise) that Λ must be a linear map. We can say more.

Lemma 3.1 *Let D be a 4×4 real matrix whose associated quadratic form $x \mapsto x^t Dx$ vanishes on the light cone C. Then $D = \lambda G$.*

Proof. This is left as an exercise for the reader. □

Lemma 3.2 *Let $\Lambda : \mathbb{R}^{1,3} \to \mathbb{R}^{1,3}$ be a linear map such that $\Lambda(C) \subseteq C$. Then there exists a constant $\lambda \in \mathbb{R}$ such that $\Lambda^t G \Lambda = \lambda G$.*

Proof. Let $D = \Lambda^t G \Lambda$. Given $x \in C$, we have $\Lambda x \in C$, and so

$$x^t Dx = x^t(\Lambda^t G \Lambda)x = (\Lambda x)^t G(\Lambda x) = 0 .$$

This shows that the quadratic form of D vanishes on the light cone, and therefore $D = \lambda G$ for some λ, by Lemma 3.1. □

Thus, our change of frames $\Lambda = \overline{\mathcal{T}} \circ \mathcal{T}^{-1}$ is a special matrix, in that it satisfies $\Lambda^t G \Lambda = \lambda G$ for some λ. Now, it is possible to *dilate* one of the reference frames mappings, say $\mathcal{T}$, by a suitable multiple of the identity so that the resulting Λ will satisfy the equality with $\lambda = 1$. Performing such a dilation has no physical effect: it simply means changing the unit of measurement in that reference frame.

Summarizing, we can think of a change of inertial frames in special relativity as given by a *Lorentz transformation*, a linear map that preserves the Minkowski inner product.

Definition 3.3 *A Lorentz transformation is a linear map $\Lambda : \mathbb{R}^{1,3} \to \mathbb{R}^{1,3}$ that preserves the Minkowski metric, i.e. such that $\Lambda^t G \Lambda = G$, or equivalently*

$$\langle \Lambda x, \Lambda y \rangle_M = \langle x, y \rangle_M ,$$

for all $x, y \in M = \mathbb{R}^{1,3}$.

[1] Actually, *a priori* we know this fact only for lines that respect the causality structure induced by the light cone on Minkowski space, but this need not bother us here.

It is an easy exercise to verify that the set of all Lorentz transformations is a group under composition. This group is called the *Lorentz group*. Note that if $\{e_0, e_1, e_2, e_3\}$ is the canonical basis of M, then a linear map Λ as above is Lorentz if and only if $\langle \Lambda e_i, \Lambda e_j \rangle_M = g_{ij}$, where $g_{ij} = \langle e_i, e_j \rangle_M$ are the components of the Minkowski metric tensor. This is simply another way of saying that $\Lambda^t G \Lambda = G$. In particular, taking determinants on both sides of this last equation, we deduce that $\det \Lambda = \pm 1$.

Example 3.4 *Here are two special types of Lorentz transformations that are worth writing down.*

(i) Time-preserving transformations. These are Lorentz transformations such that $(\Lambda x)^0 = x^0$ for all x. Because Λ preserves the Minkowski inner product, it follows that Λ leaves the orthogonal decomposition $\mathbb{R}^{1,3} = \mathbb{R} \oplus \mathbb{R}^3$ invariant. Hence its restriction to the spatial coordinate 3-space is an orthogonal transformation. Thus, the matrix Λ has the form

$$\Lambda = \begin{pmatrix} 1 & 0 & 0 & 0 \\ 0 & & & \\ 0 & & [a_{ij}] & \\ 0 & & & \end{pmatrix},$$

where $A = [a_{ij}] \in O(3)$.

(ii) Lorentz boosts. A boost or simple Lorentz transformation along the axis of a given spatial coordinate is a Lorentz transformation that leaves the other two coordinates unchanged. For instance, a Lorentz boost along the x^1 axis has the form

$$\Lambda = \begin{pmatrix} a_{00} & a_{01} & 0 & 0 \\ a_{10} & a_{11} & 0 & 0 \\ 0 & 0 & 1 & 0 \\ 0 & 0 & 0 & 1 \end{pmatrix}.$$

From $\Lambda^t G \Lambda = G$, we deduce that

$$(a_{00})^2 - (a_{10})^2 = 1$$
$$(a_{11})^2 - (a_{01})^2 = 1$$
$$a_{00}a_{01} = a_{10}a_{11} .$$

From these relations, it is not difficult to see (exercise) that there exists a real number θ such that $a_{00} = a_{11} = \cosh\theta$ and $a_{01} = a_{10} = \sinh\theta$;

in other words,

$$\Lambda = B_\theta = \begin{pmatrix} \cosh\theta & \sinh\theta & 0 & 0 \\ \sinh\theta & \cosh\theta & 0 & 0 \\ 0 & 0 & 1 & 0 \\ 0 & 0 & 0 & 1 \end{pmatrix} .$$

Now, there exists a unique number v *such that* $\tanh\theta = -v/c$ *(note that* $|v| < c$*, necessarily). Using this number, we can write* $\cosh\theta = \gamma$ *and* $\sinh\theta = -\gamma v/c$*, where*

$$\gamma = \frac{1}{\sqrt{1 - v^2/c^2}} .$$

With this notation, the Lorentz boost transformation $\overline{x} = \Lambda x$ *can be written in coordinates as follows:*

$$\overline{x}^0 = \gamma\left(x^0 - \frac{v}{c}x^1\right) \; ; \; \overline{x}^1 = \gamma\left(x^1 - \frac{v}{c}x^0\right) \; ; \; \overline{x}^2 = x^2 \; ; \; \overline{x}^3 = x^3 .$$

The number v *is the relative velocity between the two frames: the spatial origin* $(\overline{x}^1, \overline{x}^2, \overline{x}^3) = (0, 0, 0)$ *in the frame* $\overline{O}$ *satisfies the equation* $x^1 - vt = 0$ *in the frame* $\overline{O}$*.*

Finally, we briefly discuss the relation of causality in Minkowski space. Let us consider the quadratic form associated with the Minkowski inner product, namely

$$Q(x) = x_0^2 - x_1^2 - x_2^2 - x_3^2 .$$

Definition 3.5 *A vector* x *in Minkowski space* $M = \mathbb{R}^{1,3}$ *is called* timelike, spacelike, *or* lightlike *depending on whether* $Q(x) > 0$, $Q(x) < 0$, *or* $Q(x) = 0$, *respectively.*

The set of all timelike vectors in M minus the origin has two connected components, each of which is a cone in M. We let

$$C^+ = \{x \in M : Q(x) > 0 \text{ and } x_0 > 0\}$$

be the *positive light cone* in Minkowski spacetime. Using this positive cone, we can define a partial order relation $\ll$ on spacetime.

Definition 3.6 *Given two vectors* $x, y \in M$*, we say that* x *causally precedes* y*, and write* $x \ll y$*, if* $y - x \in C^+$*.*

This order relation is called *causality*. Now, it is an easy matter to see that a Lorentz transformation either preserves or reverses causality. The set of all Lorentz transformations that preserve causality is a group, called the *proper*

Lorentz group, usually denoted $L_+^\uparrow$. All other Lorentz transformations are called *improper*. Special relativity calls for a preservation of causality by all laws of physics.

Remark 3.7 It was proved by Zeeman in 1964 that any causality-preserving transformation of Minkowski's spacetime (without any assumption of continuity) must be of the form $T \circ D \circ \Lambda$, where Λ is a Lorentz transformation, D is a dilation, and T is a translation. In particular, every such causality-preserving transformation is *linear*. See [Na1] for a complete proof of Zeeman's theorem. As will be seen in Appendix B, Section B.6, a mathematically rigorous way of incorporating causality into a quantum theory of fields is a central quality of a branch of knowledge known as *algebraic quantum field theory*.

3.2 Relativistic kinematics

Note from Example 3.4 (ii) that the composition of two Lorentz boots B_θ and B_φ along the x^1-axis (or any other fixed axis) is also a boost along the same axis, namely $B_{\theta+\varphi} = B_\theta B_\varphi$. This immediately yields the relativistic law of addition of velocities. Suppose O, O_1, O_2 are three inertial frames, and assume that O_1 has velocity v_1 with respect to O, moving along O's x^1-axis, say, and that O_2 moves along O_1's x^1-axis with velocity v_2. Then the velocity v with which O_2 moves with respect to O is given by

$$v = \frac{v_1 + v_2}{1 + v_1 v_2 / c^2} \,. \tag{3.2}$$

Let us now discuss the relativistic kinematics of a particle in a bit more detail. Such a particle's motion is described by a parameterized curve $x(\lambda) = (x^\mu(\lambda))$ in Minkowski space. This curve is called the particle's *world-line*. The motion should respect the causality relation defined above. In other words, for each λ the tangent 4-vector to the world-line at $x(\lambda)$, namely

$$\frac{dx}{d\lambda} = \left(\frac{dx^\mu}{d\lambda} \right) ,$$

must lie inside the solid light cone C^+, i.e. must be timelike. Thus,

$$0 < \left\langle \frac{dx}{d\lambda}, \frac{dx}{d\lambda} \right\rangle_M = g^{\mu\nu} \frac{dx^\mu}{d\lambda} \frac{dx^\nu}{d\lambda} = \left(\frac{dt}{d\lambda} \right)^2 \left[c^2 - \sum_{j=1}^{3} \left(\frac{dx^j}{dt} \right)^2 \right] .$$

Hence the expression between brackets on the right-hand side must be always positive, and this tells us that the particle's velocity in the given Lorentzian frame satisfies

$$v^2 = \boldsymbol{v} \cdot \boldsymbol{v} = \sum_{j=1}^{3} \left(\frac{dx^j}{dt} \right)^2 < c^2 .$$

This shows that causality entails that a particle's velocity can never exceed the speed of light.

The above parameterization of the particle's world-line uses as parameter any monotone function of $t = x^0/c$. There is, however, a special choice for λ that is very natural: we can use (normalized) Minkowski length. More precisely, let s denote the Minkowski arc length along the particle's world-line and set $\tau = s/c$. Infinitesimally, with respect to any other parameterization, this is tantamount to writing

$$d\tau = \frac{1}{c} \sqrt{g^{\mu\nu} \frac{dx^\mu}{d\lambda} \frac{dx^\nu}{d\lambda}}\, d\lambda = \frac{dt}{\gamma} .$$

The parameter τ defined is this way is called the *Lorentz proper time*. From a physical standpoint, τ measures the time as told by a clock placed at the particle's instantaneous location in space. Using proper time, we define the particle's velocity 4-vector $V = (V^\mu)$ by

$$V^\mu = \frac{dx^\mu}{d\tau} = \gamma \frac{dx^\mu}{dt} .$$

In other words, we have $V = \gamma(c, \boldsymbol{v})$. Note also that (V^μ) has constant Minkowski length, namely

$$\langle V, V \rangle_M = V^\mu V_\mu = c^2 .$$

3.3 Relativistic dynamics

The *momentum* 4-*vector*, or 4-*momentum* of a relativistic particle, denoted $P = (P^\mu)$, is defined as $P = mV$, where m is the particle's *rest mass*. The rest mass is defined as the usual inertial mass, i.e. a measure of the particle's resistance to motion relative to an inertial frame with respect to which the particle is at rest. The relativistic 4-*force* $F = (F^\mu)$ acting on the particle is given by analogy with Newton's second law,

$$F^\mu = \frac{dP^\mu}{d\tau} = m \frac{dV^\mu}{d\tau} .$$

Note that we are guiding ourselves by the first postulate: the laws of mechanics must remain the same across all inertial frames. Because

$$\langle P, P \rangle_M = P^\mu P_\mu = m^2 V^\mu V_\mu = m^2 c^2 ,$$

we deduce that

$$\left\langle P, \frac{dP}{d\tau} \right\rangle_M = P^\mu \frac{dP^\mu}{d\tau} = 0 .$$

Using the fact that $P = mV = (m\gamma c, m\gamma \boldsymbol{v})$, this last equality yields

$$m\gamma c \frac{dP^0}{d\tau} - m\gamma \boldsymbol{F} \cdot \boldsymbol{v} = 0 ,$$

where we have written $F = (F^0, \boldsymbol{F})$, and the dot denotes the standard inner product in $\mathbb{R}^3$. Because $\boldsymbol{v} = d\boldsymbol{x}/d\tau$, we get

$$c \frac{dP^0}{d\tau} = \boldsymbol{F} \cdot \frac{d\boldsymbol{x}}{d\tau} ,$$

or $c dP^0 = \boldsymbol{F} \cdot d\boldsymbol{x}$. But $dW = \boldsymbol{F} \cdot d\boldsymbol{x}$ is the infinitesimal work effected on our particle by the 3-force $\boldsymbol{F}$. The law of conservation of energy tells us that $dW = dE$; i.e. this infinitesimal work is equal to the infinitesimal change of kinetic energy. Thus, $dE = c dP^0$, and we are justified in writing $E = cP^0$ for the particle's relativistic kinetic energy. Therefore

$$E = m\gamma c^2 = \frac{mc^2}{\sqrt{1 - v^2/c^2}} .$$

When $v = 0$ we get $E = mc^2$, which is Einstein's famous formula for the *rest energy* of a particle whose rest mass is m.

3.4 The relativistic Lagrangian

The reader will not have failed to notice that, in relativistic mechanics, invariance under Galilean transformations is replaced with invariance under Lorentz transformations, which as we have seen is the group of isometries of the Minkowski metric

$$ds^2 = c^2 dt - dx^2 - dy^2 - dz^2 = dx_0^2 - dx_1^2 - dx_2^2 - dx_3^2 .$$

Just as in the case of classical mechanics, one can derive the basic dynamical laws from a variational principle. For simplicity, we shall do this for the relativistic *free* particle (upon which no forces act).

The relevant action happens to be the following. Given a path γ in M, we write

$$S(\gamma) = -mc \int_\gamma ds .$$

Here m is the *rest mass* of our particle. This is an intrinsic definition of the action, in the sense that it does not depend on a particular choice of inertial frame. Once we fix such a Lorentzian frame, however, we can write

$$S(\gamma) = -mc^2 \int dt \sqrt{1 - \frac{v^2}{c^2}} \; .$$

This makes it clear that the *relativistic Lagrangian* for a free particle is

$$L = -mc^2 \sqrt{1 - \frac{v^2}{c^2}} \; .$$

Following the Lagrangian formalism to the script, the momentum components are thus

$$p_i = \frac{\partial L}{\partial v_i} = \frac{mv_i}{\sqrt{1 - v^2/c^2}} \; .$$

In particular, we have

$$\boldsymbol{p} = \frac{m\boldsymbol{v}}{\sqrt{1 - v^2/c^2}} \; .$$

The Hamiltonian in this relativistic context, being the Legendre transform of L, becomes

$$H = \sum p_i v_i - L = \frac{mv^2}{\sqrt{1 - v^2/c^2}} + mc^2 \sqrt{1 - v^2/c^2} \; .$$

This expression simplifies to

$$H = \frac{mc^2}{\sqrt{1 - v^2/c^2}} \; .$$

The motion of our free particle is such that H is constant along it. This can be recast in the following way: the kinetic energy of a free particle, namely

$$E = \frac{mc^2}{\sqrt{1 - v^2/c^2}}$$

is conserved along the motion. We therefore recover Einstein's formula from a variational principle. We can rewrite this last formula in yet another way, in which only the rest mass m, the energy E, and the scalar momentum p appear:

$$E = c\sqrt{p^2 + m^2c^2} \; . \tag{3.3}$$

In other words, $E^2 = p^2c^2 + m^2c^4$. From this last equality, we see that the particle's so-called *rest energy* E_0 (corresponding to $v = 0$) satisfies $E_0 = mc^2$, which is Einstein's famous equation. From now on in this book, we shall work with units for which $c = 1$, so that (3.3) becomes

$$E^2 = p^2 + m^2 .$$

3.5 Dirac's equation

In the light of special relativity, and despite its important role in the standard formulation of quantum mechanics, the Schrödinger equation for the wave function ψ, namely

$$i\frac{\partial \psi}{\partial t} = H\psi \ , \ \text{with } H = -\Delta + V \ ,$$

has a major shortcoming: it is not Lorentz invariant. Dirac's attempt to bring together quantum mechanics and relativity resulted in a new, more fundamental equation, known as Dirac's equation, and the notion of a spinor. The equation that Dirac found is Lorentz invariant, but there seemed to be a price to be paid: one had to allow eigenstates having negative energy. Dirac used this seemingly paradoxical fact to predict the existence of antiparticles (more precisely, the positron). This incredible prediction was confirmed experimentally just a few years later, and brought Dirac universal fame.

The first attempt at a relativistic (Lorentz) invariant equation that could fill in for Schrödinger's equation was the Klein–Gordon equation. The introduction of this equation is motivated by the so-called correspondence principle alluded to in Chapter 2. If we consider a system having a single free particle, then its classical non-relativistic energy is simply the kinetic energy given – up to an additive constant – by

$$E = \frac{p^2}{2m} .$$

As we saw in Chapter 2, the correspondence principle says that, upon canonical quantization, $\boldsymbol{p}$ should be replaced by $i\nabla$, so that $p^2 = \boldsymbol{p} \cdot \boldsymbol{p}$ becomes $-\nabla^2 = -\Delta$, and E should be replaced by $i\partial/\partial t$. This yields Schrödinger's equation for a free particle ($V = 0$).

In the case of a free *relativistic* particle, however, the equation relating energy and momentum is, as we have seen, $E^2 = p^2 + m^2$. Here, as before, m is the particle's rest mass. Hence, if we proceed by analogy guided by the correspondence principle, we arrive at the Klein–Gordon equation,

$$\left(\frac{\partial^2}{\partial t^2} - \Delta + m^2 \right) \psi = 0 .$$

This equation has some important features, both good and bad from a physical standpoint:

(i) It is relativistic, i.e. invariant under Lorentz transformations (good).
(ii) It is of second order in time – unlike the Scrödinger (or Heisenberg) equation, which is of first order in time and therefore an evolution equation – and therefore less amenable to dynamical interpretation (bad).
(iii) It allows negative-energy eigenstates; in particular, the spectrum of the Klein–Gordon operator is not bounded from below (bad).
(iv) Unlike the solutions to Schrödinger's equation, which as wave functions give rise to probability densities, the solutions to the Klein–Gordon equation admit no such probabilistic interpretation (bad).

We leave to the reader the task of examining (i) and (ii) above. Let us say a few words about (iii) and (iv).

The Klein–Gordon equation admits plane-wave solutions of the form

$$\psi(t, \boldsymbol{x}) = \exp\{-i(Et - \boldsymbol{p} \cdot \boldsymbol{x})\} ,$$

as long as $\boldsymbol{p} \in \mathbb{R}^3$ is a fixed (momentum) vector and E is a real constant such that $E^2 = \boldsymbol{p}^2 + m^2$. Here it does not matter whether E is positive or negative; as far as the Klein–Gordon equation goes, positive and negative values of E are on an equal footing. Moreover, the possible values are clearly not bounded from below. This is the elaboration of (iii).

As for (iv), it is worth noting that associated to every solution of the Klein–Gordon equation, there is something that plays the role of a density, albeit not a positive one. Indeed, we can define a 4-vector $(j^\mu) = (\rho, \boldsymbol{j})$, where

$$\rho = \frac{i}{2m}\left(\psi^* \frac{\partial \psi}{\partial t} - \frac{\partial \psi^*}{\partial t}\psi\right)$$

and

$$\boldsymbol{j} = \frac{-i}{2m}\left(\psi^* \nabla \psi - \nabla \psi^* \psi\right) .$$

In these expressions, we follow the physicist's notation ψ^* for the complex conjugate of ψ. Now, the 4-vector (j^μ) is divergence-free. To see this, note that

$$\frac{\partial \rho}{\partial t} = \frac{i}{2m}\left(\psi^* \frac{\partial^2 \psi}{\partial t^2} - \frac{\partial^2 \psi^*}{\partial t^2}\psi\right) = \frac{i}{2m}\left(\psi^* \Delta \psi - (\Delta \psi)^* \psi\right) .$$

Likewise, we have

$$\nabla \cdot \boldsymbol{j} = \frac{-i}{2m}\left(\psi^* \Delta \psi - (\Delta \psi)^* \psi\right) .$$

Putting these facts together, we deduce that $\nabla \cdot (j^\mu) = 0$, or more explicitly,

$$\frac{\partial \rho}{\partial t} + \nabla \cdot \boldsymbol{j} = 0 .$$

This is a *continuity* equation akin to Bernoulli's equation in fluid dynamics. However, ρ is not necessarily positive, so it cannot be interpreted as a probability density.

Dirac sought after a *first-order* equation that would not suffer from these difficulties. The equation he found eliminates the bad points (ii) and (iv) above (keeping Lorentz invariance), but it still allows (iii). Let us reproduce Dirac's reasoning (in modern mathematical notation). The idea is to "extract the square-root of the wave operator"; in other words, one seeks a first-order differential operator D such that

$$D^2 = \Box = \frac{\partial^2}{\partial t^2} - \Delta \, .$$

Writing $D = i\gamma^\mu \partial_\mu$, where $\partial_0 = \partial/\partial t$ and $\partial_j = \partial/\partial x_j$ for $j = 1, 2, 3$ and the coefficients γ^μ are to be determined, we see by pure algebraic computation that

$$D^2 = \frac{1}{2}\{\gamma^\mu, \gamma^\nu\}\, \partial_\mu \partial_\nu \, ,$$

where the brackets represent Dirac's anti-commutator

$$\{\gamma^\mu, \gamma^\nu\} = \gamma^\mu \gamma^\nu + \gamma^\nu \gamma^\mu \, .$$

Because we want $D^2 = \Box$, it follows that

$$\{\gamma^\mu, \gamma^\nu\} = 2g^{\mu\nu} \, ,$$

where $g^{\mu\nu}$ are the components of the Minkowski metric tensor, namely

$$(g^{\mu\nu}) = \begin{pmatrix} 1 & 0 & 0 & 0 \\ 0 & -1 & 0 & 0 \\ 0 & 0 & -1 & 0 \\ 0 & 0 & 0 & -1 \end{pmatrix} .$$

It is worth writing the above relations more explicitly, as follows:

$$(\gamma^0)^2 = 1 \, , \quad (\gamma^1)^2 = (\gamma^2)^2 = (\gamma^3)^2 = -1 \, , \qquad (3.4)$$
$$\gamma^\mu \gamma^\nu = -\gamma^\nu \gamma^\nu \; (\mu \neq \nu) \, .$$

These relations define what is known as a *Clifford algebra*. This can be realized as a matrix algebra. We can take each γ^μ to be a 4×4 matrix,

$$\gamma^\mu = \begin{pmatrix} 0 & -\sigma_\mu \\ \sigma_\mu & 0 \end{pmatrix} ,$$

where the σ_μ's are the so-called Pauli 2×2 matrices

$$\sigma_0 = \begin{pmatrix} 1 & 0 \\ 0 & 1 \end{pmatrix}; \quad \sigma_1 = \begin{pmatrix} 0 & 1 \\ 1 & 0 \end{pmatrix}$$

$$\sigma_2 = \begin{pmatrix} 0 & i \\ -i & 0 \end{pmatrix}; \quad \sigma_3 = \begin{pmatrix} -1 & 0 \\ 0 & 1 \end{pmatrix}.$$

The reader can not only check that the Dirac matrices γ^μ constructed in this fashion indeed satisfy the relations (3.4), but also prove as an exercise that four is the smallest possible order for which such a matrix representation of the Clifford algebra is possible.

Thus, the Dirac operator D must act not on ordinary wave functions, but on 4-vector-valued functions called *spinors*. If we look at a spinor solution to Dirac's equation

$$(i\gamma^\mu \partial_\mu - m)\,\psi = 0\,, \tag{3.5}$$

we realize, applying the conjugate operator $i\gamma^\mu \partial_\mu + m$ to both sides of this equation, that each component of ψ satisfies the Klein–Gordon equation.

Now, Dirac's equation (3.5) is free from two of the three bad features of the Klein–Gordon equation, although still preserving its good feature – Lorentz invariance. The Pauli matrices $\sigma_1, \sigma_2, \sigma_3$ used above yield the generators of $\mathfrak{su}(2)$, the Lie algebra of $SU(2)$, and it is a fact (exercise) that the Lie algebra of the Lorentz group is $\mathfrak{su}(2) \oplus \mathfrak{su}(2)$. This points toward the fact that Lorentz transformations will leave Dirac's equation invariant. The computation verifying that this indeed happens is left to the reader as yet another exercise.

What about the bad point (iv) raised against the Klein–Gordon equation? We claim that it goes away in the case of Dirac's equation. This can be seen as follows. First we take the Hermitian conjugate of (3.5), taking into account that $(\gamma^0)^\dagger = \gamma^0$ and $(\gamma^j)^\dagger = -\gamma^j$ $(j = 1, 2, 3)$. We obtain the equation

$$\psi^\dagger\,(-i\gamma^0 \partial_0^\dagger + \gamma^j \partial_j^\dagger - m) = 0\,,$$

where $\partial_\mu^\dagger$ simply means the differential operator ∂_μ acting on the *left*. If we multiply both sides of this last equation on the right by γ^0 and take into account that $\gamma^0\gamma^j = -\gamma^j\gamma^0$, we get

$$\bar{\psi}\,(i\gamma^\mu \partial_\mu^\dagger + m) = 0\,,$$

where now $\bar{\psi} = \psi^\dagger\gamma^0$ is the so-called *adjoint spinor* to ψ. Using this adjoint spinor, we define a current 4-vector (j^μ) whose components are given by

$$j^\mu = \bar{\psi}\gamma^\mu\psi\,.$$

An easy computation now shows that

$$\nabla \cdot (j^\mu) = \partial_\mu j^\mu = (\partial_\mu \bar{\psi})\gamma^\mu \psi + \bar{\psi}\gamma^\mu(\partial_\mu \psi) = 0 .$$

Therefore the current is conserved, just as in the case of the Klein–Gordon equation, but this time we see that

$$j^0 = \bar{\psi}\gamma^0\psi = \psi^\dagger(\gamma^0)^2\psi = \psi^\dagger\psi = |\psi_0|^2 + |\psi_1|^2 + |\psi_2|^2 + |\psi_3|^2 ,$$

which is obviously non-negative and therefore can play the role of a (probabilistic) density.

But Dirac's equation still suffers from the bad point raised in (iii): negative energy states. These could spell trouble because an electron could in principle fall into a state with arbitrarily large negative energy, thereby emitting an arbitrarily large amount of energy in the process. Dirac's ingenious idea, however, was to regard these infinitely many negative energy states as already occupied by a "sea" of electrons; he used the *exclusion principle* discovered by Pauli to justify this picture. Holes in this infinite sea would appear as positive energy, positively charged particles. If an electron fell into such a vacant spot, the hole and the electron could be thought of as annihilating each other, with energy being produced in place of their combined mass according to Einstein's formula $E = mc^2$. Dirac's prediction of such holes, or antiparticles as they are now called (positrons in this case), was confirmed experimentally shortly afterwards by Anderson. In the face of the fact that particles could be created and destroyed, physicists after Dirac were forced to give up the idea of systems having a fixed number of particles. They soon started to study physical processes in terms of *fields*, describing particles as properties of fields. For a conceptual description of these ideas and much more, see the beautiful exposition by R. Penrose in [Pen].

Exercises

3.1 Let B_θ denote the Lorentz boost along the x^1-axis in Minkowski space, with parameter θ.
- (a) Using the law of addition for hyperbolic sine and cosine, show that $B_{\theta+\varphi} = B_\theta B_\varphi$ for all $\theta, \varphi \in \mathbb{R}$.
- (b) Deduce from this fact the law of addition of velocities presented in (3.2).

3.2 Let γ^μ, $\mu = 0, 1, 2, 3$, denote the Dirac matrices, and let us write $\gamma^5 = i\gamma^0\gamma^1\gamma^2\gamma^3$.
- (a) Show that γ^5 is Hermitian and that $(\gamma^5)^2 = 1$.
- (b) Show that $\{\gamma^5, \gamma^\mu\} = 0$, for $\mu = 0, 1, 2, 3$.

4

Fiber bundles, connections, and representations

This chapter represents a predominantly mathematical intermezzo. Here we present in a condensed and systematic way the language of fiber bundles, cocycles, and connections. This language is absolutely crucial in the formulation of modern field theories (classical or quantum), as we discuss at the end of the chapter, and as will be exemplified in Chapter 5.

4.1 Fiber bundles and cocycles

Let us start with the following basic definition. Suppose E, F, M are smooth manifolds.

Definition 4.1 *A* fiber bundle *with fiber F, base M, and total space E is a submersion $\pi : E \to M$ with the following property. There exist an open covering $\{U_i\}$ of the base M and diffeomorphisms $\phi_i : \pi^{-1}(U_i) \to U_i \times F$ such that $\pi_1 \circ \phi_i = \pi$, where π_1 denotes the projection onto the first factor.*

It follows that for each $x \in M$, the fiber above x, $E_x = \pi^{-1}(x)$, is diffeomorphic to F. Moreover, there exist maps $\rho_{ij} : U_i \cap U_j \to \mathrm{Diff}(F)$ such that the map

$$\phi_j \circ \phi_i^{-1} : (U_i \cap U_j) \times F \to (U_i \cap U_j) \times F$$

is given by

$$\phi_j \circ \phi_i^{-1}(x, y) = (x, \rho_{ij}(x)(y)) .$$

The reader can check that $\rho_{ij}(x) \circ \rho_{jk}(x) = \rho_{ik}(x)$ for all $x \in U_i \cap U_j \cap U_k$.

Definition 4.2 *A* section *of the fiber bundle $\xi = (E, \pi, M)$ is a differentiable map $s : M \to E$ such that $\pi \circ s = \mathrm{id}_M$. We shall denote by $\Gamma(E)$ the space of sections of ξ.*

We remark here that if the fiber F comes equipped with a structure, e.g. as a vector space, group, or algebra, and this structure is preserved under each of the maps ρ_{ij}, then each fiber E_x inherits this structure. It follows that the space of sections $\Gamma(E)$ also possesses this structure. For example, if F is a vector space and $\rho_{ij} : U_i \cap U_j \to GL(F)$ then we say that $\xi = (E, \pi, M)$ is a vector bundle, and in this case $\Gamma(E)$ is an infinite-dimensional vector space (in fact, it is a module over the ring of C^∞ functions on M).

A section $s \in \Gamma(E)$ defines a family of functions $s_i : U_i \to F$ such that $s(x) = \phi_i(x, s_i(x))$ for all $x \in U_i$. This family of functions satisfies $s_j = \rho_{ij} \circ s_i$. Conversely, each family $\{s_i\}$ satisfying this condition defines a section of our fiber bundle.

Definition 4.3 *A* morphism *between two fiber bundles $\xi_1 = (E_1, \pi_1, F_1)$ and $\xi_2 = (E_2, \pi_2, F_2)$ is a pair of differentiable maps $f, \tilde{f}$ such that the following diagram commutes:*

$$\begin{array}{ccc} E_1 & \xrightarrow{\tilde{f}} & E_2 \\ \pi_1 \downarrow & & \downarrow \pi_2 \\ M_1 & \xrightarrow[f]{} & M_2 \end{array}$$

It follows from this definition that $\tilde{f}$ maps the fiber $(E_1)_x$ into the fiber $(E_2)_{f(x)}$. If the fiber bundles come with some additional structure, morphisms are assumed to preserve such structure.

Definition 4.4 *Let M be a manifold and let G be a Lie group. A* cocycle *of G in M is an open covering $\{U_i\}$ of M together with a family of differentiable maps $\gamma_{ij} : U_i \cap U_j \to G$ such that $\gamma_{ij} \cdot \gamma_{jk} = \gamma_{ik}$ for all i, j, k (in particular, $\gamma_{ii} = e$, the identity element of G).*

Definition 4.5 *A* representation *of G in* Diff(F) *is a group homomorphism $\rho : G \to \mathrm{Diff}(F)$ such that the map $G \times F \to F$ given by $(g, y) \to \rho(g)y$ is differentiable. This map is called a left action of G in F.*

The following proposition shows that a fiber bundle with fiber F can be constructed from a given family of transition functions and a representation of a given Lie group G into the group of diffeomorphisms of F.

Proposition 4.6 *Let $\rho : G \to \mathrm{Diff}(F)$ be a representation of G, let $\{U_i\}$ be an open covering of M, and let $\gamma_{ij} : U_i \cap U_j \to G$ be a cocycle. Then there exists a fiber bundle $\xi = (E, \pi, M)$ with fiber F whose transition functions $\rho_{ij} : U_i \cap U_j \to \mathrm{Diff}(F)$ are given by $\rho_{ij} = \rho \circ \gamma_{ij}$.*

Let us now discuss two important classes of examples of fiber bundles.

Example 4.7 *The first example is our old friend the tangent bundle. Let M be a smooth real n-dimensional manifold, let $F = \mathbb{R}^n$, and let $\{\varphi_i : U_i \to \mathbb{R}^n\}$ be an atlas in M. We define $\gamma_{ij} : U_i \cap U_j \to GL(\mathbb{R}^n)$ by*

$$\gamma_{ij}(x) = d(\varphi_j \circ \varphi_i^{-1})(\varphi_i(x)) .$$

The chain rule tells us that this family of maps is a cocycle of $G = GL(\mathbb{R}^n)$ in M. We take as our representation $\rho : GL(\mathbb{R}^n) \to \mathrm{Diff}(\mathbb{R}^n)$ the inclusion homomorphism. Then the tangent bundle of M is the fiber bundle TM obtained from Proposition 4.6 for these choices of F, $\{\gamma_{ij}\}$, and ρ. The sections of TM are called vector fields *and the space of sections $\Gamma(TM)$ is denoted by $\mathcal{X}(M)$.*

Example 4.8 *Our second example is the tensor bundle $\mathcal{T}^{r,s}(M)$ of tensors over a differentiable manifold M that are r-covariant and s-contravariant. The starting point of the construction is the tensor product space*

$$T^{r,s} = (\mathbb{R}^n)^{\otimes r} \otimes (\mathbb{R}^{n*})^{\otimes s} ,$$

which is (isomorphic to) the vector space of multilinear transformations $(\mathbb{R}^{n*})^r \times (\mathbb{R}^n)^s \to \mathbb{R}$. For each tensor $\tau \in T^{r,s}$ and each invertible linear map $\varphi \in GL(\mathbb{R}^n)$ the pull-back *of τ by φ is defined as*

$$\varphi^*\tau\,(\lambda_1, \ldots, \lambda_r; v_1, \ldots, v_s) = \tau(\lambda_1 \circ \varphi, \ldots, \lambda_r \circ \lambda_r; \varphi(v_1), \ldots, \varphi(v_s)) .$$

Thus, for each $\varphi \in GL(\mathbb{R}^n)$, we have a well-defined linear map $\varphi^* : T^{r,s} \to T^{r,s}$, satisfying

$$(\varphi \circ \psi)^* = \psi^* \circ \varphi^*$$
$$(\varphi^*)^{-1} = (\varphi^{-1})^* .$$

This shows that if we are given a representation ρ of a Lie group G into $GL(\mathbb{R}^n)$ then the map $\rho_* : G \to GL(T^{r,s})$ given by $\rho_* g = \rho(g^{-1})^*$ is also a representation. In particular, taking $G = GL(\mathbb{R}^n)$ and $\rho = \mathrm{id}$, we get the tensor bundle $\mathcal{T}^{r,s}(M)$ using the representation ρ_* and the cocycle $\{\gamma_{ij}\}$ of Example 4.7 and applying Proposition 4.6. It is clear from these definitions that $\mathcal{T}^{1,0}(M)$ coincides with the tangent bundle TM, and that $\mathcal{T}^{0,1}(M)$ coincides with the cotangent bundle T^*M. The space of sections of T^*M consists of differential 1-forms, and it will be denoted by $\Omega^1(M)$. Note that $\sigma \in \Gamma(\mathcal{T}^{r,s}(M))$ if and only if

$$\sigma : \mathcal{X}(M) \times \cdots \times \mathcal{X}(M) \times \Omega^1(M) \times \cdots \times \Omega^1(M) \to \mathbb{R}$$

is a multilinear map and for each $f \in C^\infty(M, \mathbb{R})$ we have

$$\begin{aligned} &\sigma(X_1, \ldots, fX_i, \ldots, X_r; \alpha_1, \ldots, \alpha_s) \\ &= f\sigma(X_1, \ldots, X_i, \ldots, X_r; \alpha_1, \ldots, \alpha_s) , \end{aligned}$$

as well as

$$\sigma(X_1, \ldots, X_r; \alpha_1, \ldots, f\alpha_j, \ldots, \alpha_s)$$
$$= f\sigma(X_1, \ldots, X_r; \alpha_1, \ldots, \alpha_j \ldots, \alpha_s) .$$

Finally, because the subspace $\wedge^k(\mathbb{R}^n) \subset T^{0,k}(\mathbb{R}^n)$ of alternating multilinear forms is invariant under the representation ρ_* (exercise), we have a sub-bundle $\wedge^k(T^*M)$ of $T^{r,s}(M)$. A section of $\wedge^k(T^*M)$ is called a differential k-form *on M. The vector space of such sections is denoted by $\Omega^k(M)$.*

There are several ways of constructing fiber bundles using other fiber bundles as building blocks. For instance, given two vector bundles ξ, η over the same base manifold M, one can define their *direct sum* $\xi \oplus \eta$ as the vector bundle over M whose fibers above each point $x \in M$ are the direct sums $F_x^\xi \oplus F_x^\eta$ of the corresponding fibers of ξ, η above x.

Another universal construction is the *pull-back*. This once again uses Proposition 4.6. Let $\xi = (E, \pi, M)$ be a fiber bundle with structural group G, fiber F, and representation ρ. Let $\{U_i\}$ be an open covering of M and associated cocycle $\{\gamma_{ij}\}$. Given another smooth manifold N and a smooth map $f : N \to M$, consider the covering of N by the open sets $V_i = f^{-1}U_i$ and let $\tilde{\gamma}_{ij} : V_i \cap V_j \to G$ be given by $\tilde{\gamma}_{ij} = \gamma_{ij} \circ f$. Then $\{\tilde{\gamma}_{ij}\}$ is a cocycle in N. Finally, let $\tilde{\rho} = \rho$; i.e. keep the same representation of G. Let $f^*\xi$ be the fiber bundle over N obtained from these data (cocycle and representation) by applying Proposition 4.6. This new fiber bundle is called the pull-back of ξ by f. Its total space is denoted by f^*E. Note that f induces a morphism from $f^*\xi$ to ξ. The top map $\tilde{f} : f^*E \to E$ of this morphism has a local expression (via transition charts) given by the maps

$$\begin{aligned} V_i \times F &\to U_i \times F \\ (x, y) &\mapsto (f(x), y) . \end{aligned}$$

4.2 Principal bundles

Now we move to a class of fiber bundles that is so important (in mathematics as well as in physics) that it deserves separate treatment. These bundles are called *principal bundles*.

Here and throughout, we let G be a Lie group, with Lie algebra $\mathcal{G} = TG_e$.

Definition 4.9 *A* principal bundle with structure group G *(or* principal G-bundle*) is a fiber bundle (P, π, M) whose fiber is G and whose representation $\rho : G \to$* Diff(G) *is such that $\rho(g) : G \to G$ is given by left multiplication; i.e. $\rho(g)h = g \cdot h$.*

Note that here we appeal once again to Proposition 4.6. Let us recall that construction in the context of principal bundles. Let $\{U_i\}$ be an open covering of the base manifold M, and let $\{\gamma_{ij} : U_i \cap U_j \to G\}$ be a G-valued cocycle in M. Take the disjoint union $\hat{P}$ of the products $U_i \times G$, on which G acts on the right in an obvious way, and factor it by the equivalence relation $\sim$ identifying $(x, g) \in U_i \times G$ with $(x', g') \in U_j \times G$ if $x = x' \in U_i \cap U_j$ and $g' = g \cdot \gamma_{ij}(x)$. The quotient space $P = \hat{P}/\sim$ is a manifold on which G also acts on the right. The projection map π is the quotient map on P of the projections $\pi_i : U_i \times G \to M$ into the base. The quotient action $P \times G \to P$ preserves the fibers $\pi^{-1}(\cdot)$ and is transitive in each such fiber. The action of a given group element g on $x \in P$ will be written $x \cdot g$. This right action defines an anti-homomorphism $R : G \to \mathrm{Diff}(P)$, with $R(g) = R_g : P \to P$ given by $R_g(x) = x \cdot g$, satisfying $R_{g_1 g_2} = R_{g_2} \circ R_{g_1}$ for all $g_1, g_2 \in G$.

Conversely, if P is a manifold and G is a compact Lie group and $P \times G \to P$ is a right smooth action of G in P that is free (no fixed points), then the orbit space M is a smooth manifold and the quotient projection map defines a principal G-bundle over M.

A *local trivialization* of a principal bundle (P, π, M) over a neighborhood $U \subseteq M$ is a smooth map $\psi : U \times G \to \pi^{-1}(U)$ such that $\pi \circ \psi(x, g) = x$ for all $x \in U$, $g \in G$. Given a cocycle as above, we can build a family of local trivializations (or charts) $\psi_i : U_i \times G \to \pi^{-1}(U_i)$ for the bundle in such a way that the chart transitions are

$$\psi_j^{-1} \circ \psi_i : (x, g) \mapsto (x, g \cdot \gamma_{ij}(x)) .$$

A *local section* $\sigma : U \to \pi^{-1}(U)$ is a smooth map such that $\pi \circ \sigma = \mathrm{id}_U$. There is a natural one-to-one correspondence between local trivializations and local sections: given σ, let ψ_σ be given by $\psi_\sigma(x, g) = g \cdot \sigma(x)$. Note also that if we are given an open cover $\{U_i\}$ as before and local sections $\sigma_i : U_i \to \pi^{-1}(U_i)$, then we get an associated cocycle given by $\gamma_{ij}(x) = \sigma_i(x)\sigma_j(x)^{-1}$ for all $x \in U_i \cap U_j$.

Example 4.10 *Our first example of a principal bundle is the frame bundle of an n-dimensional manifold M. The group is $G = GL(\mathbb{R}^n)$. The relevant cocycle is the same as given in Example 4.7. A point in the total space P corresponds to a point $x \in M$ together with a basis (reference frame) $\{v_1(x), \ldots, v_n(x)\}$ of TM_x. The right action $GL(\mathbb{R}^n) \ni \tau \mapsto \mathcal{R}_\tau : P \to P$ is given by*

$$\mathcal{R}_\tau(x, \{v_1(x), \ldots, v_n(x)\}) = (x, \{\tau^{-1}(v_1(x)), \ldots, \tau^{-1}(v_n(x))\}) .$$

Example 4.11 *Our second example is the orthonormal frame bundle of an oriented, n-dimensional Riemannian manifold M. Here, the group is $SO(n)$,* *the group of orthogonal $n \times n$ matrices having determinant equal to one. To construct the relevant cocycle, let $\{\varphi_i : U_i \to \mathbb{R}^n\}$ be an oriented atlas on M, and let $\{v_1^{(i)}(x), \ldots, v_n^{(i)}(x)\}$ be the basis of TM_x whose vectors are given by*

$$d\varphi_i(x)\, v_j^{(i)}(x) = \frac{\partial}{\partial x^j} \, .$$

Applying the Gram–Schmidt orthonormalization procedure to this basis (using the Riemannian inner product on TM_x), we get an orthonormal basis $\{e_1^{(i)}(x), \ldots, e_n^{(i)}(x)\}$ of TM_x. Hence, to define $\gamma_{ij} : U_i \cap U_j \to SO(n)$, simply let $\gamma_{ij}(x)$ be the orthogonal matrix that makes the change of basis $\{e_1^{(i)}(x), \ldots, e_n^{(i)}(x)\} \to \{e_1^{(j)}(x), \ldots, e_n^{(j)}(x)\}$.

Example 4.12 *An interesting and non-trivial example of a principal bundle that is relevant for our purposes is the Hopf bundle. Here, the group is $G = U(1) = \{z \in \mathbb{C} : |z| = 1\}$ (which is topologically the unit circle $\mathbb{S}^1$), the total space P is the unit 3-sphere, which we view as a subset $\mathbb{S}^3 \subset \mathbb{C}^2$, namely*

$$\mathbb{S}^3 = \left\{(z_1, z_2) \in \mathbb{C}^2 :\ |z_1|^2 + |z_2|^2 = 1\right\} \, ,$$

and the base is the 2-sphere, which we view as the complex projective space $\mathbb{C}P^1$. The projection map $\pi : \mathbb{S}^3 \to \mathbb{C}P^1$ is given by $\pi(z_1, z_2) = [z_1 : z_2]$. The abelian group $U(1)$ acts on $\mathbb{S}^3$, in fact in the whole of $\mathbb{C}^2$, in the obvious way: $(z_1, z_2) \cdot e^{i\theta} = (z_1 e^{i\theta}, z_2 e^{i\theta})$. The Lie algebra of $U(1)$ is $u(1) = \{z \in \mathbb{C} : \mathrm{Re}\, z = 0\}$, the imaginary axis.

Example 4.13 *Another example of a principal bundle, of special interest in Yang–Mills theory, is the quaternionic Hopf bundle. Here the relevant Lie group is $SU(2)$, which is topologically the 3-sphere $\mathbb{S}^3$. We view it as a subgroup of the group $\mathbb{H}$ of quaternions, consisting of all quaternions of norm one. Recall that*

$$\mathbb{H} = \{q = x_0 + x_1 \boldsymbol{i} + x_2 \boldsymbol{j} + x_3 \boldsymbol{k} \ :\ x_0, x_1, x_2, x_3 \in \mathbb{R}\} \, ,$$

where $\boldsymbol{i}, \boldsymbol{j}, \boldsymbol{k}$ satisfy the relations

$$\boldsymbol{i}^2 = \boldsymbol{j}^2 = \boldsymbol{k}^2 = -1\, ;\ \boldsymbol{i} \cdot \boldsymbol{j} = -\boldsymbol{j} \cdot \boldsymbol{i} = \boldsymbol{k}\, ;\ \boldsymbol{j} \cdot \boldsymbol{k} = -\boldsymbol{k} \cdot \boldsymbol{j} = \boldsymbol{i}\, ;\ \boldsymbol{k} \cdot \boldsymbol{i} = -\boldsymbol{i} \cdot \boldsymbol{k} = \boldsymbol{j} \, .$$

The norm of a quaternion q is given by $\|q\| = x_0^2 + x_1^2 + x_2^2 + x_3^2$. Our group then is $\mathbb{S}^3 = \{q \in \mathbb{H} : \|q\| = 1\}$, a subgroup of $\mathbb{H}$. The base space of our principal bundle is the 4-sphere $\mathbb{S}^4$, which is naturally identified with the quaternionic projective space $\mathbb{H}P^1$, just as the 2-sphere was identified with complex projective space in the previous example. The total space P in this case is the 7-sphere

$$\mathbb{S}^7 = \left\{(q_1, q_2) \in \mathbb{H} \times \mathbb{H} :\ \|q_1\|^2 + \|q_2\|^2 = 1\right\}$$

and the projection $\pi : \mathbb{S}^7 \to \mathbb{H}P^1 \simeq \mathbb{S}^4$ *is given by*

$$\pi(q_1, q_2) = [q_1 : q_2] = \{(\lambda q_1, \lambda q_2) \; : \; \lambda \in \mathbb{H} \setminus \{0\}\} .$$

Finally, the right action $\mathcal{R} : \mathbb{S}^7 \times \mathbb{S}^3 \to \mathbb{S}^7$ *is given by*

$$\mathcal{R}((q_1, q_2); \lambda) = (q_1 \cdot \lambda, q_2 \cdot \lambda) .$$

The first example we gave (Example 4.8) can be generalized: given any vector bundle (E, π, M), one can consider the principal bundle of *frames* for E as the space P whose fiber over a point $x \in M$ is the collection of all frames for E_x. The structure group is $\text{Aut}(V)$, where V is the vector space on which the vector bundle is modeled. Conversely, given a principal bundle (P, π, M) with structure group G and a representation $\rho : G \to \text{Aut}(V)$, where V is some vector space, we see that there is a right action of G on $P \times V$ defined by $(p, v) \cdot g = (p \cdot g, \rho(g^{-1})v)$. The quotient space $P \times_\rho V$ of $P \times V$ by this action is a vector bundle over M whose fibers are isomorphic to V. This bundle is called the *associated vector bundle*, or vector bundle associated to (P, π, M) via the representation ρ.

4.3 Connections

A *connection* on the principal bundle (P, π, M) is a 1-form ω on the total space P with values in the Lie algebra $\mathcal{G}$ that is equivariant with respect to the right action on P, in the sense that $R_g^*\omega = \text{Ad}_{g^{-1}} \circ \omega$ for all $g \in G$. Here $\text{Ad} : G \to L(\mathcal{G}, \mathcal{G})$ is the *adjoint representation* of G, given at each g by the derivative at the identity of the inner automorphism $h \mapsto g^{-1}hg$. Geometrically, a connection defines a family of *horizontal subspaces* $H_x = \ker \omega_x$ that are invariant under the action, in the sense that $(R_g)_* H_x = H_{x \cdot g}$ for all $x \in P$ and all $g \in G$. We have a decomposition $TP_x = H_x \oplus V_x$ of the tangent space at each point as a direct sum of horizontal spaces H_x with corresponding vertical spaces $V_x = T(\pi^{-1}(\pi(x)))_x$. Accordingly, the tangent bundle of P splits as a direct sum of two sub-bundles, horizontal and vertical: $TP = HP \oplus VP$.

Given a connection on P and a smooth curve $\gamma : [0, 1] \to M$ connecting the points x and y, for each point v in the fiber over x there exists a unique lift of γ to a curve $\tilde{\gamma} : [0, 1] \to P$ that projects onto γ and has a horizontal tangent vector at each point. We say that $\tilde{\gamma}(1)$ is the *parallel transport* of v along γ. The parallel transport along a curve γ defines a diffeomorphism from the fiber over x onto the fiber over y that is equivariant with respect to the right action of G on P.

A connection ω in P gives rise to a *covariant exterior derivative*, which carries k-forms with values in the Lie algebra into $(k + 1)$-forms with values in the Lie algebra. This differential operator $d^\omega : \wedge^k(P) \otimes \mathcal{G} \to \wedge^{k+1}(P) \otimes \mathcal{G}$ is given by

$$d^\omega \alpha(X_1, X_2, \ldots, X_{k+1}) = d\alpha(X_1^h, X_2^h, \ldots, X_{k+1}^h)$$

for all $\alpha \in \wedge^k(P)$ and all vector fields $X_1, \ldots, X_{k+1}$ on P, where X_i^h is the horizontal component of X_i.

In particular, we define the *curvature* of a connection ω to be the 2-form $\Omega \in \wedge^2(P) \otimes \mathcal{G}$ given by

$$\Omega(X, Y) = d^\omega \omega(X, Y) = d\omega(X^h, Y^h) .$$

Theorem 4.14 (Cartan's formula) *We have* $\Omega = d\omega + \omega \wedge \omega$.

When expressed in terms of local sections, the connection ω gives rise to a family of local $\mathcal{G}$-valued 1-forms on the base manifold. These local forms are defined as follows. Suppose $\{U_i\}$ is an open cover of M and let $\sigma_i : U_i \to \pi^{-1}(U)$ be local sections. Take $\mathcal{A}_i = \sigma_i^* \omega$; this is a 1-form on U_i with values in $\mathcal{G}$.

Likewise, the curvature Ω gives rise to a family of local $\mathcal{G}$-valued 2-forms on M, namely $\mathcal{F}_i = \sigma_i^* \Omega$. Now, using the cocycle associated to the family of sections σ_i, we have the following fundamental fact.

Proposition 4.15 *The families of local connection forms* $\{\mathcal{A}_i\}$ *and local curvature forms* $\{\mathcal{F}_i\}$ *transform in the following way:*

$$\mathcal{A}_j = \theta_{ij}^{-1} \mathcal{A}_i \theta_{ij} + \theta_{ij}^{-1}\, d\theta_{ij}$$

$$\mathcal{F}_j = \theta_{ij}^{-1} \mathcal{F}_i \theta_{ij} \, ,$$

where $\theta_{ij}(x) = \mathrm{ad}(\gamma_{ij}(x))$ *and* ad *is the adjoint representation of the group* G.

Note that neither of these families defines a form on the base M.

However, these formulas show that the curvature is a 2-form on M with values in an associated vector bundle: the *adjoint* bundle. This is the vector bundle that corresponds to the adjoint representation $\mathrm{ad}\colon G \to \mathrm{Aut}(G)$. A section of this bundle is locally given by a function $\sigma_i : U_i \to \mathcal{G}$ and $\sigma_j(x) = (\theta_{ij}(x))^{-1} \circ \sigma_i(x) \circ \theta_{ij}(x)$, where $\theta_{ij}(x) = \mathrm{ad}(\gamma_{ij}(x))$. Hence, if X and Y are vector fields on the base manifold M, then the curvature of the connection associates to these vector fields a section of the adjoint bundle. Hence the curvature may be interpreted as a 2-form on M with values in the adjoint bundle. Notice that the connection is *not* a 1-form with values in the adjoint bundle, because of the second term in its transformation law, but the difference of two connections *is* indeed such a form. So the space of connections on a principal bundle is an affine space modeled in the space of sections of the bundle $\mathrm{Ad}(P) \otimes \Omega^1(M)$ and a curvature of a connection is a section of $\mathrm{Ad}(P) \otimes \Omega^2(M)$.

A connection on a principal bundle induces a *covariant derivative* in each associated bundle $\pi : E \to M$. A covariant derivative is a map that associates to each vector field X on M a linear map $\nabla_X : \Gamma(E) \to \Gamma(E)$ of the space of sections of E that satisfy the Leibnitz rule: $\nabla_X(f\sigma) = X(f)\sigma + f\nabla_X(\sigma)$ for any section σ and any smooth function f. Here $X(f)(x) = df(x) \cdot X(x)$. In a local trivialization, the section σ is represented by a function $\sigma_i : U_i \to V$ and $(\nabla_X(\sigma))_i = d\sigma(x) \cdot X(x) + A_i(x)(\sigma_i(x))$, where $A_i : U_i \to \mathrm{End}(V)$. To define a covariant derivative, the A_i's must transform as $A_j(x) = \rho_{ij}(x)^{-1} \circ A_i(x) \circ \rho_{ij}(x)$, where $\rho_{ij} : U_i \cap U_j \to \mathrm{Aut}(V)$ are the transition functions of the bundle. Because $\rho_{ij}(x) = \rho(\gamma_{ij}(x))$, where γ_{ij} are the transition functions of the principal bundle and $\rho : G \to \mathrm{Aut}(V)$ is the representation of the associated bundle, it is clear from this formula that a connection on the principal bundle induces a unique covariant derivative in each associated vector bundle.

Example 4.16 *Let us go back to our first example, the Hopf bundle. One simple way to define a non-trivial connection on this bundle is to consider a 1-form in $\mathbb{C}^2$ that is $U(1)$-invariant (and $u(1)$-valued) and take its pull-back to $\mathbb{S}^3$ by the inclusion map $j : \mathbb{S}^3 \to \mathbb{C}^2$. For example, consider*

$$\overline{\omega} = i \ \mathrm{Im}\ (\overline{z}_1\, dz_1 + \overline{z}_2\, dz_2)$$

and then take $\omega = j^*\overline{\omega} \in \Omega^1(\mathbb{S}^3) \otimes u(1)$. We leave it as an easy but instructive exercise for the reader to compute the curvature of this connection.

4.4 The gauge group

Let P be a principal bundle over a manifold M with group G. We will construct an infinite-dimensional group that will act in the space of connections on P and on the space of sections of associated vector bundles. This group will play a crucial role in Yang–Mills theory. An element γ of the gauge group $\mathbb{G}$ is represented locally by a function $\gamma_i : U_i \to G$ that transforms as $\gamma_j(x) = \gamma_{ij}^{-1}(x)\gamma_i(x)\gamma_{ij}(x)$, where $\gamma_{ij} : U_i \cap U_j \to G$ are the transition functions of P. Such an element is a section of the adjoint bundle corresponding to the adjoint representation $\mathrm{Ad} : G \to \mathrm{Aut}(G)$ that associates to each $g \in G$ the inner automorphism $h \in G \mapsto ghg^{-1}$. This is an infinite-dimensional group whose Lie algebra is the space of sections of the adjoint bundle given by the adjoint representation of G in the Lie algebra $\mathcal{G}$. Given a connection A and an element γ of the gauge group, we have another connection $\gamma \cdot A$, which is locally defined as

$$(\gamma \cdot A)_i = \gamma_i(x) \cdot A_i(x) \cdot (\gamma_i(x))^{-1} + (\gamma_i)^{-1} d\gamma_i \,.$$

Here, we use the notation $v \in \mathcal{G} \mapsto g \cdot v \cdot g^{-1} \in \mathcal{G}$ to indicate the endomorphism of the Lie algebra associated to $g \in G$ under the adjoint representation. The reader can verify that this indeed defines a connection and an action of the gauge group in the space of connections. The curvature of the connection $\gamma \cdot A$ is locally given by $\gamma_i(x) \cdot F_i(x) \cdot \gamma_i(x)^{-1}$. The reader can also verify that given an associate vector bundle E, given by a representation $\rho\colon G \to \mathrm{Aut}(V)$, there is a natural homomorphism of the group of gauge transformations of P into the group of sections of the bundle $\mathrm{Aut}(E)$, the bundle of automorphisms of E. This bundle is a sub-bundle of the bundle $\mathrm{End}(E)$ of endomorphisms of E (associated to the representation $G \to \mathrm{End}(V)$; $\mathrm{End}(V) \ni M \mapsto \rho(g)M\rho(g)^{-1}$). Hence, the group of gauge transformations acts in the space of sections of associated bundles. Many important action functionals in physics are invariant under the action of the group of gauge transformations, as we will see.

Let us write down the local expression of a connection A, curvature F, and the covariant derivative in local trivializations over an open set $U \subset M$. We start by choosing a basis $\{T^a\}$ for the Lie algebra $\mathcal{G}$ with $[T^a, T^b] = f_c^{ab} T^c$, where f_c^{ab} are the *structure constants* of the Lie algebra. We also choose a frame $\{\frac{\partial}{\partial x_\mu}\}$, i.e. vector fields on U that at each point give a basis for the tangent space. Then we may write

$$A\left(\frac{\partial}{\partial x_\mu}\right) = A_\mu^a T^a \quad \text{and} \quad F\left(\frac{\partial}{\partial x_\mu}, \frac{\partial}{\partial x_\nu}\right) = F_{\mu\nu}^a T^a \,,$$

where T_μ^a and $F_{\mu\nu}^a$ are functions on U. Then the curvature, which the physicists call *field strength*, is given by

$$F_{\mu\nu}^a = \frac{\partial}{\partial x_\nu} A_\mu - \frac{\partial}{\partial x_\mu} A_\nu + f_{bc}^a A_\mu^b A_\nu^c \,. \tag{4.1}$$

Finally, if we choose a basis $\{v^a\}$ for the vector space V, which is the fiber of the associate fiber bundle E, then a section $\psi = \psi_a v^a$ and the corresponding covariant derivative is

$$(\nabla_{\partial/\partial x_\mu} \psi)^a = (\nabla_\mu \psi)^a = \frac{\partial}{\partial x_\mu} \psi^a + A_{\mu\, b}^{a} \psi^b \,. \tag{4.2}$$

In the above equation, $A_{\mu\, b}^{a}(x)$ is the matrix in the basis v^a of the linear map that corresponds to $A_\mu(x) \in \mathcal{G}$ under the representation $\mathcal{G} \to \mathrm{End}(V)$.

4.5 The Hodge ⋆ operator

Let V be a real finite-dimensional vector space, and let $g : V \times V \to \mathbb{R}$ be an inner product on V, i.e. a symmetric and non-degenerate bilinear form on V. Then b induces a linear isomorphism between V and its dual V^*, given by $v \mapsto g(v, \cdot)$.

More generally, consider the space $\wedge^k(V^*)$ of exterior k-forms on V, where $0 \le k \le n = \dim V$. An element of $\wedge^k(V^*)$ can be written as

$$\alpha = \sum_{(i_1, i_2, \ldots, i_k)} a_{i_1 i_2 \cdots i_k} \lambda_{i_1} \wedge \lambda_{i_2} \cdots \wedge \lambda_{i_k} ,$$

where $\{\lambda_1, \lambda_2, \ldots, \lambda_n\}$ is a basis of V^*. The orthogonal group $O_g(V)$ of linear transformations that preserve g acts on $\wedge^k(V^*)$ by pull-back. We claim that there exists a unique symmetric non-degenerate bilinear form $\hat{g}$ on $\wedge^k(V^*)$ that is invariant under the action of $O_g(V)$. Indeed, we define

$$g(\alpha_1 \wedge \cdots \wedge \alpha_k; \beta_1 \wedge \cdots \wedge \beta_k) = \sum_{\sigma} (-1)^{\mathrm{sign}(\sigma)} g(\alpha_1, \beta_{\sigma(1)}) \cdots g(\alpha_k, \beta_{\sigma(k)}),$$

the sum being over all permutations σ on k elements, and then extend $\hat{g}$ to all of $\wedge^k(V^*)$ by k-linearity (uniqueness is left as an exercise). The resulting bilinear form,

$$\hat{g} : \wedge^k(V^*) \times \wedge^k(V^*) \to \mathbb{R} ,$$

is easily seen to be symmetric and non-degenerate (i.e., an inner product). Hence, as before, we have a linear isomorphism $\wedge^k(V^*) \simeq (\wedge^k(V^*))^*$, for all $0 \le k \le n$.

Now, the Hodge operator $\star : \wedge^k(V^*) \to \wedge^{n-k}(V^*)$ can be defined as follows. Let $\mu \in \wedge^n(V^*)$ be a normalized volume form, and note that if we take the wedge product of a k-form with an $(n-k)$-form we get a multiple of μ (because $\dim \wedge^n(V^*) = 1$). Hence, given $\beta \in \wedge^k(V^*)$, let $\star\beta$ be the unique $(n-k)$-form such that, for all $\alpha \in \wedge^k(V^*)$,

$$\alpha \wedge \star\beta = \hat{g}(\alpha, \beta)\, \mu .$$

Suppose we are given an (oriented) orthonormal basis $\{e^1, e^2, \ldots, e^n\}$ of V under the inner product, such that $g_{ij} = g(e^i, e^j) = \pm\delta_{ij}$. The determinant of the resulting matrix is equal to 1 or -1, and it is called the signature of the inner product. The signature is independent of which basis we choose. Denoting by $\{e_1, e_2, \ldots, e_n\}$ the dual basis of V^*, we see that the normalized volume form in $\wedge^n(V^*)$ is $\mu = e_1 \wedge e_2 \wedge \cdots e_n$. Given a set of distinct indices $0 \le i_1, \ldots, i_k \le n$, and letting $j_1, j_2, \ldots, j_{n-k}$ be the complementary indices, one can easily check (exercise) that

$$\star(e_{i_1} \wedge e_{i_2} \wedge \cdots e_{i_k}) = \epsilon_{i_1 \cdots i_k; j_1 \cdots j_{n-k}} \epsilon_{i_1} \cdots \epsilon_{i_k} e_{j_1} \wedge e_{j_2} \wedge e_{j_{n-k}} .$$

Here, we used the so-called Levi–Civita symbol $\epsilon_{i_1 \cdots i_k; j_1 \cdots j_{n-k}}$, which is equal to $+1$ if $\{i_1 \cdots i_k, j_1 \cdots j_{n-k}\}$ is an even permutation of $\{1, 2, \ldots, n\}$, -1 if it is odd, and zero otherwise. Moreover, $\epsilon_i = g_{ii} = \pm 1$.

The Hodge operator is an involution up to a sign. Indeed, it is easy to see, for example working with the help of an orthonormal basis as above, that

$$\star\star = (-1)^{k(n-k)} s \text{ id},$$

where s is the signature of the given inner product, and id : $\wedge^k(V^*) \to \wedge^k(V^*)$ is the identity.

This definition of *Hodge operator* carries over to differential forms on any (pseudo) Riemannian manifold M in a straightforward manner. Indeed, all we have to do is to apply the Hodge $\star$ operator just defined in each tangent space. More precisely, given a differential k-form α on M, define $\star\alpha \in \Omega^k(M)$ by $(\star\alpha)_x = \star\alpha_x \in \wedge^{n-k}(T^*M_x)$ for each $x \in M$. The resulting operator allows us to define an inner product in the space of differential k-forms on M: if we are given two differential k-forms α, β, then $\alpha \wedge \star\beta \in \Omega^n(M)$ is a volume form, and we can integrate it over M to get a number (α, β). In other words,

$$(\alpha, \beta) = \int_M \alpha \wedge \star\beta .$$

The Hodge operator can be generalized still further, to differentiable forms in M with values in vector bundles over M. It is a basic ingredient in the formulation of the Yang–Mills action. The space of fields is the space of connections on a principal bundle P over a manifold M endowed with a pseudo-Riemannian (Minkowski) metric g. The Hodge operator gives an isomorphism between sections of $\text{ad}(P) \otimes \Omega^2(M)$ and sections of $\text{ad}(P) \otimes \Omega^{d-2}(M)$, where d is the dimension of M. Another ingredient is the bundle map Tr induced by the trace map Tr: $\text{End}((G)) \to \mathbb{R}$ that gives a linear map from the space of sections of the adjoint bundle to the space of functions on M. So if F is the curvature of a connection A then $\text{Tr}(F \wedge *F)$ is a d-form on M. Thus we may write the pure Yang–Mills action as

$$S(A) = \int \text{Tr}(F \wedge *F) .$$

If the group is non-commutative, the curvature is quadratic in the gauge field (the connection) and so the Yang–Mills Lagrangian is not quadratic but has also terms of degrees 3 and 4 in the gauge field. In particular, the Euler–Lagrange equations are non-linear!

4.6 Clifford algebras and spinor bundles

4.6.1 Clifford algebras

Let V be a finite-dimensional vector space over $k = \mathbb{R}$ or $\mathbb{C}$, and let $B : V \times V \to k$ be a bilinear form. For simplicity of exposition we will assume that B is positive definite (in the real case) or Hermitian (in the complex case). Everything we will do in this section can be adapted, *mutatis mutandis*, to the general case of a symmetric non-degenerate bilinear form, including the Minkowski case that will be considered later. Consider the full-tensor algebra of V, namely

$$\mathcal{T}(V) = k \oplus \bigoplus_{n=1}^{\infty} V^{\otimes n} .$$

Let $\mathcal{I}(V, B)$ be the ideal in $\mathcal{T}(V)$ generated by the elements of the form

$$v \otimes w + w \otimes v - 2B(v, w) ,$$

with $v, w \in V$.

Definition 4.17 *The Clifford algebra of V (with respect to B) is the quotient algebra* $\mathrm{Cl}(V, B) = \mathcal{T}(V)/\mathcal{I}(V, B)$.

We shall write simply ab for the product of two elements $a, b \in \mathrm{Cl}(V, B)$. If we are given an orthonormal basis $\{e_1, e_2, \ldots, e_n\}$ of V (with respect to B), then we have the relations $e_i e_j = -e_j e_i$ and $e_i^2 = 1$. Therefore each element θ of $\mathrm{Cl}(V, B)$ can be written uniquely as

$$\theta = \alpha 1 + \sum_i \alpha_i e_i + \sum_{i<j} \alpha_{ij} e_i e_j + \cdots + \sum_{i_1<i_2<\cdots<i_s} \alpha_{i_1 i_2 \cdots i_s} e_{i_1} e_{i_2} \cdots e_{i_s}$$
$$+ \cdots + \alpha_{12\cdots n} e_1 e_2 \cdots e_n .$$

From now on, we shall write simply $\mathrm{Cl}(V)$ instead of $\mathrm{Cl}(V, B)$. The above expression clearly shows that, as a vector space, $\mathrm{Cl}(V)$ has dimension 2^n.

Note that if $v \in V$ has unit norm ($B(v, v) = 1$), then $v^2 = 1$. Hence every unit vector v is invertible, and equal to its inverse: $v^{-1} = v$. We define $\mathrm{Pin}(V)$ to be the group generated by all the unit vectors of V.

Lemma 4.18 *If $v \in V$ is a unit vector and $w \in V$ is arbitrary, then $-vwv^{-1}$ belongs to V and is the reflection of w across the orthogonal complement of v in V.*

Proof. The proof is left as an exercise. □

This lemma shows that we have a well-defined action $\mathrm{Pin}(V) \times V \to V$ given by $w \mapsto \theta w \theta^{-1}$. Moreover, for each $\theta \in \mathrm{Pin}(V)$, the map $w \mapsto \theta w \theta^{-1}$ is a composition of orthogonal reflections and therefore it is an orthogonal map. This yields a group homomorphism

$$\mathrm{Pin}(V) \xrightarrow{\pi} O(V) \supset SO(V).$$

Because every orthogonal transformation of V is a composition of a finite number of reflections (a classical result), this group homomorphism is surjective.

Definition 4.19 *The subgroup* $\mathrm{Spin}(V) \subset \mathrm{Pin}(V)$*, called the* spin group *of* V*, is the pre-image of* $SO(V)$ *under the above group homomorphism.*

It follows that every element of $\mathrm{Spin}(V)$ can be written as a product of an even number of unit vectors in V.

Lemma 4.20 *The restricted homomorphism*

$$\pi\Big|_{\mathrm{Spin}(V)} : \mathrm{Spin}(V) \to SO(V)$$

has kernel equal to $\{\pm 1\}$*. Moreover,* $\mathrm{Spin}(V)$ *is a simply connected group. In particular,* $\mathrm{Spin}(V)$ *is the universal cover of* $SO(V)$*.*

Proof. Again, an exercise for the reader. □

When V is the Euclidean three-dimensional space $\mathbb{R}^3$, we deduce from this lemma that $\mathrm{Spin}\left(\mathbb{R}^3\right) = SU(2)$.

4.6.2 Representations of Spin(V) and the Dirac operator

Now, let S be a vector space that is also a left module over the Clifford algebra $\mathrm{Cl}(V)$. This is the same as saying that we have an algebra homomorphism $\tilde{\rho} : \mathrm{Cl}(V) \to \mathrm{End}(S)$. This homomorphism restricts to a representation

$$\rho : \mathrm{Spin}(V) \to \mathrm{Aut}(S).$$

This representation allows us to define the so-called Dirac operator on the smooth S-valued vector functions on V, as follows.

Definition 4.21 *The* Dirac operator *is the first-order operator* $D : C^\infty(V, S) \to C^\infty(V, S)$ *given by*

$$D\varphi = \sum_{i-1}^{n} e_i \partial_{e_i} \varphi,$$

where $\{e_i\}_{1 \leq i \leq n}$ *is an orthonormal basis for* V *and* ∂_{e_i} *is the directional derivative in the direction of* e_i.

We leave it as an exercise for the reader to show that the definition of $D\varphi$ is independent of which orthonormal basis one chooses, and that

$$D^2\varphi = \sum_{i=1}^{n} \partial_{e_i}^2 \varphi \; ;$$

in other words, $D^2 = \Delta$, the Euclidean Laplacian operator. This generalizes the discussion of Dirac's equation at the end of Chapter 3.

4.6.3 Spin bundles over a spacetime manifold

Next, we will show how to construct Dirac operators acting on spaces of sections of vector bundles over a manifold M that admits a special structure known as a *spin structure*. This structure is given by a spin bundle over M, i.e. a principal bundle over M with structure group Spin (V) for a given V.

This somewhat vague description will now be made precise. In the discussion to follow, the underlying vector space will be $V = \mathbb{R}^{1,n-1}$, the n-dimensional Minkowski space, with inner product given by the bilinear functional

$$B(v, w) = v^1 w^1 - \sum_{i=2}^{n} v_i w_i \; .$$

Let M be an n-dimensional pseudo-Riemannian manifold modeled on V, i.e. endowed with a Minkowski metric. Let us consider the orthonormal frame bundle over M

This is a principal bundle with structure group $\mathcal{L}$, the Lorentz group, which is the group of all linear transformation of $\mathbb{R}^{1,n-1}$ that leave the Minkowski inner product invariant. As in the previous discussion for the Euclidean case, there is a universal covering homomorphism (two-to-one) $\tilde{\mathcal{L}} \to \mathcal{L}$, where $\tilde{\mathcal{L}}$ is the spin group contained in the Clifford algebra Cl $\left(\mathbb{R}^{1,n-1}\right)$.

Roughly speaking, we say that the manifold M has a *spin structure* if the above principal bundle has a double cover. More precisely, a spin structure on M consists of a double covering $\tilde{P} \to P$ with the property that

$$\begin{array}{ccc} \tilde{\mathcal{L}} & \longrightarrow & \tilde{P} \\ & & \downarrow \\ & & M \end{array}$$

is a principal bundle over M, so that

$$\begin{array}{ccc} \tilde{\mathcal{L}} & \longrightarrow & \tilde{P} \\ {\scriptstyle (2-1)}\downarrow & & \downarrow{\scriptstyle (2-1)} \\ \mathcal{L} & \longrightarrow & P \\ & & \downarrow \\ & & M \end{array}$$

Now let S be a vector space, and let $\tilde{\rho} : \mathrm{Cl}\left(\mathbb{R}^{1,n-1}\right) \to \mathrm{End}(S)$ be a representation of algebras. As before, this restricts to a representation

$$\rho : \mathrm{Spin}\left(\mathbb{R}^{1,n-1}\right) \to \mathrm{Aut}(S) .$$

This representation yields an associated vector bundle

$$\begin{array}{ccc} S & \longrightarrow & E \\ & & \downarrow \\ & & M \end{array}$$

Given a connection A on the principal bundle $(P, M, \mathrm{Spin}\left(\mathbb{R}^{1,n-1}\right))$, let ∇ be the associated covariant derivative on the vector bundle (E, M, S). Recall what this means: if $X \in \mathcal{X}(M)$ is a vector field on M, then ∇_X is a first-order differential operator on sections of E satisfying the Leibnitz rule. The Dirac operator D_A acting on sections of E associated with the connection A will be defined using ∇.

Take a local trivialization over an open set $U \subseteq M$, and a (smoothly varying) orthonormal frame $\{e_1(x), e_2(x), \ldots, e_n(x)\}$ on TM_x for all $x \in U$. These ingredients allow us to identify TM_x with $\mathbb{R}^{1,n-1}$ and E_x with S. Through these identifications and the representation $\tilde{\rho}$, we get a representation $\tilde{\rho}_x : \mathrm{Cl}\,(TM_x) \to \mathrm{End}(E_x)$. Therefore, for each section $\psi : M \to E$ we define

$$D_A\psi(x) = \sum_{i=1}^{n} e_i(x) \left(\nabla_{e_i}\psi\right)(x) .$$

It can be shown (exercise) that this definition is independent of the choices of local trivialization and orthonormal frames used in the construction.

4.6.4 Abstract actions

Let us show how the above machinery can be used to write down natural actions on spaces of *fields*. There are fields of two types: *force* fields, which are connections on principal bundles over spacetime M, and *matter* fields, which are sections of suitable associated vector bundles over M. For physically relevant examples, see Chapter 5.

To construct action functionals, we need the following ingredients:

(1) a manifold M endowed with a Minkowski metric and a principal bundle over M;

(2) an associated vector bundle (E, π, M), together with its dual bundle (E^*, π^*, M), and a bundle isomorphism

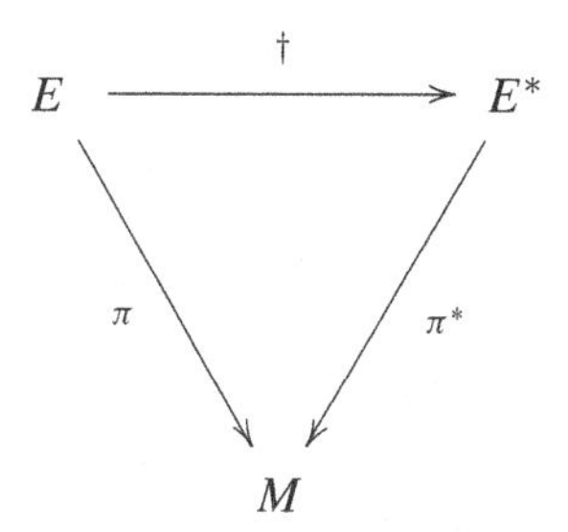

(the image of a section ψ of E being denoted by $\psi^\dagger$), as well as a pairing map

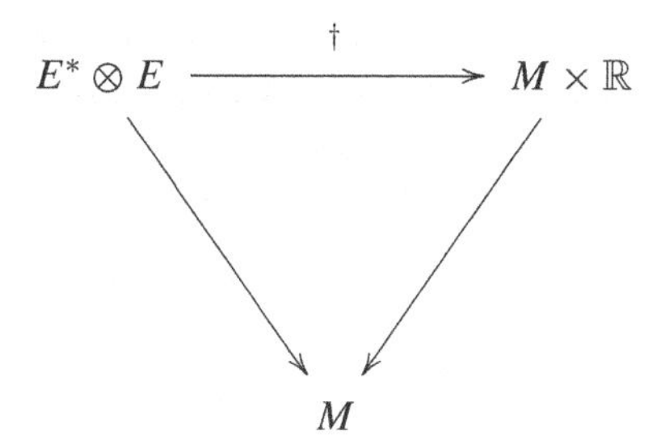

(3) as fields, we take a connection A, a section $\psi : M \to E$, and a scalar function $\phi : M \to \mathbb{R}$;

(4) the Hodge $\star$-operator, defined using the Minkowski structure on M.

With these ingredients at hand, we can write down an action functional

$$S(A, \phi, \psi) = \int_M \mathrm{Tr}\,(\mathcal{F}_A \wedge \star\mathcal{F}_A) + \int_M \Big(\psi^\dagger D_A \psi + \frac{1}{2}(\nabla\phi)^2 + \mu\psi^\dagger\psi + \frac{1}{2}m^2\phi^2 + \lambda\phi^4 + \psi^\dagger\psi\phi\Big)\, dV\,,$$

where dV is the Minkowski volume form, ∇ denotes the Minkowski gradient, $\mathcal{F}_A = dA + A \wedge A$ is the curvature of the connection A, and D_A is the Dirac operator associated to A.

4.7 Representations

The theory of Lie group representations is a very broad subject, a significant portion of which was developed because of quantum physics. Here we content ourselves with presenting just a few basic results leading to Wigner's definition of a (quantum) particle.

4.7.1 The Lorentz and Poincaré groups

We consider four-dimensional spacetime $\mathbb{R}^{1,3} = \mathbb{R} \times \mathbb{R}^3$ endowed with the Lorentz (or Minkowski) metric coming from the inner product

$$\langle x, y \rangle = x_0 y_0 - \sum_{j=1}^{3} x_j y_j \, . \tag{4.3}$$

The (full) Lorentz group is the group of linear isometries of this metric. We are interested in the connected component of the identity in this group. This subgroup is the *restricted* Lorentz group $L_+^\uparrow = SO(1,3)$ of linear maps of $\mathbb{R}^{1,3}$ that preserve the bilinear form (4.3) and also leave invariant the positive cone $\{x \in \mathbb{R}^{1,3} : x_0 > 0 \text{ and } \langle x, y \rangle > 0\}$.

It turns out that the Lorentz group $L_+^\uparrow$ is doubly covered by $SL(2, \mathbb{C})$. This can be seen as follows. There is a natural identification between spacetime $\mathbb{R}^{1,3}$ and the space of 2×2 complex Hermitian matrices, given by

$$\mathbb{R}^{1,3} \ni x \mapsto \begin{pmatrix} x_0 - x_3 & x_1 + ix_2 \\ x_1 - ix_2 & x_0 + x_3 \end{pmatrix} \in i\mathfrak{su}(2) \, .$$

This identification can also be written as $x \mapsto \sum_{j=0}^{3} x_j \sigma_j$, where σ_j, $j = 0, \ldots, 3$ are the Pauli matrices

$$\sigma_0 = \begin{pmatrix} 1 & 0 \\ 0 & 1 \end{pmatrix} \; ; \; \sigma_1 = \begin{pmatrix} 0 & 1 \\ 1 & 0 \end{pmatrix} \; ; \; \sigma_2 = \begin{pmatrix} 0 & i \\ -i & 0 \end{pmatrix} \; ; \; \sigma_3 = \begin{pmatrix} -1 & 0 \\ 0 & 1 \end{pmatrix} .$$

Using this identification, we define an action of $SL(2, \mathbb{C})$ on Lorentzian spacetime through the map $SL(2, \mathbb{C}) \times i\mathfrak{su}(2) \to i\mathfrak{su}(2)$ given by $(\Lambda, X) \mapsto \Lambda X \Lambda^*$. Note that, when $X = \sum_{j=0}^{3} x_j \sigma_j$, we have

$$\det(X) = x_0^2 - x_1^2 - x_2^2 - x_3^2 = \|x\|^2 \, .$$

Because we also have

$$\det(\Lambda X \Lambda^*) = \det(X) \, ,$$

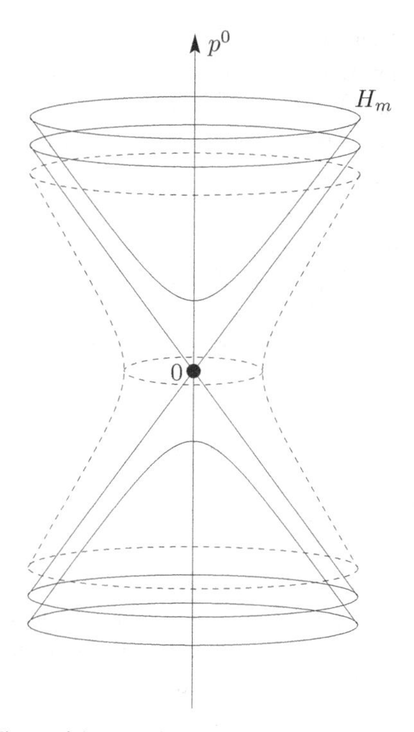

Figure 4.1 Hyperboloids in momentum space.

we see that the action just defined preserves the Lorentz metric. In other words, $SL(2,\mathbb{C})$ acts isometrically on spacetime, and we have a well-defined map $SL(2,\mathbb{C}) \xrightarrow{\Phi} SO(1,3)$, a group homomorphism in fact. Now, it is an easy exercise to see that if $\Lambda \in SL(2,\mathbb{C})$ is such that $\Lambda X \Lambda^* = X$ for all Hermitian matrices X, then $\Lambda = \pm I$. From this it follows that Φ is a double covering map, and we have $L_+^\uparrow = SO(1,3) \cong SL(2,\mathbb{C})/\{\pm I\} = PSL(2,\mathbb{C})$.

The orbit structure of the action of $L_+^\uparrow$ on momentum space is fairly simple. Each hyperboloid (see Figure 4.1) of the form

$$H_m = \left\{ p \in \mathbb{R}^{1,3} : (p^0)^2 - (p^1)^2 - (p^2)^2 - (p^3)^2 = m^2 \right\}$$

is invariant under the group action, including the degenerate case $m = 0$ (the light cone). When $m^2 < 0$, we have a one-sheeted hyperboloid. When $m^2 > 0$, we have a two-sheeted hyperboloid, and each sheet is invariant. In either case, the group action is transitive in each sheet.

4.7.2 Induced representations of unitary type

Let G be a Lie group, and let H be a closed subgroup of G. Suppose we are given a representation of H into a Hilbert space. Does it somehow yield a representation of the larger group? We shall see how to construct such an induced representation provided we are given the following ingredients.

(i) A principal H-bundle (P, π, M), where P and M are smooth manifolds.

(ii) A representation $\rho : H \to \mathrm{Aut}(V)$, where V is a complex Hilbert space, through *unitary* automorphisms.

(iii) A left action $G \times P \xrightarrow{\alpha} P$ that is smooth and sends fibers $\pi^{-1}(x) \subseteq P$ onto fibers. In particular, there is a quotient action $G \times M \xrightarrow{\beta} M$ on the base. We have a commutative diagram

$$\begin{array}{ccc} G \times P & \xrightarrow{\alpha} & P \\ {\scriptstyle i\times\pi}\downarrow & & \downarrow{\scriptstyle \pi} \\ G \times M & \xrightarrow[\beta]{} & M \end{array}$$

We assume also that the left G-action commutes with the right H-action on P.

(iv) A Borel measure μ on M that is G-invariant ($g_*\mu = \mu$ for all $g \in G$).

Using these ingredients, we will show how to build an induced representation $\hat{\rho} : G \to U(\mathbb{H})$, where $\mathbb{H}$ is a Hilbert space arising as a subspace of the space of sections of the vector bundle associated to the representation ρ. The most important special case of this construction happens when $P = G$ and M is the homogeneous space G/H. In this case the third ingredient above comes for free (take α to be the standard left action by translations, and let β be the obvious quotient action).

Example 4.22 *The main physical example is the case where* $P = G = SL(2, \mathbb{C})$ *(the double cover of the Lorentz group), the little group is* $H = SU(2)$ *(the double cover of* $SO(3)$*, the isotropy group of the point* $(m, 0, 0, 0)$ *in Lorentz spacetime), and* $M = SL(2, \mathbb{C})/SU(2)$ *is the hyperboloid* $\{x \in \mathbb{R}^{1,3} : x_0^2 - x_1^2 - x_2^2 - x_3^2 = m^2\}$ *(the orbit of* $(m, 0, 0, 0)$ *under the Lorentz group). The manifold* M *with the metric induced from the Minkowski metric on* $\mathbb{R}^{1,3}$ *is isometric to hyperbolic* 3*-space with the hyperbolic metric. Hence the measure* μ *is the hyperbolic volume form transported by this isometry (so* μ *is obviously invariant under* $SL(2, \mathbb{C})$*). The representations* $\rho : SU(2) \to U(V)$ *are all finite-dimensional (*$V = \mathbb{C}^N$ *for some* N*) and will be described later.*

Let us go back to the general situation. In order to construct the induced representation $\hat{\rho}$ of the larger group G, let (E, π_E, M) be the associated vector bundle corresponding to the representation $\rho : H \to \mathrm{Aut}(V)$ of the smaller group H. Each fiber $V_x = \pi_E^{-1}(x) \subseteq E$ is isomorphic to V. The first step is the following.

Lemma 4.23 *The space of sections $\Gamma(E)$ is isomorphic to the space W of H-equivariant maps $f : P \to V$, i.e. maps satisfying $f(p \cdot h) = \rho(h^{-1})f(p)$ (for all $p \in P$ and $h \in H$).*

Proof. Recall that $E = P \times V/H$, where the H-action on $P \times V$ is given by $(p, v) \cdot h = (ph^{-1}, \rho(h)v)$. Let $\psi \in \Gamma(E)$ be a section of E. For each $p \in P$ there exists $v \in V$ such that $\psi(\pi(p)) = [(p, v)]$ (here $[\cdot]$ denotes an orbit of the H-action). We claim that v is uniquely determined by p. Indeed, if $w \in V$ is such that $[(p, w)] = [(p, v)]$, then there exists $h \in H$ such that $p \cdot h^{-1} = p$ and $\rho(h)v = w$. Because H acts freely on each fiber of P, we must have $h = e$, and therefore $w = v$. Hence we have $v = f_\psi(p)$ for a well-defined function $f_\psi : P \to V$. This function clearly satisfies $f_\psi(ph^{-1}) = \rho(h)v = \rho(h)f_\psi(p)$, so it is H-equivariant. This defines a linear map $L : \Gamma(E) \to W$, given by $L(\psi) = f_\psi$. The construction of f_ψ from ψ can be reversed to show that L is surjective. Moreover, if $f_\psi = 0$ then ψ must be the zero section, so L is injective as well. □

The second step is to define a left action $G \times W \to W$ in the obvious way: if $g \in G$ and $f \in W$, let $g \cdot f$ be given by $g \cdot f(p) = f(g^{-1} \cdot p)$. This is well defined, because the left G-action and the right H-action on P commute, and therefore

$$\begin{aligned} g \cdot f(p \cdot h) &= f(g^{-1} \cdot (p \cdot h)) \\ &= f((g^{-1} \cdot p) \cdot h) \\ &= \rho(h^{-1}) f(g^{-1} \cdot p) \\ &= \rho(h^{-1})(g \cdot f)(p)\,, \end{aligned}$$

showing that, indeed, $g \cdot f \in W$. Now, this action of G on W can be transported, via the isomorphism L of Lemma 4.23, to an action $\gamma : G \times \Gamma(E) \to \Gamma(E)$ according to the following diagram:

$$\begin{array}{ccc} G \times W & \longrightarrow & W \\ {\scriptstyle i_G \times L^{-1}} \big\downarrow & & \big\downarrow {\scriptstyle L^{-1}} \\ G \times \Gamma(E) & \underset{\gamma}{\longrightarrow} & \gamma(E) \end{array}$$

Thus, $g \cdot \psi = \gamma(g, \psi) = L^{-1}(g \cdot f_\psi)$, for all $g \in G$ and all $\psi \in \Gamma(E)$.

The third step is to define the Hilbert space of L^2-sections of the vector bundle (E, π_E, M), on which G will act. To define an inner product on sections, first we take their scalar product on each fiber, and then we integrate over the base manifold M using our fourth ingredient, the Borel measure μ. More precisely, if $\phi, \psi \in \Gamma(E)$ and $x \in M$, let

$$\langle \phi(x), \psi(x) \rangle_{V_x} = \langle f_\phi(p), f_\psi(p) \rangle ,$$

where $p \in P$ is any point such that $\pi(p) = x$. This is well defined, for if $q \in P$ is any other point with $\pi(q) = x$, then $q = p \cdot h$ for some $h \in H$, and therefore, by H-equivariance,

$$\begin{aligned} \langle f_\phi(q), f_\psi(q) \rangle &= \langle \rho(h^{-1}) f_\phi(p), \rho(h^{-1}) f_\psi(p) \rangle \\ &= \langle f_\phi(p), f_\psi(p) \rangle , \end{aligned}$$

where we have used that $\rho(h^{-1}) : V \to V$ is unitary. The inner product on sections is given by

$$\langle \phi, \psi \rangle = \int_M \langle \phi(x), \psi(x) \rangle_{V_x} \, d\mu(x) .$$

Now let $L^2(\Gamma(E), \mu) \subseteq \Gamma(E)$ be the subspace consisting of those $\psi \in \Gamma(E)$ such that $\langle \psi, \psi \rangle < \infty$. This is easily seen to be a Hilbert space. We shall denote this Hilbert space by $\mathbb{H}$.

Lemma 4.24 *The restricted action $\gamma : G \times \mathbb{H} \to \mathbb{H}$ is well defined and unitary.*

Proof. The definitions given so far assure us that $(g \cdot \psi)(x) = [(p, g \cdot f_\psi(p))]$, for all $\psi \in \Gamma(E)$ and all $g \in G$, where $p \in P$ is any point with $\pi(p) = x$. Hence, given $\phi, \psi \in \Gamma(E)$, we have

$$\begin{aligned} \langle g \cdot \phi(x), g \cdot \psi(x) \rangle_{V_x} &= \langle g \cdot f_\phi(p), g \cdot f_\psi(p) \rangle \\ &= \langle f_\phi(g^{-1} \cdot p), f_\psi(g^{-1} \cdot p) \rangle \\ &= \langle \phi(g^{-1} \cdot x), \psi(g^{-1} \cdot x) \rangle_{V_{g^{-1} \cdot x}} . \end{aligned}$$

Here, we have used that $\pi(g^{-1} \cdot p) = g^{-1} \cdot \pi(p) = g^{-1} \cdot x$. This shows that

$$\langle g \cdot \phi, g \cdot \psi \rangle = \int_M \langle \phi(g^{-1} \cdot x), \psi(g^{-1} \cdot x) \rangle_{V_{g^{-1} \cdot x}} \, d\mu(x) . \tag{4.4}$$

Because $g_*^{-1}\mu = \mu$ (for the measure μ is G-invariant), the change of variables $y = g^{-1} \cdot x$ in (4.4) at last yields $\langle g \cdot \phi, g \cdot \psi \rangle = \langle \phi, \psi \rangle$. In other words, each $g \in G$ acts on $\mathbb{H}$ as a unitary isometry. □

Finally, using the action $\gamma : G \times \mathbb{H} \to \mathbb{H}$ just constructed, we define a unitary representation

$$\hat{\rho} : G \to U(\mathbb{H})$$

quite naturally by $\hat{\rho}(g) = \gamma(g, \cdot) : \mathbb{H} \to \mathbb{H}$. This is the induced unitary representation of G (induced by $\rho : H \to \mathrm{Aut}(V)$) that we were looking for.

4.7.3 Wigner's classification of particles

In a seminal work, Wigner proposed the following mathematical notion of an elementary particle. For a discussion of the physical motivation behind this definition and of Wigner's work, see [St, pp. 148–50].

Definition 4.25 *A quantum mechanical particle is a projective, irreducible unitary representation of the Poincaré group.*

One can be a bit more restrictive here: the above representations may be required to satisfy additional conditions, whose nature and relevance are dictated by physical context. Moreover, instead of projective representations of the Poincaré group $\mathcal{P}$, one can consider representations of its universal (double) covering group

$$\tilde{\mathcal{P}} = SL(2, \mathbb{C}) \rtimes \mathbb{R}^{1,3} , \tag{4.5}$$

that is to say, the semi-direct product of the group $SL(2, \mathbb{C})$ with the translation group in Minkowski space. This will be the point of view adopted here.

Our goal in this section is to present Wigner's classification of particles in a nutshell. Wigner's classification theorem provides the (correct) mathematical framework for the study of elementary particles, and has stimulated a great deal of research in the theory of group representations. The classification amounts to finding all unitary irreducible representations of the group $\tilde{\mathcal{P}}$. The general problem of finding irreducible, unitary representations of semi-direct products such as (4.5) was thoroughly investigated by Mackey (but also by Wigner in the specific case at hand). Such representations are typically infinite-dimensional. In order to simplify our discussion here, we shall ignore the translation group factor $\mathbb{R}^{1,3}$ in the semi-direct product (4.5). This amounts to studying the unitary irreducible representations of the Lorentz group, or of its double cover $SL(2, \mathbb{C})$. Using the results we proved in Section 4.7.2, it suffices to classify the irreducible unitary representations of the isotropy groups of points in Minkowski space with respect to the underlying action of the Lorentz group in that space.

The isotropy groups are isomorphic either to the special orthogonal group $SO(3)$ or to the group of Euclidean motions of the plane, $E(2)$. We shall deal explicitly here with the case of $SO(3)$, which is compact. The Euclidean group $E(2)$ is not compact, but a further reduction can be used to study its irreducible representations (we note that $E(2)$ contains $SO(2)$ as a maximal compact subgroup).

The fact that $SO(3)$ is compact makes our job easier, because of the following classical theorem.

Theorem 4.26 (Peter–Weyl) *Every irreducible representation of a compact Lie group is finite-dimensional.*

Proof. The proof can be found in many references, among them [St, Appendix E]. □

Now, we have seen already that $SO(3)$ is doubly covered by $SU(2)$. Hence it suffices to determine the irreducible unitary representations of this last group.

It is not difficult to exhibit countably many (unitary) representations of $SU(2)$, the double covering of $SO(3)$. The idea is very simple. First note that $SU(2)$ acts in $\mathbb{C}^2$, through skew-Hermitian linear transformations. Therefore we have a regular representation $r : SU(2) \to \mathrm{Aut}(\mathcal{A})$, where $\mathcal{A} = C(\mathbb{C}^2, \mathbb{C})$ is the algebra of all complex-valued continuous functions on $\mathbb{C}^2$, given by $r(A)f = f \circ A^{-1}$. Let $V_n \subset \mathcal{A}$ be the subspace of all homogeneous degree n polynomials (in the complex variables z and w, say). An element $p \in V_n$ can be written in the form

$$p(z, w) = \sum_{j=0}^{n} \alpha_j z^j w^{n-j} .$$

In other words, the monomials $z^n, z^{n-1}w, \ldots, zw^{n-1}, w^n$ are a basis of V_n. If $A \in SU(2)$ has matrix

$$A = \begin{pmatrix} a & b \\ -\bar{b} & \bar{a} \end{pmatrix} ,$$

then

$$p \circ A^{-1}(z, w) = \sum_{j=0}^{n} \alpha_j \, (\bar{a}z - bw)^j \left(\bar{b}z + aw\right)^{n-j} ,$$

which is still, as the reader can easily check, a homogenous polynomial of degree n. This shows that r restricts to a representation in V_n for each n. It is also not difficult to see what inner product on V_n one should pick so that this representation is unitary (this is left as yet another exercise).

We claim that these restricted representations (one for each value of n) are irreducible, and they are *all* irreducible unitary representations of $SU(2)$. There are several equivalent ways of proving these facts, for example using one of the several alternative formulations of the Peter–Weyl theorem. We prefer to prove them using the Lie algebra $\mathfrak{su}(2)$. Recall that if $\rho : G \to \mathrm{Aut}(V)$ is a (say, finite-dimensional) representation of a Lie group G, one has a corresponding Lie algebra representation $\dot{\rho} : \mathrm{Lie}(G) \to \mathrm{End}\, V$, given by

$$\dot{\rho}(X) = \frac{d}{dt}\bigg|_{t=0} \rho(\exp(tX)) .$$

If ρ is reducible, the same will happen to $\dot{\rho}$. Hence, after we prove Theorem 4.27 below, we will have established that the representations of $SU(2)$ that we constructed above are indeed all the irreducible representations.

Furthermore, any representation of a Lie algebra $\mathfrak{g}$ into a (complex) vector space extends to a representation of the complexified Lie algebra $\mathfrak{g}_{\mathbb{C}} = \mathbb{C} \otimes \mathfrak{g}$, with the same invariant subspaces. In particular, if the representation of $\mathfrak{g}$ is irreducible, so will be the extended representation of $\mathfrak{g}_{\mathbb{C}}$. These facts are left as straightforward exercises to the reader.

We will need to use the fact that the complexified Lie algebra $\mathfrak{su}_{\mathbb{C}}(2)$ agrees with $\mathfrak{sl}(2, \mathbb{C})$, and as such it has a basis τ_1, τ_2, τ_3 (over $\mathbb{C}$) satisfying the commutation relations

$$[\tau_1, \tau_2] = \tau_3 ; \quad [\tau_2, \tau_3] = \tau_1 ; \quad [\tau_3, \tau_1] = \tau_2 .$$

These are easily constructed from the Pauli matrices (another exercise). Let us consider

$$L = \tau_2 + i\tau_1 ; \quad \theta_3 = i\tau_3 ; \quad R = \tau_2 - i\tau_1 .$$

These elements also form a basis of $\mathfrak{su}(2)$, and one easily sees that

$$\begin{aligned}
[\theta_3, L] &= i\tau_3(\tau_2 + i\tau_1) - (\tau_2 + i\tau_1)(i\tau_3) \\
&= -i[\tau_3, \tau_2] - [\tau_3, \tau_1] \\
&= -\tau_2 + i\tau_1 = -L .
\end{aligned}$$

Likewise, one sees that $[\theta_3, R] = R$.

Theorem 4.27 *For each non-negative half-integer s, there exists an irreducible, skew-Hermitian finite-dimensional representation ρ_s of the Lie algebra $\mathfrak{sl}(2, \mathbb{C})$ into a complex vector space V_s of dimension $2s + 1$. Each such representation is unique up to unitary equivalence. Moreover, there exists a basis $\{v_0, v_1, \ldots, v_{2s}\}$ of V_s such that*

(i) The linear map $\rho_s(\theta_3)$ is diagonalizable in this basis, and $\rho_s(\theta_3)v_j = (-s + j)v_j$, for all $0 \leq j \leq 2s$;

(ii) The operators $\rho_s(L)$ and $\rho_s(R)$ act as shift operators; in other words, $\rho_s(L)v_j = v_{j-1}$ and $\rho_s(R)v_j = v_{j+1}$ (for the appropriate values of j).

Furthermore, these are all the irreducible skew-hermitian representations of $\mathfrak{sl}(2, \mathbb{C})$ (and hence also of $\mathfrak{su}(2)$).

Proof. We perform a *ladder reasoning* similar to the one used in the analysis of the harmonic oscillator, or of spectra of the angular momentum operators (Chapter 2). Given a finite-dimensional irreducible representation of $\mathfrak{sl}(2, \mathbb{C})$, let V be the complex vector space. Let us keep representing by L, R, θ_3 the images of the Lie algebra generators under the given representation. Note that all eigenvalues of θ_3 must be real (can you see why?). Suppose λ is the smallest eigenvalue of θ_3, and let $v_0 \in V$ be an eigenvector of θ_3 with eigenvalue λ. Then the commutation relations imply that

$$\theta_3(Rv_0) = L\theta_3(v_0) + Rv_0 = (\lambda + 1)Rv_0 \, .$$

Hence $v_1 = Rv_0$ is an eigenvector of θ_3 with eigenvalue $\lambda + 1$, unless of course $Lv = 0$. We can continue inductively as long as we do not hit the zero vector, obtaining a sequence of eigenvectors v_0, $v_1 = Rv_0$, $v_2 = R^2 v_0, \ldots, v_k = R^{k-1} v_0$ with eigenvalues $\lambda, \lambda + 1, \ldots, \lambda + k - 1$ that must terminate because V is finite-dimensional. In the end we have $R^k v_0 = 0$. The subspace generated by these k eigenvectors is invariant under θ_3, under R, and (by a similar ladder reasoning going downwards) under L also. But because the representation is assumed to be irreducible, this can only happen if $k = n$. We have deduced in particular that θ_3 is diagonalizable, and that L and R act as shift operators. More importantly, because θ_3 must have zero trace, we get

$$\sum_{j=0}^{n-1} (\lambda + j) = n\lambda + \frac{n(n-1)}{2} = 0 \, .$$

This shows that

$$\lambda = -\frac{n-1}{2} \, .$$

Hence, the eigenvalues of θ_3 are

$$-\frac{n-1}{2}, -\frac{n-3}{2}, \ldots, \frac{n-3}{2}, \frac{n-1}{2} \, ,$$

a sequence that may or may not go through 0, depending on whether n is odd or even, respectively. We have proved both (a) and (b), provided we take, of course, $s = (n-1)/2$. □

Therefore, the irreducible unitary representations of $SU(2)$ corresponding to *massive* particles are labeled by a non-negative half-integer s, called the *spin* of the particle,

$$s = 0, \ \frac{1}{2}, \ 1, \ \frac{3}{2}, \ldots \, .$$

Particles with integer spin are called *bosons*. Those with half-integer spin are called *fermions* (an example of which is the electron).

On the other hand, the irreducible unitary representations of $SO(2)$ are all one-dimensional (because the group is abelian) and correspond to *massless* particles such as photons. They are labeled by the eigenvalues $h \in \mathbb{Z}$ of the angular momentum J_z, which is the generator of the group $SO(2)$ of rotations in the $(x, y, 0)$ plane around the vertical z-axis. The number h is called the *helicity* of the particle.

4.7.4 Spinor representations of $SL(2, \mathbb{C})$

As we have seen, the Lorentz group is doubly covered by $SL(2, \mathbb{C})$. In light of the previous discussion, it is a matter of considerable interest in quantum physics to find all finite-dimensional irreducible representations of $SL(2, \mathbb{C})$. It is easy to give an enumeration of all such irreducible representations, although not so easy to prove that the given list indeed exhausts all possible irreducible representations.

Let us do the easy part. For each pair (s, t) of half-integers, let $V^{s,t}$ denote the complex vector space of all polynomials $p(z, \bar{z})$ of degree at most $2s$ in z and at most $2t$ in $\bar{z}$, i.e. polynomials of the form

$$p(z, \bar{z}) = \sum_{0 \leq j \leq 2s} \sum_{0 \leq k \leq 2t} a_{jk} z^j \bar{z}^k ,$$

where the coefficients a_{jk} are complex. The monomials $z^j \bar{z}^k$ clearly form a basis of $V^{s,t}$, so $\dim V^{s,t} = (2s + 1)(2t + 1)$. Now, given $A \in SL(2, \mathbb{C})$, say

$$A = \begin{pmatrix} a & b \\ c & d \end{pmatrix} ,$$

we define $D^{s,t}(A) : V^{s,t} \to V^{s,t}$ by

$$D^{s,t}(A)(p(z, \bar{z})) = (cz + d)^{2s} (\bar{c}\bar{z} + \bar{d})^{2t} \, p(w, \bar{w}) ,$$

where

$$w = \frac{az + b}{cz + d} .$$

is the fractional linear transformation associated to A. This is clearly well defined and linear in p. It is an easy exercise to verify that $D^{s,t}(I)$ is the identity and that $D^{s,t}(AB) = D^{s,t}(A) \circ D^{s,t}(B)$. Hence $A \mapsto D^{s,t}(A)$ is a finite-dimensional representation of $SL(2, \mathbb{C})$. Such representation is called a *spinor representation* of $SL(2, \mathbb{C})$. Now we have the following fundamental result.

Theorem 4.28 *For all non-negative half-integers s, t, the spinor representation*

$$D^{s,t} : SL(2,\mathbb{C}) \to V^{s,t}$$

is irreducible. Moreover, every finite-dimensional irreducible representation of $SL(2,\mathbb{C})$ is equivalent to one of these.

We will not prove this theorem here. A complete proof can be found in [GMS]. Nevertheless, we invite the reader to compare this result with Theorem 4.27. The derived representation $\dot{D}^{s,t}$ at the level of the Lie algebra $\mathfrak{sl}(2,\mathbb{C})$ is made up by taking the tensor product of *two* of the representations appearing in that theorem.

Exercises

4.1 Let ω be the connection of Example 4.10 (the Hopf bundle), and let A be the corresponding (local) connection 1-form on the base ($\mathbb{S}^2$).

(a) Show that in local coordinates (given by stereographic projection) we can write

$$A = \frac{i}{2}\mathrm{Im}\left(\frac{\overline{z}\,dz}{1+|z|^2}\right) .$$

(b) Deduce that the local curvature 2-form is, in local coordinates, given by

$$F = dA = \frac{1}{2}\frac{d\overline{z}\wedge dz}{(1+|z|^2)^2} .$$

4.2 Work out the analogue of Example 4.10 for the quaternionic Hopf bundle presented in Example 4.8. More precisely, let $\overline{\omega}$ be the 1-form in $\mathbb{H}^2$ given by

$$\overline{\omega} = \mathrm{Im}\,(\overline{q_1}\,dq_1 + \overline{q_2}\,dq_2) .$$

Consider the inclusion $j : \mathbb{S}^7 \to \mathbb{H}^2$, and let $\omega = j^*\overline{\omega}$ be the connection defined on the quaternionic Hopf bundle. Again, denote by A the corresponding local 1-form on the base ($\mathbb{S}^4$) defined via stereographic projection.

(a) Show that, in these local quaternionic coordinates, we have

$$A = \mathrm{Im}\left(\frac{\overline{q}\,dq}{1+|q|^2}\right) .$$

(b) Show that the corresponding curvature 2-form F is given by

$$F = \frac{d\overline{q}\wedge dq}{(1+|q|^2)^2} .$$

5
Classical field theory

The concept of *field* is central in modern physics. In this chapter, we study the basic classical fields, such as the electromagnetic field, presenting them from a unified mathematical perspective. The expression “classical field” is used here in contradistinction to “quantum field” (a concept to be defined in Chapter 6), and is taken to mean “field before quantization.” Thus, we treat fermionic fields as classical, even though strictly speaking fermions are *bona fide* quantum objects, with no actual classical counterpart.

5.1 Introduction

As we shall see, all such fields arise as sections of certain bundles over the spacetime manifold. The basic paradigm was introduced by Yang and Mills in their fundamental paper (see [tH2]). The central idea of Yang–Mills theory is that there is a background field (such as the electromagnetic field) that is given by a connection A defined on a principal bundle over spacetime. The structural group of this bundle represents the internal symmetries of the background field. The possible interactions – also called couplings – of the background field with, say, particles such as photons or electrons are dictated by the representations of the group. Each particle field turns out to be a section of the associated vector bundle constructed from the principle bundle with the help of a given representation of the group. These fields, say $\varphi, \psi, \ldots$, together with the background connection, should satisfy a variational principle. In each case, we have a Lagrangian $\mathcal{L} = \mathcal{L}(A, \varphi, \psi, \ldots)$ defined on the product of the spaces of sections of the bundles corresponding to each field, and taking values in the space of volume forms in the spacetime M. Integrating this Lagrangian, we get an action

$$S = \int_M \mathcal{L}(A, \varphi, \psi, \ldots) .$$

The physically relevant fields are the critical points of this action. These are solutions to the Euler–Lagrange equations, coming from setting the first variation of the action functional equal to zero, i.e.

$$\frac{\delta S}{\delta \partial_\mu A} = 0 ; \quad \frac{\delta S}{\delta \partial_\mu \varphi} = 0 ; \ldots .$$

The resulting equations are linear PDEs in the case of electromagnetic fields and non-linear in the case of gravitational fields. We will construct a generalization of electromagnetism, Yang–Mills fields, which will also lead to non-linear PDEs.

5.2 Electromagnetic field

Let us start with the best understood of all physical fields, the electromagnetic field. What follows is elementary, and appears in one form or another in many physics books.

5.2.1 Maxwell's equations

The study of electric and magnetic phenomena goes back to the 18th century with the pioneering work of Coulomb on electric charges, but it only became systematic after Faraday, Gauss, Ampere, and Maxwell in the 19th century. Their work culminated in a great synthesis with the famous Maxwell equations. In the vacuum, these equations read as follows:

$$\nabla \cdot \boldsymbol{E} = 0 , \quad \nabla \wedge \boldsymbol{B} - \frac{\partial \boldsymbol{E}}{\partial t} = 0$$

$$\nabla \cdot \boldsymbol{B} = 0 , \quad \nabla \wedge \boldsymbol{E} + \frac{\partial \boldsymbol{B}}{\partial t} = 0 .$$

Here, $\boldsymbol{E} = (E_1, E_2, E_3)$ is the electric field and $\boldsymbol{B} = (B_1, B_2, B_3)$ is the magnetic field. These equations make it clear that electric and magnetic phenomena are not independent, but deeply intertwined. For instance, the first line says that, in the absence of electric sources, the electric field is divergence-free (first equation), whereas a non-zero magnetic field must be present as soon as we have a time-varying electric field (second equation). The equations in the second line have analogous physical interpretations.

5.2.2 The scalar and vector potentials

Maxwell realized that the above equations could be deduced from the assumption that both the electric and the magnetic fields were in some sense conservative, i.e. derived from potentials. He introduced a scalar potential ϕ and a vector potential $\boldsymbol{A}$ and arrived at the following expressions:

$$\boldsymbol{E} = -\nabla\phi - \partial_t \boldsymbol{A} \tag{5.1}$$

$$\boldsymbol{B} = \nabla \wedge \boldsymbol{A} \,. \tag{5.2}$$

It is easy to derive the four Maxwell equations from these two. For example, because the divergence of the rotational of any vector field is identically zero, taking the divergence on both sides of $\boldsymbol{B} = \nabla \wedge \boldsymbol{A}$ yields $\nabla \cdot \boldsymbol{B} = 0$, the third of Maxwell's equations. Because the rotational of the gradient of any function is identically zero, we also have, combining the two expressions above,

$$\frac{\partial \boldsymbol{B}}{\partial t} = \nabla \wedge \left(\frac{\partial \boldsymbol{A}}{\partial t}\right) = \nabla \wedge (-\nabla\phi - \boldsymbol{E}) = -\nabla \wedge \boldsymbol{E} \,,$$

which gives us the fourth of Maxwell's equations.

It was observed by Lorentz at the end of the 19th century that Maxwell's equations are invariant under a large group of linear transformations of space-time $\mathbb{R}^4$. This group is the Lorentz group we met in Chapter 3, namely, the group of isometries of $M = \mathbb{R}^4$ under the Lorentz (or Minkowski) metric given by

$$ds^2 = -dx^2 - dy^2 - dz^2 + c^2\, dt^2 \,,$$

where c stands for the speed of light. Lorentz invariance lies at the heart of Einstein's relativity theory.

Using coordinates $x^0 = ct$, $x^1 = x$, $x^2 = y$, $x^3 = z$ in spacetime, we see that the Lorentz metric tensor $g_{\mu\nu}$, $\mu, \nu = 0, 1, 2, 3$, is given by the diagonal 4×4 matrix with diagonal entries $g_{00} = 1$, $g_{11} = g_{22} = g_{33} = -1$.

5.2.3 The field strength tensor

A synthetic way to express Maxwell's equations is obtained using the modern language of differential forms. Let us consider the 2-form F in spacetime $M = \mathbb{R}^4$ given by

$$\begin{aligned} F = {} & E_1\, dx^1 \wedge dx^0 + E_2\, dx^2 \wedge dx^0 + E_3\, dx^3 \wedge dx^0 \\ & + B_1\, dx^2 \wedge dx^3 + B_2\, dx^3 \wedge dx^1 + B_3\, dx^1 \wedge dx^2 \,. \end{aligned}$$

There is another 2-form associated to F, called the Hodge dual to F (see Section 4.5), which is denoted by $\star F$ and is given by

$$\begin{aligned} \star F = {} & B_1\, dx^1 \wedge dx^0 + B_2\, dx^2 \wedge dx^0 + B_3\, dx^3 \wedge dx^0 \\ & + E_1\, dx^2 \wedge dx^3 + E_2\, dx^3 \wedge dx^1 + E_3\, dx^1 \wedge dx^2 \,. \end{aligned}$$

In terms of these forms, Maxwell's equations in the vacuum become simply

$$dF = 0\,, \quad d\star F = 0\,. \tag{5.3}$$

Note that we can write

$$F = \frac{1}{2}\, F_{\mu\nu}\, dx^\mu \wedge dx^\nu\,,$$

where $(F_{\mu\nu})$ is the skew-symmetric tensor given by the matrix

$$\begin{bmatrix} 0 & -E_1 & -E_2 & -E_3 \\ E_1 & 0 & B_3 & -B_2 \\ E_2 & -B_3 & 0 & B_1 \\ E_3 & B_2 & -B_1 & 0 \end{bmatrix}.$$

This tensor is called the *field strength* tensor.

What is the relationship between the field strength and the scalar and vector potentials introduced by Maxwell? Let us consider first the 1-form $A = A_\mu dx^\mu$, where $A_0 = -\phi$ is the negative of the scalar potential and A_1, A_2, A_3 are the spatial components of the vector potential. Then,

$$dA = d(A_\mu dx^\mu) = \left(\partial_\nu A_\mu\, dx^\nu\right) \wedge dx^\mu = \frac{1}{2}(\partial_\mu A_\nu - \partial_\nu A_\mu)\, dx^\mu dx^\nu\,.$$

Using the equations (5.1), we see after some simple computations that

$$F_{\mu\nu} = \partial_\mu A_\nu - \partial_\nu A_\mu\,.$$

In other words, we have precisely $F = dA$. Thus, the field strength 2-form is exact, and therefore it is closed as well: $dF = 0$.

5.2.4 The electromagnetic Lagrangian

Can Maxwell's equations be derived from a variational principle? The answer is yes. Note that the exterior product of F with $\star F$ yields a volume form in spacetime, which can be integrated. Thus, we may consider the functional given by

$$S_{em}(A) = \int_M F \wedge \star F\,. \tag{5.4}$$

This is the electromagnetic action (in the vacuum). The Euler–Lagrange equations for this functional, obtained by imposing the condition that its first variation vanishes, i.e.

$$\frac{\delta S_{em}}{\delta A} = 0\,,$$

are precisely Maxwell's equations (5.3)! (The proof of this fact uses Stokes's theorem and is left as an exercise.) We may therefore regard $L(A) = F \wedge \star F$ as the Lagrangian of electromagnetism (in the absence of charges).

Let us express the Lagrangian (or the corresponding action) using a notation that is more familiar to physicists. Using the Lorentz metric, we can "raise the indices" of the field strength tensor $F_{\mu\nu}$ to get another tensor $F^{\mu\nu}$ whose components are given by

$$F^{\mu\nu} = F_{\alpha\beta}\, g^{\alpha\mu} g^{\beta\nu} .$$

In this new notation, it is an easy exercise to check that

$$S_{em}(A) = -\frac{1}{4}\int_M F_{\mu\nu} F^{\mu\nu}\, d^4x ,$$

where $d^4x = dx^0 \wedge dx^1 \wedge dx^2 \wedge dx^3$ is the standard volume form in $M = \mathbb{R}^4$.

5.2.5 Gauge invariance

An extremely important property of the Maxwell Lagrangian is its invariance under gauge transformations. A gauge transformation has the form

$$A_\mu \mapsto A_\mu + \partial_\mu \Theta ,$$

where $\Theta : \mathbb{R}^4 \to \mathbb{R}$ is an arbitrary (smooth) function. It is clear that the Lagrangian (5.4) does not change when the vector potential (connection) A is changed in this fashion. Hence we have considerable freedom when choosing A. For instance, given A, we can always make $A_0 = 0$ after a suitable gauge transformation. Indeed, let

$$\Theta(t, \boldsymbol{x}) = \int^t A_0(s, \boldsymbol{x})\, ds .$$

Then $\partial_0 \Theta = A_0$, and therefore the gauge-transformed field

$$A'_\mu = \partial_\mu \int^t A_0(s, \boldsymbol{x})\, ds$$

has $A'_0 = 0$. We can further gauge transform A' in many ways to get yet another connection A'' that still has first component equal to zero, provided we use as gauge a function that is independent of t. One way to do this is the following. Let

$$\Psi(x) = -\frac{1}{4\pi}\int \nabla_y \cdot \boldsymbol{A}'(t, \boldsymbol{y})\, \frac{d^3y}{|\boldsymbol{x} - \boldsymbol{y}|} .$$

Maxwell's equations show that, because $A'_0 = 0$, we have $\partial_0(\nabla \cdot \boldsymbol{A}') = 0$. This shows that Ψ is indeed independent of time (i.e. $\Psi(x) = \Psi(0, \boldsymbol{x})$). Now let

$$A''_\mu = A'_\mu + \partial_\mu \Psi \, .$$

Note that we still have $A''_0 = 0$. Using the identity on distributions

$$\nabla_x^2 \left(\frac{1}{4\pi |\boldsymbol{x} - \boldsymbol{y}|} \right) = -\delta^3(\boldsymbol{x} - \boldsymbol{y}) \, ,$$

we see that $\nabla \cdot \boldsymbol{A}'' = 0$ (exercise). In other words, we have $\partial^\mu A''_\mu = 0$; this choice of gauge is called the *Lorentz gauge*. Using the equations of motion $\partial_\mu F_{\mu\nu} = 0$, we deduce that the connection components A''_μ satisfy the massless Klein–Gordon equation

$$\Box A''_\mu = 0 \, . \tag{5.5}$$

These Klein–Gordon equations admit plane-wave solutions of the form

$$A''_\mu(x) = \epsilon_\mu(k) e^{-ik\cdot x} + \epsilon^*_\mu(k) e^{ik\cdot x}$$

for each $k \in \mathbb{R}^4$, where the coefficients $\epsilon_\mu(k) \in \mathbb{C}^4$ (and their complex conjugates $\epsilon^*_\mu(k)$) are called *polarization vectors*. Because the Klein–Gordon equation is linear, the superposition principle tells us that these plane waves can be combined to yield the general solution. Later (Chapter 6), when we quantize the electromagnetic field, we will see that these equations (5.5) describe massless particles, namely photons.

5.2.6 Maxwell Lagrangian with an external current

We have presented Maxwell's equations *in vacuo*, which are homogeneous, but of course it is also necessary to consider the electromagnetic field in the presence of a charge distribution. This is represented by a 4-vector $(J^\mu) = (\rho, \boldsymbol{J})$, where ρ denotes the charge density, and $\boldsymbol{J}$ the associated current. The inhomogeneous Maxwell equations corresponding to this situation are

$$\nabla \cdot \boldsymbol{E} = \rho \, , \quad \nabla \times \boldsymbol{B} - \frac{\partial \boldsymbol{E}}{\partial t} = \boldsymbol{J}$$

$$\nabla \cdot \boldsymbol{B} = 0 \, , \quad \nabla \times \boldsymbol{E} + \frac{\partial \boldsymbol{B}}{\partial t} = 0 \, .$$

These equations turn out to be the Euler–Lagrange equations for the action functional with Lagrangian density given by

$$\mathcal{L} = -\frac{1}{4} F_{\mu\nu} F^{\mu\nu} - J^\mu A_\mu \, .$$

5.3 Conservation laws in field theory

Just as in classical mechanics, in field theory there is a close relationship between symmetry and conservation laws. This relationship is expressed through Noether's theorem. Rather than present the most general formulation of this theorem, we will deal here with the case of internal symmetries only. A symmetry is called a *spacetime* symmetry if it acts on the spacetime variables, and an *internal* symmetry otherwise. For more details we recommend [Fr, Chapter 20], as well as [BL, Chapter 3] and [Ma, Chapter 3].

5.3.1 The Euler–Lagrange equations

Let us try to free the discussion from dependence on coordinates. We shall examine the case of a field Lagrangian depending on a single field ϕ. We make the following assumptions.

(i) The spacetime M is a pseudo-Riemannian n-manifold (typically a Lorentzian 4-manifold) with a fixed pseudo-Riemannian metric g (typically the Minkowski metric). The volume element of M can be written, in local coordinates, as

$$dV_M = \sqrt{|\det g|}\, dx^1 \wedge dx^2 \wedge \cdots \wedge dx^n$$

(we ignore a sign here, which is determined by a choice of orientation of M).

(ii) The relevant fields are sections of a vector (or spinor) bundle $E \to M$ with N-dimensional fibers. These sections will be denoted $\phi = (\phi^\alpha)_{\alpha=1,\ldots,N}$. The vector bundle E is provided with a fixed connection, i.e. a fixed covariant derivative ∇. In local coordinates $x : U \to \mathbb{R}^n$ ($U \subset M$ being a coordinate patch), this covariant derivative has an expression of the form

$$\nabla_j \phi^\alpha = \partial_j \phi^\alpha + \omega^\alpha_{j,\beta} \phi^\beta ,$$

where the $\omega^\alpha_{j,\beta}$ are smooth functions on U.

(iii) The relevant action functional $S : \Gamma(E) \to \mathbb{R}$ is given by integration of a suitable Lagrangian density, namely

$$S(\phi) = \int_M \mathcal{L}(\phi, \nabla\phi)\, dV_M . \tag{5.6}$$

Here $\mathcal{L}(\phi, \nabla\phi) \in C^\infty(M)$ for each $\phi \in \Gamma(E)$. Thus, the Lagrangian density is assumed to depend explicitly only on the field ϕ and on its covariant derivative.

We shall presently derive the Euler–Lagrange equations for the critical points (fields) of the action functional. Following the recipe dictated by the calculus of variations, let us work out the variation δS corresponding to a given variation $\delta\phi \in \Gamma(E)$ of the field ϕ. Because the connection is fixed and $\delta\partial_j = \partial_j\delta$ for all j, we have

$$\delta(\nabla_j\phi^\alpha) = \nabla_j(\delta\phi^\alpha)\,.$$

Using this fact and taking into account that the metric on M is also fixed, we see that

$$\begin{aligned}\delta S(\phi) &= \int_M \delta\mathcal{L}(\phi, \nabla\phi)\, dV_M \\ &= \int_M \left\{\frac{\partial\mathcal{L}}{\partial\phi^\alpha}(\delta\phi^\alpha) + \frac{\partial\mathcal{L}}{\partial(\nabla_j\phi^\alpha)}\,\nabla_j(\delta\phi^\alpha)\right\} dV_M\,. \end{aligned} \tag{5.7}$$

Let us pause here to understand the invariant tensorial meaning of the terms in the integrand on the last line of (5.7). Because the expression in brackets is assumed to be a function on M, the products appearing as summands must be *pairings* of the tensor fields involved. In each pairing, a section of a given bundle must be paired against a section of the corresponding dual bundle. Thus, because $\delta\phi \in \Gamma(E)$, we see that

$$\left(\frac{\partial\mathcal{L}}{\partial\phi^\alpha}\right) \in \Gamma(E^*)\,,$$

i.e. a section of the dual bundle E^*. Likewise, because the covariant derivative maps sections of E to sections of $E \otimes T^*M$, we see that

$$\left(\frac{\partial\mathcal{L}}{\partial(\nabla_j\phi^\alpha)}\right) \in \Gamma(E^* \otimes TM)\,.$$

Now, going back to the calculation of δS, we note that

$$\nabla_j\left(\frac{\partial\mathcal{L}}{\partial(\nabla_j\phi^\alpha)}\delta\phi^\alpha\right) = \frac{\partial\mathcal{L}}{\partial(\nabla_j\phi^\alpha)}\,\delta(\nabla_j\phi^\alpha) + \nabla_j\left(\frac{\partial\mathcal{L}}{\partial(\nabla_j\phi^\alpha)}\right)\delta\phi^\alpha\,. \tag{5.8}$$

Here there is an abuse of notation. All covariant derivatives here are being denoted by the same symbol, but they live in different vector bundles. Thus, the covariant derivative on the left-hand side of (5.8) is the pseudo-Riemannian covariant derivative on TM, whereas on the right-hand side we have the given covariant derivative ∇ acting on ϕ in the first summand, and the induced covariant derivative acting on the section of $\Gamma(E^* \otimes TM)$ appearing in the second summand. Using (5.8) in (5.7), we get

$$\delta S(\phi) = \int_M \left\{ \frac{\partial \mathcal{L}}{\partial \phi^\alpha} - \nabla_j \left(\frac{\partial \mathcal{L}}{\partial (\nabla_j \phi^\alpha)} \right) \right\} \delta\phi^\alpha \, dV_M + \int_M \nabla_j \left(\frac{\partial \mathcal{L}}{\partial (\nabla_j \phi^\alpha)} \delta\phi^\alpha \right) dV_M \, . \tag{5.9}$$

Now, suppose the spacetime manifold is compact and has a smooth boundary ∂M (with the induced orientation from M). Then, applying the divergence (i.e. Stokes) theorem to this last integral, we get

$$\delta S(\phi) = \int_M \left\{ \frac{\partial \mathcal{L}}{\partial \phi^\alpha} - \nabla_j \left(\frac{\partial \mathcal{L}}{\partial (\nabla_j \phi^\alpha)} \right) \right\} \delta\phi^\alpha \, dV_M + \int_{\partial M} \left(\frac{\partial \mathcal{L}}{\partial (\nabla_j \phi^\alpha)} \delta\phi^\alpha \right) N_j \, dA_M \, , \tag{5.10}$$

where dA_M is the area-form on ∂M and $\boldsymbol{n} = (N_j)_{j=1}^n$ is the unit normal to the boundary of M. Hence, if ϕ is a critical point for the action functional S, in other words if the first variation $\delta S(\phi)$ vanishes for all field variations $\delta\phi$ vanishing at ∂M, then we must have

$$\frac{\delta \mathcal{L}}{\delta \phi^\alpha} = \frac{\partial \mathcal{L}}{\partial \phi^\alpha} - \nabla_j \left(\frac{\partial \mathcal{L}}{\partial (\nabla_j \phi^\alpha)} \right) = 0 \, .$$

The symbol on the left-hand side of the above equality stands for the so-called *functional derivative* of the Lagrangian. Note that this object is a section of the dual bundle E^*. It is also customary to write

$$\operatorname{div} \left(\frac{\partial \mathcal{L}}{\partial \nabla \phi} \right) = \nabla_j \left(\frac{\partial \mathcal{L}}{\partial (\nabla_j \phi^\alpha)} \right) , \tag{5.11}$$

so that the functional derivative of $\mathcal{L}$ with respect to ϕ can be written without reference to components as

$$\frac{\delta \mathcal{L}}{\delta \phi} = \frac{\partial \mathcal{L}}{\partial \phi} - \operatorname{div} \left(\frac{\partial \mathcal{L}}{\partial \nabla \phi} \right) .$$

5.3.2 Noether's theorem for internal symmetries

Let us now assume that the Lagrangian $\mathcal{L}$ is invariant under a one-parameter group of symmetries. In other words, let G be the structure group of the vector (or spinor) bundle on which the field ϕ lives, and suppose we have a one-parameter subgroup $t \mapsto g_t \in G$. This induces a one-parameter group of fiber-preserving self-maps of the vector bundle E, and also a one-parameter group of self-maps of $\Gamma(E)$, all of which we denote by the same symbol g_t. In particular, we have a motion of sections $\phi_t = g_t\phi$. The meaning of the symmetry in question is that $\mathcal{L}$ is invariant under such motion, i.e.

$$\frac{\partial}{\partial t}\bigg|_{t=0} \mathcal{L}(\phi_t, \nabla\phi_t) = 0 .$$

Now, we can write $g_t = e^{tY}$, where Y is an element of the Lie algebra of G. For a coordinate neighborhood $U \subset M$, each field ϕ is represented by a column vector $(\phi^\alpha)_{\alpha=1}^N$, each g_t is represented by a matrix, and

$$\phi_t^\alpha = g_{t,\beta}^\alpha \phi^\beta = (e^{tY})_\beta^\alpha \phi^\beta .$$

The first variation $\delta\phi$ of ϕ along this symmetry is given by

$$\delta\phi^\alpha = \frac{\partial\phi_t^\alpha}{\partial t}\bigg|_{t=0} = Y_\beta^\alpha \phi^\beta .$$

We call $\delta\phi$ a *variation by symmetries of the Lagrangian*. The theorem of Noether tying symmetries of the Lagrangian to conservation laws is the following.

Theorem 5.1 *Let the field ϕ be an extremal of the Lagrangian $\mathcal{L}$. If $\delta\phi$ is a variation by symmetries of the Lagrangian, then*

$$\operatorname{div}\left(\frac{\partial\mathcal{L}}{\partial\nabla\phi^\alpha}\delta\phi^\alpha\right) = 0 .$$

In particular, the vector field J given by

$$J^\mu = \frac{\partial\mathcal{L}}{\partial\nabla_\mu\phi^\alpha} Y_\beta^\alpha \phi^\beta$$

has divergence zero.

Proof. Because ϕ is extremal for the action functional with Lagrangian $\mathcal{L}$, we know from (5.9) and the fact that $\delta\phi$ is a variation by symmetries that

$$0 = \delta S = \int_V \nabla_j \left(\frac{\partial\mathcal{L}}{\partial(\nabla_j\phi^\alpha)}\delta\phi^\alpha\right) dV_M$$

for every compact n-dimensional submanifold of M with boundary. Because V is arbitrary, the integrand above must vanish identically. Remembering (5.11), this proves the theorem. □

Remark 5.2 Suppose we are dealing with a *scalar field* ϕ, defined over flat Minkowski spacetime $M = \mathbb{R}^4$, and let us denote the corresponding Noether current by j^μ. We know from Theorem 5.1 that $\partial_\mu j^\mu = 0$. Separating the time component j^0 of j from its spatial components, i.e. writing $j = (j^0, \boldsymbol{j})$, we see that the equation $\partial_\mu j^\mu = 0$ becomes $\partial_0 j^0 = -\nabla \cdot \boldsymbol{j}$. This is similar to the situation we already encountered in Chapter 3 when we discussed Dirac's equation: current conservation. Indeed, let $V \subset \mathbb{R}^3$ be a region in 3-space with smooth boundary ∂V. Defining

$$Q_V(t) = \int_V j^0(t, \boldsymbol{x})\, d^3x \ ,$$

it follows from the divergence theorem that $\partial_t Q_V(t)$ is equal to the *flux* of $\boldsymbol{j}$ across the boundary ∂V. This justifies calling Q_V the *charge* and j the *current* associated to the symmetry $\delta\phi$. The reader can check, as an instructive exercise, that when $\mathcal{L}$ is the Maxwell Lagrangian with a source term, and the present discussion is adapted to the vector potential A, the charge and current introduced above are precisely the electric charge and electric current that are so familiar in electromagnetism. The remainder of this chapter will be devoted to further important examples of free or interacting fields and their symmetries and associated conservation laws.

5.4 The Dirac field

5.4.1 The free Dirac field

The *free* (or *pure*) Dirac Lagrangian describes a fermion, which mathematically is represented by a spinor field, in the absence of interactions. The spinor field ψ is a section of a spinor bundle over spacetime M (whose fibers are identified with a vector space V isomorphic to either $\mathbb{R}^4$ or $\mathbb{C}^2$, as described in Chapter 4). For simplicity, let us work in Minkowski's flat space. Using the standard Dirac matrices γ^μ, the Lagrangian density can be written as

$$\mathcal{L}_D = \overline{\psi}(i\gamma^\mu \partial_\mu - m)\psi \ , \tag{5.12}$$

where $\overline{\psi} = \psi^\dagger \gamma^0$ is the adjoint spinor to ψ (here $\psi^\dagger$ is the Hermitian conjugate to ψ). See our discussion of the Dirac equation in Chapter 3. The Dirac Lagrangian is clearly invariant under the $U(1)$ symmetry $\psi \mapsto e^{i\theta}\psi$ (a one-parameter group), and therefore we can apply Noether's theorem to recover the result that the associated *Dirac current*

$$j^\mu = \overline{\psi}\gamma^\mu\psi$$

is conserved, as we saw in in Chapter 3. The Euler–Lagrange equations for the Dirac Lagrangian yield, not surprisingly, the *Dirac equation*

$$(i\gamma^\mu \partial_\mu - m)\psi = 0 \ .$$

In the exercises at the end of this chapter, we analyze the plane-wave solutions to this equation and their superpositions to yield a general solution formula in Minkowski space.

5.4.2 Coupling with the electromagnetic field

Let us now describe how the Dirac field can be (minimally) coupled with the electromagnetic field. The idea is to replace the ordinary derivative ∂_μ in (5.12) with a *covariant* derivative

$$\nabla_\mu = \partial_\mu + iqA_\mu ,$$

where A_μ is the vector potential, and q is called the *charge*. This presupposes a connection on a suitable principal $U(1)$ bundle over spacetime, and a suitable representation $U(1) \to \mathrm{Aut}(V)$, where V is a vector space (real four-dimensional or complex two-dimensional) representing the fiber of the vector bundle on which ψ lives. Here we shall ignore the interaction of the electromagnetic field with itself. The Lagrangian is thus the same as (5.12) with ∇_μ replacing ∂_μ, and we get

$$\mathcal{L}_D^A = \overline{\psi}(i\gamma^\mu\partial_\mu - m)\psi - q\overline{\psi}\gamma^\mu A_\mu\psi . \tag{5.13}$$

The Euler–Lagrange equations in this case yield the modified Dirac equation

$$(i\gamma^\mu\partial_\mu - m)\psi = q\gamma^\mu A_\mu\psi . \tag{5.14}$$

Remark 5.3 We want to record here an important symmetry enjoyed by the above Dirac equation. If we have a solution ψ to (5.14), then we can take the complex conjugate of both sides, getting

$$\left[(\gamma^\mu)^*(-i\partial_\mu - qA_\mu) - m\right]\psi^* = 0 .$$

Recall that the components A_μ of the electromagnetic field are real. Now, the Dirac matrices as we defined them (in the so-called chiral representation) are such that $(\gamma^2)^* = -\gamma^2$, whereas $\gamma^0, \gamma^1, \gamma^3$ are real. Multiplying both sides of this last equation by γ^2 and using the anti-commutation relations satisfied by these matrices, we get

$$(i\gamma^\mu\partial_\mu - m)(\gamma^2\psi^*) = q\gamma^\mu(-A_\mu)(\gamma^2\psi^*) .$$

But this equation is again the Dirac equation, with the new electromagnetic field $-A_\mu$. In physical terms, if ψ is a positive-energy solution, say, of the Dirac equation for a particle with charge q in the electromagnetic field A_μ, then $\psi^C = -i\gamma^2\psi^*$ is a negative-energy solution to the Dirac equation in the charge conjugate field $A_\mu^C = -A_\mu$. The operation $\psi \mapsto \psi^C$, $A_\mu \mapsto A_\mu^C$ is called *charge conjugation*. The choice of factor $-i$ in the definition of ψ^C

is made so that the resulting operation is an involution, i.e. equal to its own inverse. If we take into account the interpretation of negative-energy solutions of Dirac's equation as representing antiparticles, we can say, informally, that charge conjugation is an operator that replaces matter with antimatter and vice versa.

To end this section, we would like to say a few words about the solution to (5.14). It is based on the Green's function method widely used by physicists. This will in fact lead us to the Feynman propagator for Dirac's equation. The method has two steps.

(i) First we solve

$$(i\gamma^\mu \partial_\mu - m)G(x, x') = \delta^{(4)}(x - x') . \tag{5.15}$$

Here, $\delta^{(4)}$ denotes the four-dimensional Dirac delta distribution. From a physical standpoint, by analogy with the standard wave equation, the Green function $G(x, x')$ should be thought of as representing the effect at $x \in \mathbb{R}^4$ of a wave originated by placing a unit source at $x' \in \mathbb{R}^4$. Note that this Green function is a (4×4) matrix-valued function (so in fact the Dirac delta distribution on the right-hand side of the equation should be thought of as multiplied by the identity matrix).

(ii) Having obtained $G(x, x')$, we form its convolution with the right-hand side of (5.14) and get the integral equation

$$\psi(x) = q \int_{\mathbb{R}^4} G(x, x')\gamma^\mu A_\mu(x')\psi(x')\, d^4x' .$$

This integral equation can be solved, in principle, by an iterative procedure. Taking ψ_0 to be, say, any solution of the homogeneous Dirac equation, we define inductively a sequence of spinor fields ψ_n by

$$\psi_{n+1}(x) = q \int_{\mathbb{R}^4} G(x, x')\gamma^\mu A_\mu(x')\psi_n(x')\, d^4x' .$$

If the limit of this sequence exists, it will be a solution to (5.14).

These steps and the Green function G itself lie at the basis of *perturbative field theory*; see Chapter 7. The translation invariance of equation (5.15) tells us that $G(x, x')$ depends only on the difference $x - x'$, so we write $G(x, x') = G_F(x - x')$. In order to write G_F explicitly, let us denote by S_F its Fourier transform to momentum space, so that

$$G_F(x - x') = \frac{1}{(2\pi)^4} \int_{\mathbb{R}^4} S_F(p)\, e^{-ip\cdot(x-x')}\, d^4p . \tag{5.16}$$

We call S_F the *Feynman propagator* associated to Dirac's equation. Note that the scalar product in the exponential is Minkowski's. Recall the usual notational conventions about 4-vectors in Minkowski's space, according to which $p = (p_0, \boldsymbol{p})$, $x = (t, \boldsymbol{x})$, and so forth. Recall also that the Dirac delta-distribution has the following Fourier representation:

$$\delta^{(4)}(x - x') = \frac{1}{(2\pi)^4} \int_{\mathbb{R}^4} e^{-ip\cdot(x-x')} \, d^4 p \,. \tag{5.17}$$

Putting (5.16) and (5.17) back into (5.15), we get

$$\frac{1}{(2\pi)^4} \int_{\mathbb{R}^4} (\gamma^\mu p_\mu - m) S_F(p) \, e^{-ip\cdot(x-x')} \, d^4 p = \frac{1}{(2\pi)^4} \int_{\mathbb{R}^4} e^{-ip\cdot(x-x')} \, d^4 p \,.$$

From this equality, we deduce that

$$(\gamma^\mu p_\mu - m) S_F(p) = 1 \,.$$

The reader should keep in mind that this is an identity involving matrices, not complex numbers. It is now an easy exercise to invert the matrix $(\gamma^\mu p_\mu - m)$, and from this we get the expression

$$S_F(p) = \frac{\gamma^\mu p_\mu + m}{p^2 - m^2} \,.$$

Going back to (5.16), we obtain the following formula for the Green's function:

$$G_F(x - x') = \frac{1}{(2\pi)^4} \int_{\mathbb{R}^4} \frac{\gamma^\mu p_\mu + m}{p^2 - m^2} \, e^{-ip\cdot(x-x')} \, d^4 p \,.$$

We will perform a partial evaluation of this integral, reducing it to an integral over three-dimensional momentum space. Let us write the 4-tuple of gamma matrices (γ^μ) as $(\gamma^0, \boldsymbol{\gamma})$, so that

$$\gamma^\mu p_\mu = \gamma^0 p_0 - \boldsymbol{\gamma} \cdot \boldsymbol{p} \,.$$

Recall the relativistic energy–momentum relationship $E^2 = \boldsymbol{p}^2 + m^2$ (we let $c = 1$, as usual). Then we have

$$p^2 - m^2 = p_0^2 - \boldsymbol{p}^2 - m^2 = p_0^2 - E^2 \,.$$

Therefore we can write

$$G_F(x - x') = \frac{1}{(2\pi)^4} \int_{\mathbb{R}^3} e^{i\boldsymbol{p}\cdot(\boldsymbol{x}-\boldsymbol{x}')} \left[\int_{-\infty}^{\infty} e^{-ip_0\cdot(t-t')} \frac{\gamma^0 p_0 - \boldsymbol{\gamma} \cdot \boldsymbol{p} + m}{(p_0 - E)(p_0 + E)} \, dp_0 \right] d^3 \boldsymbol{p} \,.$$

The integral appearing inside brackets can be evaluated by means of the residue theorem. The appropriate contour depends on the sign of $t - t'$. If $t > t'$, then we choose the lower contour in Figure 5.1, enclosing the pole at $p_0 = E$. This

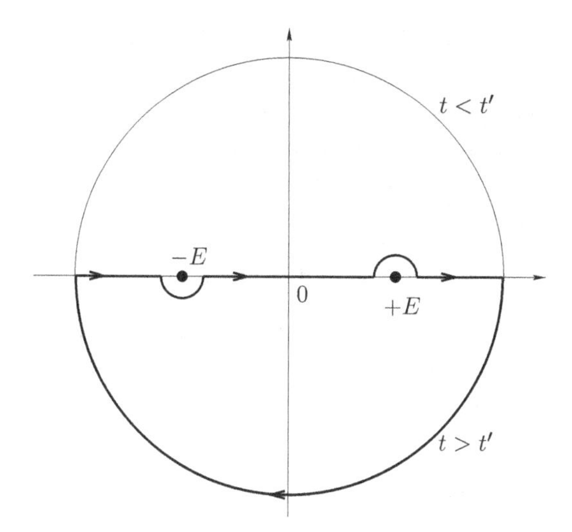

Figure 5.1 Contours for Dirac's Green function.

yields

$$\int_{-\infty}^{\infty} e^{-ip_0\cdot(t-t')}\,\frac{\gamma^0 p_0 - \boldsymbol{\gamma}\cdot\boldsymbol{p} + m}{(p_0-E)(p_0+E)}\,dp_0 = -2\pi i\,\mathrm{Res}\,(p_0 = E)$$

$$= -2\pi i\, e^{-iE(t-t')}\,\frac{\gamma^0 E - \boldsymbol{\gamma}\cdot\boldsymbol{p} + m}{2E}\,,$$

and therefore

$$G_F(x-x') = \frac{-i}{(2\pi)^3}\int_{\mathbb{R}^3} e^{i\boldsymbol{p}\cdot(\boldsymbol{x}-\boldsymbol{x}')}\,e^{-iE(t-t')}\,(\gamma^0 E - \boldsymbol{\gamma}\cdot\boldsymbol{p} + m)\frac{d^3\boldsymbol{p}}{2E}\,.$$

Similarly, if $t < t'$, then we choose the upper contour in Figure 5.1, enclosing the pole at $p_0 = -E$. Applying the residue theorem as before, we arrive at

$$G_F(x-x') = \frac{-i}{(2\pi)^3}\int_{\mathbb{R}^3} e^{i\boldsymbol{p}\cdot(\boldsymbol{x}-\boldsymbol{x}')}\,e^{iE(t-t')}\,(-\gamma^0 E - \boldsymbol{\gamma}\cdot\boldsymbol{p} + m)\frac{d^3\boldsymbol{p}}{2E}\,.$$

Remark 5.4 We have performed these calculations using the contours of Figure 5.1, which cleverly avoid the poles of the propagator on the real line. Alternatively, we could have kept the (limiting) contour in the real line, shifting the poles of S_F instead. In other words, we could have used as propagator

$$S_F(p) = \frac{\gamma^\mu p_\mu + m}{p^2 - m^2 + i\epsilon}\,.$$

This viewpoint turns out to be very useful when one performs quantization via path integrals and Feynman diagrams; see Chapter 7.

5.4.3 The full QED Lagrangian

Now is the time to take into account the interaction of the electromagnetic field with itself. This introduces an extra, self-interacting term into the Lagrangian, which is obtained from the electromagnetic field strength tensor. Mathematically, the field strength tensor is the curvature of the connection determined by (A_μ), and is given by

$$F_{\mu\nu} = \partial_\mu A_\nu - \partial_\nu A_\mu \ ,$$

as we have seen in Section 5.2.3. Therefore, the full Lagrangian of the Dirac field coupled with this field is

$$\mathcal{L}_{QED} = -\frac{1}{4} F_{\mu\nu} F^{\mu\nu} + \overline{\psi}(i\gamma^\mu \nabla_\mu - m)\psi \ . \qquad (5.18)$$

This Lagrangian is called the *QED Lagrangian* (the acronym QED stands for *quantum electrodynamics*). It describes the theory before quantization, of course. Fully expanded out, $\mathcal{L}_{QED}$ can be written as

$$\mathcal{L}_{QED} = -\frac{1}{4} F_{\mu\nu} F^{\mu\nu} + \overline{\psi}(i\gamma^\mu \partial_\mu - m)\psi - q\overline{\psi}\gamma^\mu A_\mu \psi \ . \qquad (5.19)$$

We see very clearly from this expression that $\mathcal{L}_{QED}$ is obtained as the sum of the Maxwell with the Dirac Lagrangian, plus an interaction Lagrangian.

Remark 5.5 We want to point out that the actual Lagrangian of quantum electrodynamics (before quantization) corresponds to the case where $q = -e$ (the electron charge). The above Lagrangian corresponds to the slightly more general case of a fermion with charge q.

Remark 5.6 One can go a bit further and add also an external charge–current distribution J^μ to (5.19) (as in the case of the inhomogeneous Maxwell Lagrangian), and the result is

$$\mathcal{L}^J_{QED} = -\frac{1}{4} F_{\mu\nu} F^{\mu\nu} + \overline{\psi}(i\gamma^\mu \partial_\mu - m)\psi - J^\mu A_\mu - q\overline{\psi}\gamma^\mu A_\mu \psi \ . \qquad (5.20)$$

If one takes J^μ to be the charge–current distribution of an atomic nucleus, this Lagrangian turns out to be precisely the one that explains most of the chemistry of the periodic table.

5.5 Scalar fields

Scalar fields are useful mathematical representations of certain types of bosons, as well as of certain types of mesons.

5.5.1 The Klein–Gordon field

Let us consider first the (hypothetical) case of a scalar field $\phi : M \to \mathbb{R}$ defined over Minkowski's spacetime $M = \mathbb{R}^{1,3}$. The relevant connection here is the flat connection $\nabla_\mu = \partial_\mu$. The Lagrangian is

$$\mathcal{L}(\phi, \partial_\mu\phi) = \frac{1}{2}(\partial_\mu\phi)(\partial^\mu\phi) - \frac{1}{2}m^2\phi^2 ,$$

where m is a constant (mass). Here and throughout, ∂^μ is simply ∂_μ raised by the metric: $\partial^\mu = g^{\mu\nu}\partial_\nu$, where $g^{\mu\nu}$ is the Minkowski metric tensor. An immediate computation yields

$$\frac{\partial\mathcal{L}}{\partial\phi} = -m^2\phi , \quad \frac{\partial\mathcal{L}}{\partial(\partial_\mu\phi)} = \partial^\mu\phi .$$

Thus, the Euler–Lagrange equations applied to this situation give rise to the dynamical field equation

$$-m^2\phi - \partial_\mu(\partial^\mu\phi) = 0 .$$

This can be written as

$$(\partial_\mu\partial^\mu + m^2)\phi = 0 ,$$

or

$$(\Box + m^2)\phi = 0 ,$$

where $\Box = \partial_\mu\partial^\mu = \partial_0^2 - \partial_1^2 - \partial_2^2 - \partial_3^2$ is the *D'Alembertian operator*. Thus, the extremal fields for the above Lagrangian satisfy a linear, second-order hyperbolic PDE, known as the *Klein–Gordon equation.*

5.5.2 Complex scalar fields

One can generalize the Klein–Gordon action to cover the case of a *complex* scalar field $\Phi : M \to \mathbb{C}$ obtained by assembling two real scalar fields ϕ_1 and ϕ_2 with the same mass m, namely $\Phi = (\phi_1 + i\phi_2)/\sqrt{2}$. We write $\Phi^* = (\phi_1 - i\phi_2)/\sqrt{2}$ for the corresponding complex conjugate field. The relevant Lagrangian is given by

$$\mathcal{L}(\Phi, \partial_\mu\Phi) = \partial_\mu\Phi^*\partial_\mu\Phi - m^2\Phi^*\Phi .$$

Note that this is merely the *sum* of the Lagrangian contributions of the two real scalar fields ϕ_1, ϕ_2. Just as in the real case, the Euler–Lagrange equations applied to this situation yield the Klein–Gordon equation. In fact, varying Φ^* yields the Klein–Gordon equation for Φ, namely

$$-\partial_\mu\partial^\mu\Phi - m^2\Phi = 0 ,$$

whereas if one varies Φ instead one gets the Klein–Gordon equation for Φ^*.

Remark 5.7 We could proceed as in the previous section on the Dirac field, and calculate the Klein–Gordon propagator, but we shall not do it here. See however our discussion of collision processes in Chapter 6.

The above Klein–Gordon Lagrangian describes an uncharged scalar field. One can also consider *charged* scalar fields. The Lagrangian for this situation is obtained from the Klein–Gordon Lagrangian through the formal substitution $i\partial_\mu \mapsto i\partial_\mu - qA_\mu$. The resulting Lagrangian,

$$\mathcal{L} = -\left[(i\partial_\mu + qA_\mu)\Phi^*\right]\left[(i\partial_\mu - qA_\mu)\Phi\right] - m^2\Phi^*\Phi ,$$

is Lorentz invariant, and the corresponding Euler–Lagrange equation is the genrealized Klein–Gordon equation, namely

$$\left[(i\partial_\mu - qA_\mu)(i\partial_\mu - qA_\mu) - m^2\right]\Phi = 0 .$$

Remark 5.8 The generalized Klein–Gordon equation is used to model the behavior of charged π^+ or π^- mesons, which are not elementary, but rather composite spin-zero particles. See Chapter 9.

5.6 Yang–Mills fields

The theory of Yang–Mills fields is a strong generalization of electromagnetism. In this theory, the internal group of symmetries is a non-abelian Lie group, typically $SU(2)$. The proper description of the Lagrangian and associated action for Yang–Mills fields requires the mathematical language of principal bundles introduced in Chapter 4, where we also described the action of pure Yang–Mills fields and also more general actions where the Yang–Mills field is coupled to some matter field – a section of a spinor bundle (in the case of fermions) or a vector bundle (in the case of bosons).

In pure Yang–Mills, we have a connection A on a principal bundle with structure group $SU(2)$, say, over spacetime M. Letting $\mathcal{F}_A = dA + A \wedge A$ denote the curvature of this connection, the pure Yang–Mills action functional is defined to be

$$S_{YM} = \int_M \mathrm{Tr}\,(\mathcal{F}_A \wedge \star\mathcal{F}_A) ,$$

where Tr denotes the trace in the Lie algebra, and $\star$ is the Hodge-$\star$ operator on (Lie algebra–valued) forms. The Euler–Lagrange equations for this action functional become

$$D\mathcal{F}_A = 0 \quad \text{and} \quad D\star\mathcal{F}_A = 0 .$$

They are called *Yang–Mills equations*. For non-abelian structure groups such as $SU(2)$, these equations are non-linear PDEs.

5.7 Gravitational fields

According to Einstein, a gravitational field is a metric deformation of spacetime given by a certain Minkowski metric. This special metric is obtained via a variational principle, as a critical point of an action functional (the Einstein–Hilbert action) on the space of all possible Minkowski metrics on spacetime.

To write down the Einstein–Hilbert action in precise mathematical terms, let us recall some basic differential geometric concepts. Let us consider a four-dimensional manifold M as the underlying spacetime manifold, and let $\mathcal{C}$ be the space of all Minkowski metrics g on M. Given $g \in \mathcal{C}$, we have at each point $x \in M$ a Lorentzian inner product $\langle \cdot, \cdot \rangle_x$, so that the metric tensor

$$g_{ij}(x) = \left\langle \frac{\partial}{\partial x^i}, \frac{\partial}{\partial x^j} \right\rangle_x$$

satisfies $\det g_{ij}(x) < 0$. Its associated volume element is

$$d^4x = \sqrt{-\det g_{ij}(x)}\, dx^1 \wedge dx^2 \wedge dx^3 \wedge dx^4 .$$

As we shall see, the action functional is defined using the *curvature* of such a metric.

5.7.1 Covariant derivative

Let us consider spacetime M with a metric g as above. Let $\mathcal{X}(M)$ be the space of vector fields on M, and pick $X \in \mathcal{X}(M)$. Recall that the *covariant derivative with respect to* X is a map

$$\nabla_X : \mathcal{X}(M) \to \mathcal{X}(M)$$

having the following properties:

(i) It satisfies the Leibnitz rule

$$\nabla_X(fY) = Df(X)Y + f\nabla_X Y$$

for all C^∞ functions f and all vector fields $Y \in \mathcal{X}(M)$.

(ii) It is compatible with the metric, in the sense that

$$D(\langle Y, Z \rangle) = \langle \nabla_X Y, Z \rangle + \langle Y, \nabla_X Z \rangle$$

for all $Y, Z \in \mathcal{X}(M)$.

(iii) It is symmetric, in the sense that

$$\nabla_X Y - \nabla_Y X = [X, Y] .$$

In local coordinates, we can write

$$\nabla_{\frac{\partial}{\partial x^i}} \frac{\partial}{\partial x^j} = \sum_{k=1}^{4} \Gamma_{ij}^{k} \frac{\partial}{\partial x^k} ,$$

where

$$\Gamma_{ij}^{k} = \frac{1}{2} \sum_{\nu=1}^{4} g^{k\nu} \left(\frac{\partial g_{\nu j}}{\partial x^i} + \frac{\partial g_{\nu i}}{\partial x^j} - \frac{\partial g_{ij}}{\partial x^\nu} \right)$$

are the so-called *Schwarz–Christoffel* symbols.

5.7.2 Curvature tensor

The curvature tensor R of our spacetime manifold (M, g) is defined by

$$R(X, Y)Z = \nabla_X \nabla_Y Z - \nabla_Y \nabla_X Z - \nabla_{[X,Y]} Z .$$

In local coordinates, we have

$$R\left(\frac{\partial}{\partial x^i}, \frac{\partial}{\partial x^j} \right) \frac{\partial}{\partial x^k} = \sum_{\nu=1}^{4} R_{ijk}^{\nu} \frac{\partial}{\partial x^\nu} ,$$

where the components R_{ijk}^{ν} can be computed either as functions of the Schwarz–Christoffel symbols or directly in terms of the components of the metric tensor.

5.7.3 Ricci and scalar curvatures

Contracting the curvature tensor R with respect to its contravariant component yields a second-order tensor known as the Ricci curvature tensor, namely

$$\mathrm{Ric}_{ij} = \sum_{\nu=1}^{4} R_{ij\nu}^{\nu} .$$

The Ricci curvature tensor can be combined with the metric to produce a scalar in the following way:

$$\mathcal{R} = \sum_{i,j=1}^{4} g^{ij} \,\mathrm{Ric}_{ij} .$$

This number is the scalar curvature of the metric g at the point $x \in M$.

5.7.4 The Einstein–Hilbert action

We now have all the necessary elements to define the Einstein–Hilbert action. This action is a function $S_{EH} : \mathcal{C} \to \mathbb{R}$ whose value at a given Minkowski metric is obtained by integrating the scalar curvature of the metric with respect to its volume element. In other words, we have

$$S_{EH} = \frac{1}{16\pi G} \int_M \mathcal{R} \sqrt{-\det g_{ij}(x)}\, d^4x \,.$$

The number G appearing in the normalizing factor before the integral is Newton's gravitational constant.

A lengthy calculation departing from the Euler–Lagrange equations for this functional arrives at the famous Einstein field equations

$$G^{ij} = \mathrm{Ric}^{ij} - \frac{1}{2}\, g^{ij}\, \mathcal{R} \,.$$

We can make the gravitational field interact with a Yang–Mills via the Hodge operator, which now is defined using the Minkovski metric g, which became a dynamical variable. As before, we can also incorporate some matter fields such as a fermion ψ, which is a section of an associated spinor bundle, and a boson ϕ, which is a scalar field. Thus we get the action

$$S(g, A, \phi, \psi) = S_{EH}(g) + \int_M \mathrm{Tr}\,(\mathcal{F}_A \wedge \star\mathcal{F}_A) + \int_M \Big(\psi^\dagger D_A \psi + \frac{1}{2}\,(\nabla\phi)^2 + \mu\psi^\dagger\psi + \frac{1}{2} m^2\phi^2 + \lambda\phi^4 + \psi^\dagger\psi\phi \Big)\, dV \,,$$

where dV is the Minkowski volume form, ∇ denotes the Minkowski gradient, $\mathcal{F}_A = dA + A \wedge A$ is the curvature of the connection A, and D_A is the Dirac operator associated to A.

Remark 5.9 Thus, we see that at the classical level gravity can be incorporated into a fairly complete field theory without much trouble. As soon as one tries to quantize this theory, however, the ensuing difficulties appear to be insurmountable.

Exercises

5.1 The purpose of this exercise is to show that Schrödinger's equation (Chapter 2) can also be derived from a variational principle. Consider the wave function $\psi = \psi(t, \boldsymbol{x})$ of a particle moving about in 3-space, subject to a potential V, and write down the Lagrangian density,

$$\mathcal{L} = -\frac{1}{2i}\left(\psi^* \frac{\partial\psi}{\partial t} - \frac{\partial\psi^*}{\partial t}\psi \right) - \frac{1}{2m}\,\langle \nabla\psi^*, \nabla\psi \rangle - \psi^*\, V\psi \,.$$

Here, the star $*$ denotes complex conjugation, as usual, and ∇ is the standard gradient with respect to the spatial coordinates (we are working with units in which Planck's constant is equal to 1).

(a) If $\delta\psi^*$ is a first variation of ψ^*, compute the corresponding first variation $\delta\mathcal{L}$.

(b) From this computation and the least-action principle, show that the resulting Euler–Lagrange equation is precisely Schrödinger's equation

$$i\frac{\partial\psi}{\partial t} = -\frac{1}{2m}\Delta\psi + V\psi .$$

Remark 5.10 Note that the above Lagrangian density is *not* Lorentz invariant.

5.2 Show that the Lagrangian for a charged scalar field is Lorentz invariant, as claimed in Section 5.5.

5.3 Let ψ denote a solution to Dirac's equation $(i\gamma^\mu\partial_\mu - m)\psi = 0$ representing a fermion in free space. Show that, if we use the decomposition $\psi = (\psi_L, \psi_R)^T$ into Weyl spinors, where

$$\begin{pmatrix}\psi_L\\ 0\end{pmatrix} = \frac{1}{2}(I-\gamma^5)\psi\ ; \quad \begin{pmatrix}0\\ \psi_R\end{pmatrix} = \frac{1}{2}(I+\gamma^5)\psi\ ,$$

then ψ_L, ψ_R satisfy the coupled equations

$$\begin{aligned} i\overline{\sigma}^\mu\partial_\mu\psi_L - m\psi_R &= 0 \\ i\sigma^\mu\partial_\mu\psi_R - m\psi_L &= 0\,, \end{aligned} \tag{E5.1}$$

where the σ^μ are the Pauli matrices (σ^0 being the 2×2 identity matrix), and where $\overline{\sigma}^0 = \sigma^0, \overline{\sigma}^j = -\sigma^j$ for $j = 1, 2, 3$.

5.4 In the situation of the previous exercise, let O' be an inertial frame with respect to which the fermion is at rest, so that in this frame its 3-momentum $\boldsymbol{p}' = 0$ and its energy E' satisfies $(E')^2 = m^2$. Verify that in this frame the coupled equations (E5.1) become the coupled equations $i\partial_0'\psi_L' = m\psi_R'$, $i\partial_0'\psi_R' = m\psi_L'$, and show that the solutions to these with positive energy $E' = m$ are

$$\psi_L' = we^{-imt'}\ , \quad \psi_R' = we^{-imt'}\ ,$$

where w is an arbitrary Weyl (i.e. two-component) spinor.

5.5 Consider now an inertial frame O with respect to which O' and the particle of the previous exercise are moving with velocity $\boldsymbol{v} = (0, 0, v)$ along the x^3-axis of O. Let

$$\Lambda = \begin{pmatrix} \cosh\theta & 0 & 0 & -\sinh\theta \\ 0 & 1 & 0 & 0 \\ 0 & 0 & 1 & 0 \\ -\sinh\theta & 0 & 0 & \cosh\theta \end{pmatrix}$$

be the Lorentz boost taking O to O'.

(a) Show that $m \cosh\theta = m\gamma = E$, where $\gamma = (1 - v^2/c^2)^{-1/2}$, and that $m \sinh\theta = mv\gamma = p$ (the relativistic momentum).

(b) Using these facts and the result of the previous exercise, show that

$$\psi_L = e^{i(-Et+px^3)} \begin{pmatrix} e^{-\theta/2} \\ 0 \end{pmatrix}, \quad \psi_R = e^{i(-Et+px^3)} \begin{pmatrix} e^{\theta/2} \\ 0 \end{pmatrix} \tag{E5.2}$$

is the solution of the coupled equations in the frame O that corresponds to the choice of initial condition $w = (1, 0)^T$ in the frame O'.

(c) Write down the analogous expressions in O for the solution that corresponds to the initial condition $w = (0, 1)^T$ in the frame O'.

5.6 The *intrinsic spin* $\boldsymbol{S}$ of a massive particle is defined to be its angular momentum operator (acting on spinor fields) in a frame with respect to which it is at rest. In such a rest frame, $\boldsymbol{S}$ is represented by the 4×4 matrix

$$\boldsymbol{\Sigma} = \frac{1}{2} \begin{pmatrix} \boldsymbol{\sigma} & 0 \\ 0 & \boldsymbol{\sigma} \end{pmatrix},$$

where $\boldsymbol{\sigma} = (\sigma^1, \sigma^2, \sigma^3)$. The *helicity* operator $\boldsymbol{h}$ is defined as

$$\boldsymbol{h} = \boldsymbol{\Sigma} \cdot \hat{\boldsymbol{p}},$$

where $\hat{\boldsymbol{p}} = \boldsymbol{p}/|\boldsymbol{p}|$ is the unit vector pointing in the direction of the particle's motion. Thus, classically, the helicity measures the projection of the particle's intrinsic spin in the direction of motion. Note that when $\boldsymbol{p} = (0, 0, p)$, we have $\boldsymbol{h} = \boldsymbol{\Sigma}_3 = \frac{1}{2}\text{diag}(\sigma^3, \sigma^3)$.

(a) Show that the positive-energy solution in (E5.2), written in normalized form as a Dirac spinor as

$$\psi_+ = \frac{1}{\sqrt{2}} e^{i(-Et+px^3)} \begin{pmatrix} e^{-\theta/2} \\ 0 \\ e^{\theta/2} \\ 0 \end{pmatrix},$$

is an eigenvector of the helicity operator $\boldsymbol{h} = \boldsymbol{\Sigma}_3$ with eigenvalue $1/2$. The normalizing constant $1/\sqrt{2}$ is chosen so that $\overline{\psi_+}\psi_+ = \psi_+^\dagger \gamma^0 \psi_+ = \psi_L^\dagger \psi_R + \psi_R^\dagger \psi_L = 1$.

(b) Similarly, show that the positive-energy solution obtained in the last item of the previous exercise, which can be written in normalized form as a Dirac spinor as

$$\psi_- = \frac{1}{\sqrt{2}} e^{i(-Et+px^3)} \begin{pmatrix} 0 \\ e^{\theta/2} \\ 0 \\ e^{-\theta/2} \\ 0 \end{pmatrix},$$

is an eigenvector of the helicity operator with eigenvalue $-1/2$.

5.7 Generalize the previous exercise to plane wave solutions to Dirac's equation with positive energy E and arbitrary 3-momentum $\boldsymbol{p}$, as follows. Consider the Pauli operator $\boldsymbol{\sigma} \cdot \hat{\boldsymbol{p}} = p_j \sigma^j$, and let $|\pm\rangle$ denote the eigenvectors of this operator, with eigenvalues ± 1, respectively.

(a) Show that the spinor field $\psi_+ = e^{i(-Et+\boldsymbol{p}\cdot\boldsymbol{x})} u_+(\boldsymbol{p})$, where

$$u_+(\boldsymbol{p}) = \frac{1}{\sqrt{2}} \begin{pmatrix} e^{-\theta/2}|+\rangle \\ e^{\theta/2}|+\rangle \end{pmatrix},$$

is an eigenvector of the helicity operator $\boldsymbol{h}$ with eigenvalue $1/2$.

(b) Similarly, show that the spinor field $\psi_- = e^{i(-Et+\boldsymbol{p}\cdot\boldsymbol{x})} u_-(\boldsymbol{p})$, where

$$u_-(\boldsymbol{p}) = \frac{1}{\sqrt{2}} \begin{pmatrix} e^{\theta/2}|-\rangle \\ e^{-\theta/2}|-\rangle \end{pmatrix},$$

is an eigenvector of the helicity operator $\boldsymbol{h}$ with eigenvalue $-1/2$.

5.8 Perform computations analogous to the ones above for *negative*-energy plane-wave solutions of Dirac's equation (i.e., those with $E < 0$) and deduce that the negative-energy eigenvectors of the helicity operator $\boldsymbol{h}$ with eigenvalues $1/2$ and $-1/2$ are respectively

$$\psi_+ = e^{i(Et-\boldsymbol{p}\cdot\boldsymbol{x})} v_+(\boldsymbol{p}), \quad \psi_- = e^{i(Et-\boldsymbol{p}\cdot\boldsymbol{x})} v_-(\boldsymbol{p}),$$

where

$$v_+(\boldsymbol{p}) = \frac{1}{\sqrt{2}} \begin{pmatrix} e^{\theta/2}|-\rangle \\ -e^{-\theta/2}|-\rangle \end{pmatrix}, \quad v_-(\boldsymbol{p}) = \frac{1}{\sqrt{2}} \begin{pmatrix} -e^{\theta/2}|+\rangle \\ e^{-\theta/2}|+\rangle \end{pmatrix}.$$

5.9 Combining the exercises above with the superposition principle for plane waves, show that the general free-space solution to Dirac's equation has the following Fourier integral representation:

$$\psi(x) = \frac{1}{(2\pi)^3} \int_{\mathbb{R}^3} \frac{1}{\sqrt{2\omega_{\boldsymbol{p}}}} \sum_{s=-,+} \left(a_{p,s} u_s(p) e^{-ip\cdot x} + b^{\dagger}_{p,s} v_s(p) e^{ip\cdot x} \right) d^3p.$$

Here, we write $x = (ct, \boldsymbol{x})$ for the 4-position vector and $p = (\omega_{\boldsymbol{p}}/c, \boldsymbol{p})$ for the 4-momentum (so $E = \omega_{\boldsymbol{p}}$ is the energy), and $d^3p = dp^1 dp^2 dp^3$ for the volume element in 3-momentum space.

5.10 Show that, in the Weyl spinor (or chiral) representation, the left and right components of the charge-conjugate field ψ^C to a Dirac field ψ are given by

$$\psi^C_L = -i\sigma^2 \psi^*_R \quad \text{and} \quad \psi^C_R = i\sigma^2 \psi^*_L.$$

6
Quantization of classical fields

In this chapter, we describe the basic ideas of the quantum theory of fields. The greater portion of this chapter is devoted to understanding the quantization of *free fields*. The theory of free quantum fields can be treated in a mathematically precise, axiomatic way, as shown by Wightman. The (constructive) theory of interacting quantum fields is at present not so well developed. But we do discuss scattering processes in some detail at the end of the chapter. From a physicist's standpoint, such processes are extremely important, because they provide a link between quantum field theory and the reality of laboratory experiments.

6.1 Quantization of free fields: general scheme

As we saw in Chapter 2, to perform the quantization of a harmonic oscillator we need to factor the Hamiltonian as a product of two mutually adjoint operators: the particle annihilator operator, which lowers the energy level of the Hamiltonian, and the particle creator operator, which raises the energy level. These two operators completely describe the spectrum of the Hamiltonian.

The basic physical idea behind the quantization of the classical free fields (Klein–Gordon, Dirac, Maxwell) is the following. Because the Lagrangians of these fields are quadratic, the corresponding Euler–Lagrange equations are linear PDEs. Therefore we can use the Fourier transform to diagonalize the quadratic form in the Lagrangian, thereby decoupling the Euler–Lagrange equations.

In the case of the Klein–Gordon equation, corresponding to scalar bosons, each Fourier mode is a harmonic oscillator that is quantized as described above. Hence we obtain a family of *pairs* of operators, one that creates a particle with a given momentum and the other that destroys a particle with that same momentum. The scalar field (corresponding to the classical solution to the Klein–Gordon equation) is then written as a Fourier integral having these operators as coefficients. These operators satisfy the commutativity relations of the harmonic oscillator. The next step is to construct the Hilbert space on which these operators act. We start with the so-called vacuum state, the state that is annihilated by all the operators that destroy particles. From the vacuum we construct states with a finite number of particles, using the operators that create particles. Taking all possible linear combinations of these states, we get a vector space with an inner product whose completion yields the desired Hilbert space.

In the case of the Dirac equation, corresponding to fermions, one proceeds along similar lines. There are some important conceptual distinctions, however. The classical analogue of the harmonic oscillator for the Dirac equation is not immediately available. The algebra of spinors imposes commutation relations different from the ones in the case of bosons. In remarkable contrast to the case of the harmonic oscillator, these Dirac commutation relations (constituting what is known as a Grassmann algebra) can be realized in a finite-dimensional Hilbert space!

In the case of Maxwell's equations, first we have to eliminate the gauge symmetry that is present (this is called gauge fixing). We then proceed in the same manner as above.

6.2 Axiomatic field theory

In this section, we present the axioms formulated by Wightman for the quantum field theory of scalar bosonic fields (bosons).

6.2.1 The Wightman axioms

Let us consider spacetime $M = \mathbb{R}^{1,3}$ with the Lorentz scalar product. The group of linear transformations that leave the corresponding metric invariant is, as we know, the Lorentz group. The Poincaré group $\mathcal{P}$ is the semi-direct product of the group of translations ($\mathbb{R}^4$) by the Lorentz group. An element of of $\mathcal{P}$ is denoted (a, Λ), where $a \in \mathbb{R}^4$ and Λ is an element of the Lorentz group. The group operation reads

$$(a, \Lambda) \cdot (a', \Lambda') = (a + \Lambda a', \Lambda\Lambda') \,.$$

Definition 6.1 *A* scalar Hermitian quantum field theory *consists of a separable Hilbert space $\mathcal{H}$ whose elements are called states, a unitary representation U of the Poincaré group $\mathcal{P}$ in $\mathcal{H}$, an operator-valued distribution φ on $\mathcal{S}(\mathbb{R}^4)$ (with values in the unbounded operators of $\mathcal{H}$), and a dense subspace $\mathcal{D} \subset \mathcal{H}$ such that the following properties hold.*

WA1 (relativistic invariance of states). *The representation $U : \mathcal{P} \to \mathcal{U}(\mathcal{H})$ is strongly continuous. Let $P_0, \ldots, P_3$ be the infinitesimal generators of the one-parameter groups $t \mapsto U(te^\mu, I)$, for $\mu = 0, \ldots, 3$.*

WA2 (spectral condition). *The operators P_0 and $P_0^2 - P_1^2 - P_2^2 - P_3^2$ are positive operators (the spectral measure in $\mathbb{R}^4$ corresponding to the restricted representation $\mathbb{R}^4 \ni a \mapsto U(a, I)$ has support in the positive light cone).*

WA3 (existence and uniqueness of the vacuum). *There exists a unique state $\psi_0 \in \mathcal{D} \subset \mathcal{H}$ such that*

$$U(a, I)\psi_0 = \psi_0 \quad \text{for all } a \in \mathbb{R}^4 .$$

This property actually implies, in combination with the first axiom, that $U(a, \Lambda)\psi_0 = \psi_0$ for all $a \in \mathbb{R}^4$ and all $\Lambda \in \mathcal{L}$. It also implies that the projection $P_{\{(0,0,0,0)\}}$ is non-trivial and its image is unidimensional.

WA4 (invariant domains for fields). *The map $\varphi : \mathcal{S}(\mathbb{R}^4) \to \mathcal{O}(\mathcal{H})$ satisfies the following:*

(a) for each $f \in \mathcal{S}(\mathbb{R}^4)$ the domains $D(\varphi(f))$ of $\varphi(f)$ and $D(\varphi(f)^*)$ of $\varphi(f)^*$ both contain $\mathcal{D}$ and the restrictions of these two operators to D agree;

(b) $\varphi(f)\mathcal{D} \subseteq \mathcal{D}$;

(c) for every fixed state $\psi \in \mathcal{D}$ the map $f \mapsto \varphi(f)\psi$ is linear.

WA5 (regularity of the field). *For every $\psi_1, \psi_2 \in \mathcal{D}$, the map $\mathcal{S}(\mathbb{R}^4) \ni f \mapsto \langle \psi_1 , \varphi(f)\psi_2 \rangle$ is a tempered distribution.*

WA6 (Poincaré invariance). *For all $(a, \Lambda) \in \mathcal{P}$ we have $U(a, \Lambda)\mathcal{D} \subseteq \mathcal{D}$ and for all test functions $f \in \mathcal{S}(\mathbb{R}^4)$ and all states $\psi \in \mathcal{D}$ we have*

$$U(a, \Lambda)\varphi(f)U(a, \Lambda)^{-1}\, \psi = \varphi((a, \Lambda)f)\, \psi .$$

WA7 (microscopic causality or local commutativity). *If the supports of two test functions $f, g \in \mathcal{S}(\mathbb{R}^4)$ are spacelike separated (i.e. if $f(x)g(y) = 0$ whenever $x - y$ does not lie in the positive light cone), then the commutator of the corresponding operators vanishes:*

$$[\varphi(f)\varphi(g) - \varphi(g)\varphi(f)] = 0 .$$

WA8 (cyclicity of the vacuum). *The set $\mathcal{D}_0 \subset \mathcal{H}$ of finite linear combinations of all vectors of the form $\varphi(f_1) \cdots \varphi(f_n)\psi_0$ is dense in $\mathcal{H}$.*

These are the Wightman axioms for a scalar field theory. These axioms require some modifications for spinorial field theories.

6.2.2 Wightman correlation functions

Here, we introduce the so-called Wightman correlation functions. We also state Wightman's reconstruction theorem, which says that, given the correlation functions, we can reconstruct the Hilbert space, the quantum fields, and the entire QFT satisfying the Wightman axioms from these correlations alone.

For simplicity, let us consider only the case of a quantum scalar field theory.

Definition 6.2 *The* nth Wightman correlation function *of a quantum field theory satisfying the Wightman axioms is a function* $\mathcal{W}_n : \mathcal{S}(\mathbb{R}^4) \times \mathcal{S}(\mathbb{R}^4) \times \cdots \mathcal{S}(\mathbb{R}^4) \to \mathbb{C}$ *given by*

$$\mathcal{W}_n(f_1, f_2, \ldots, f_n) = \langle \psi_0, \phi(f_1)\phi(f_2)\cdots\phi(f_n)\psi_0 \rangle \ ,$$

for all $f_j \in \mathcal{S}(\mathbb{R}^4)$, *where* ψ_0 *is the vacuum vector of the theory and* ϕ *is its field operator.*

We can associate to each Wightman function $\mathcal{W}_n$ a tempered distribution in a way that we proceed to describe. The association will depend on a version for tempered distributions of the so-called *nuclear theorem*, or *kernel theorem*, of Schwartz. Recall that if $f : \mathbb{R}^m \to \mathbb{C}$ and $g : \mathbb{R}^n \to \mathbb{C}$, then their tensor product $f \otimes g : \mathbb{R}^{m+n} \to \mathbb{C}$ is defined by

$$f \otimes g(x_1, \ldots, x_{m+n}) = f(x_1, \ldots, x_m)\, g(x_{m+1}, \ldots, x_{m+n}) \ .$$

The kernel theorem for tempered distributions can be stated as follows. The proof of this theorem is not elementary (Schwartz originally deduced it from Grothendieck's theory of nuclear spaces).

Theorem 6.3 *Let* $B : \mathcal{S}(\mathbb{R}^m) \times \mathcal{S}(\mathbb{R}^n) \to \mathbb{C}$ *be a separately continuous bilinear function. Then there exists a tempered distribution* $\beta \in \mathcal{S}'(\mathbb{R}^{m+n})$ *such that* $B(f, g) = \beta(f \otimes g)$ *for all* $f \in \mathcal{S}(\mathbb{R}^m)$ *and all* $g \in \mathcal{S}(\mathbb{R}^n)$.

Proof. See [FJ, pp. 70–3] for an elementary proof of a version of this theorem in which $\mathcal{S}, \mathcal{S}'$ are replaced by $\mathcal{D}, \mathcal{D}'$ (ordinary distributions). For the original proof, see [S]. For a short proof, see [Ehr]. □

Using the kernel theorem, we define the Wightman distributions as follows. The nth Wightman correlation function $\mathcal{W}_n(f_1, \ldots, f_n)$ is a multilinear function of the test functions $f_1, \ldots, f_n$. Hence by Theorem 6.3 and induction, there exists a tempered distribution $W_n \in \mathcal{S}'(\mathbb{R}^{4n})$ such that

$$\mathcal{W}_n(f_1, \ldots, f_n) = W_n(f_1 \otimes \cdots \otimes f_n) \ ,$$

for all $f_1, \ldots, f_n \in \mathcal{S}(\mathbb{R}^4)$.

Before we can state our next theorem, concerning the basic properties of the Wightman distributions, we need the following notions and notations.

(i) An involution on test functions: If $f \in \mathcal{S}(\mathbb{R}^{4n})$, we write $f^*(x_1, \ldots, x_n) = \overline{f(x_n, \ldots, x_1)}$, for all $x_j \in \mathbb{R}^4$. In particular, for

$n = 0, 1$ the $*$ involution is simply complex conjugation. Note also that if $f \in \mathcal{S}(\mathbb{R}^{4n})$ is of the form $f = f_1 \otimes \cdots \otimes f_n$ with $f_j \in \mathcal{S}(\mathbb{R}^4)$, then $f^* = f_n^* \otimes \cdots \otimes f_1^*$.

(ii) An action of the Poincaré group on test functions: If $(a, \Lambda) \in \mathcal{P}$ and $f \in \mathcal{S}(\mathbb{R}^{4n})$, we let $(a, \Lambda) \cdot f$ be given by

$$(a, \Lambda) \cdot f(x_1, \ldots, x_n) = f(\Lambda^{-1}(x_1 - a), \ldots, \Lambda^{-1}(x_n - a)) .$$

(iii) For each test function $f \in \mathcal{S}(\mathbb{R}^{4n})$, let $\hat{f} \in \mathcal{S}(\mathbb{R}^{4n-4})$ be the test function given by

$$\hat{f}(x_1, x_2, \ldots, x_{n-1}) = f(0, x_1 + x_2, \ldots, x_1 + x_2 + \cdots + x_{n-1}) .$$

(iv) A translation operator: If $a \in \mathbb{R}^4$ is a spacelike vector and $0 \leq j \leq n$, let $T_{a,j} : \mathcal{S}(\mathbb{R}^{4n}) \to \mathcal{S}(\mathbb{R}^{4n})$ be given by

$$T_{a,j} f(x_1, \ldots, x_n) = f(x_1, \ldots, x_j, x_{j+1} - a, \ldots, x_n - a) .$$

Theorem 6.4 *The Wightman distributions of a (scalar hermitian) quantum field theory enjoy the following properties.*

WA1 (positive definiteness). *Given $f_0 \in \mathbb{C}$ and $f_j \in \mathcal{S}(\mathbb{R}^{4j})$ for $j = 1, \ldots, n$, we have*

$$\sum_{j,k=0}^{n} W_{j+k}(f_k^* \otimes g_j) \geq 0 .$$

WA2 (reality). *We have $W_n(f^*) = \overline{W_n(f)}$ for each test function $f \in \mathcal{S}(\mathbb{R}^{4n})$.*

WA3 (Lorentz–Poincaré covariance). *For all $(a, \Lambda) \in \mathcal{P}$ and each test function $f \in \mathcal{S}(\mathbb{R}^{4n})$, we have $W_n((a, \Lambda) \cdot f) = W_n(f)$.*

WA4 (spectrum condition). *For each $n > 0$, there exists a tempered distribution $D_n \in \mathcal{S}'(\mathbb{R}^{4n-4})$ such that $W_n(f) = D_n(\hat{f})$ for all test functions $f \in \mathcal{S}(\mathbb{R}^{4n})$, with the property that the support of its Fourier transform $\mathcal{F}D_n$ is contained in $\overline{V_+^{n-1}}$, where $V_+ \subset \mathbb{R}^4$ is the positive light cone in Minkowski (momentum) space.*

WA5 (locality). *If f_j and f_{j+1} are spacelike separated test functions in $\mathcal{S}(\mathbb{R}^4)$, then*

$$\begin{aligned} &W_n(f_1 \otimes \cdots \otimes f_j \otimes f_{j+1} \otimes \cdots \otimes f_n) \\ &\quad = W_n(f_1 \otimes \cdots \otimes f_{j+1} \otimes f_j \otimes \cdots \otimes f_n) . \end{aligned}$$

WA6 (cluster property). *If $a \in \mathbb{R}^4$ is a spacelike vector then for all $0 \leq j \leq n$ we have*

$$\lim_{\lambda \to \infty} W_n \circ T_{\lambda a, j} = W_j \otimes W_{n-j} \,.$$

The reconstruction theorem of Wightman is a converse to the above theorem.

Theorem 6.5 (reconstruction theorem) *Let $(W_n)_{n \geq 0}$ be a sequence of tempered distributions satisfying WD1–WD6 above. Then there exists a (Hermitian scalar) quantum field theory $(\phi, \mathcal{H}, U, \mathcal{D}, \psi_0)$ satisfying the Wightman axioms, whose Wightman distributions are precisely the W_n's. In other words, for all test functions $f_1, \ldots, f_n \in \mathcal{S}(\mathbb{R}^4)$, we have*

$$W_n(f_1 \otimes \cdots \otimes f_n) = \langle \psi_0, \phi(f_1) \cdots \phi(f_n) \psi_0 \rangle \,.$$

Such quantum field theory is unique up to unitary equivalence.

Proof. For a proof, see [Gu, p. 189], or the original reference [SW]. □

6.3 Quantization of bosonic free fields

Let us now give the first important example of a free QFT. We shall construct the field theory for free bosons quite explicitly.

6.3.1 Fock spaces

Let us start with a mathematical definition. Let $\mathcal{H}$ be a complex Hilbert space.

Definition 6.6 *The Fock space of $\mathcal{H}$ is the direct sum of Hilbert spaces*

$$\mathcal{F}(\mathcal{H}) = \bigoplus_{n=0}^{\infty} \mathcal{H}^{(n)} \,,$$

where $\mathcal{H}^{(0)} = \mathbb{C}$ and

$$\mathcal{H}^{(n)} = \mathcal{H} \otimes \cdots \otimes \mathcal{H} = \bigotimes_{j=1}^{n} \mathcal{H} \,.$$

The vector $\Omega_0 = (1, 0, 0, \ldots) \in \mathcal{F}(\mathcal{H})$ is called the vacuum vector.

Given $\psi \in \mathcal{F}(\mathcal{H})$, we write $\psi^{(n)}$ for the orthogonal projection of ψ onto $\mathcal{H}^{(n)}$. The set $F_0 \subset \mathcal{F}(\mathcal{H})$ consisting of those ψ such that $\psi^{(n)} = 0$ for all sufficiently large n is a dense subspace of Fock space.

Now we define, for each n, the symmetrization operator $S_n : \mathcal{H}^{(n)} \to \mathcal{H}^{(n)}$ by

$$S_n(\psi_1 \otimes \cdots \otimes \psi_n) = \frac{1}{n!} \sum_{\sigma} \psi_{\sigma(1)} \otimes \cdots \otimes \psi_{\sigma(n)} ,$$

extending it by linearity to all of $\mathcal{H}^{(n)}$. Likewise, the anti-symmetrization operator $A_n : \mathcal{H}^{(n)} \to \mathcal{H}^{(n)}$ is defined by

$$A_n(\psi_1 \otimes \cdots \otimes \psi_n) = \frac{1}{n!} \sum_{\sigma} (-1)^{\mathrm{sign}(\sigma)} \psi_{\sigma(1)} \otimes \cdots \otimes \psi_{\sigma(n)} ,$$

extended by linearity as before. It is a very simple fact that both S_n and A_n are projections, in the sense that $S_n^2 = S_n$ and $A_n^2 = A_n$.

Using the symmetrization operators, we define the space of states for n bosons as

$$\mathcal{H}_s^{(n)} = \left\{ \psi \in \mathcal{H}^{(n)} \, : \, S_n \psi = \psi \right\} .$$

Using the anti-symmetrization operators, we define the space of states for n fermions to be

$$\mathcal{H}_a^{(n)} = \left\{ \psi \in \mathcal{H}^{(n)} \, : \, A_n \psi = \psi \right\} .$$

With these at hand, we define the bosonic Fock space as

$$\mathcal{F}_s(\mathcal{H}) = \bigoplus_{n=0}^{\infty} \mathcal{H}_s^{(n)}$$

and the fermionic Fock space as

$$\mathcal{F}_a(\mathcal{H}) = \bigoplus_{n=0}^{\infty} \mathcal{H}_a^{(n)} .$$

Example 6.7 *Let us consider a measure space M with finite measure μ. We take as our Hilbert space $\mathcal{H} = L^2(M, d\mu)$. Then*

$$\begin{aligned} \mathcal{H}^{(n)} &= L^2(M \times \cdots \times M, \, d\mu \otimes \cdots \otimes d\mu) \\ &= \Big\{ \psi : M \times \cdots \times M \to \mathbb{C} : \\ &\qquad \int_{M \times \cdots \times M} |\psi(x_1, \cdots, x_n)|^2 \, d\mu(x_1) \cdots d\mu(x_n) < \infty \Big\} . \end{aligned}$$

We have also

$$\mathcal{H}_s^{(n)} = \left\{ \psi \in \mathcal{H}^{(n)} \, : \, \psi(x_{\sigma(1)}, \ldots, x_{\sigma(n)}) = \psi(x_1, \ldots, x_n) \right\} .$$

In addition, ψ is an element of Fock space $\mathcal{F}(\mathcal{H})$ if and only if

$$\psi = (\psi^{(0)}, \psi^{(1)}, \ldots, \psi^{(n)}, \ldots)$$

with $\psi^{(n)} \in \mathcal{H}^{(n)}$ *for all n, and*

$$\sum_{n=0}^{\infty} \int_{M\times\cdots\times M} |\psi^{(n)}|^2 \, d\mu \otimes \cdots \otimes d\mu < \infty .$$

6.3.2 Operators in Fock spaces

Every unitary operator $U : \mathcal{H} \to \mathcal{H}$ in the Hilbert space $\mathcal{H}$ induces a unitary operator $\Gamma(U) : \mathcal{F}(\mathcal{H}) \to \mathcal{F}(\mathcal{H})$ in the Fock space of $\mathcal{H}$, given by

$$\Gamma(U)\big|_{\mathcal{H}^{(n)}} = U \otimes \cdots \otimes U .$$

Now, let $D \subset \mathcal{H}$ be a dense subspace and let $A : D \to \mathcal{H}$ be an essentially self-adjoint operator. Note that

$$D_A = \left\{ \psi \in F_0 \; : \; \psi^{(n)} \in \bigotimes_{j=1}^{n} D \right\}$$

is dense in $\mathcal{F}(\mathcal{H})$. We define an operator $d\Gamma(A) : \; D_A \to \mathcal{F}(\mathcal{H})$ by

$$d\Gamma(A)\big|_{D_A \cap \mathcal{H}^{(n)}} = A \otimes I \otimes \cdots \otimes I + I \otimes A \otimes \cdots \otimes I + \cdots + I \otimes I \otimes \cdots A .$$

This operator is essentially self-adjoint in D_A, and it is called the second quantization of A. An important special case is the operator $N = d\Gamma(I)$. For this operator, every vector $\psi \in \mathcal{H}^{(n)}$ is an eigenvector of N with eigenvalue n; in other words, $N\psi = n\psi$. For this reason, N is called the number operator (number of particles).

6.3.3 Segal quantization operator

We shall construct the QFT for a free scalar field following the so-called Segal quantization scheme. This scheme uses creation and annihilation operators, just as in the case of the quantum harmonic oscillator (see Chapter 2).

First we define, for each $f \in \mathcal{H}$, a map $b^-(f) : \mathcal{H}^{(n)} \to \mathcal{H}^{(n-1)}$ by

$$b^-(f)(\psi_1 \otimes \psi_2 \otimes \cdots \otimes \psi_n) = \langle f, \psi_1 \rangle \, \psi_2 \otimes \cdots \otimes \psi_n .$$

This operator extends to a bounded linear operator on Fock space, $b^-(f) : \mathcal{F}(\mathcal{H}) \to \mathcal{F}(\mathcal{H})$, with norm equal to $\|f\|$. It is adjoint to the operator $b^+(f) : \mathcal{F}(\mathcal{H}) \to \mathcal{F}(\mathcal{H})$. This is quite simple to describe: $b^+(f)$ maps $\mathcal{H}^{(n-1)}$ to $\mathcal{H}^{(n)}$ and is such that

$$b^+(f)(\psi_1 \otimes \psi_2 \otimes \cdots \otimes \psi_n) = f \otimes \psi_1 \otimes \psi_2 \otimes \cdots \otimes \psi_n ,$$

again extended to Fock space linearly. The operators $b^-(f), b^+(f)$ have the symmetric and anti-symmetric Fock spaces as invariant subspaces. We remark also that the maps $f \mapsto b^-(f)$ and $f \mapsto b^+(f)$ are complex linear and complex anti-linear, respectively.

We are now ready to define the creation and annihilation operators.

Definition 6.8 *The* particle annihilation operator *is an operator on symmetric Fock space,*

$$a^-(f) : F_0 \subset \mathcal{F}_s(\mathcal{H}) \to \mathcal{F}_s(\mathcal{H}) ,$$

given by

$$a^-(f) = \sqrt{N+1}\, b^-(f) .$$

The particle creation operator *is defined taking $a^+(f) = (a^-(f))^*$.*

Note that N is a positive operator, so $\sqrt{N+1}$ is well defined. It is not difficult to check that if $\psi, \eta \in F_0$ then

$$\left\langle \sqrt{N+1} b^-(f)\psi, \eta \right\rangle = \left\langle \psi, S b^+(f)\sqrt{N+1}\eta \right\rangle .$$

Therefore, we have

$$a^+(f)\big|_{F_0} = S b^+(f)\sqrt{N+1} .$$

Example 6.9 *Let us go back to Example 6.7, where the Hilbert space is $\mathcal{H} = L^2(M, d\mu)$. Recall that $\mathcal{H}^{(n)} = L^2(M \times \cdots \times M, d\mu \otimes \cdots \otimes d\mu)$. The annihilation operator here is given by*

$$(a^-(f)\psi)^{(n)}(x_1, \ldots, x_n) = \sqrt{n+1} \int_M \overline{f(x)} \psi^{(n+1)}(x, x_1, \ldots, x_n) d\mu(x) .$$

The creation operator can be written in the following way:

$$(a^+(f)\psi)^{(n)}(x_1, \ldots, x_n) = \frac{1}{\sqrt{n}} \sum_{j=1}^{n} f(x_j)\psi^{(n-1)}(x_1, \ldots, \widehat{x_j}, \ldots, x_n) .$$

Definition 6.10 *The* Segal quantization map

$$\mathcal{H} \ni f \mapsto \Phi_s(f) : F_0 \to \mathcal{F}_s(\mathcal{H})$$

is given by

$$\Phi_s(f) = \frac{1}{\sqrt{2}} \left(a^-(f) + a^+(f) \right) .$$

Theorem 6.11 *Let $\mathcal{H}$ be a complex Hilbert space, and let Φ_s be its Segal quantization map. Then*

(i) For each $f \in \mathcal{H}$, the operator $\Phi_s(f)$ is essentially self-adjoint on F_0.

(ii) *The vacuum vector Ω_0 is cyclic, i.e. it is in the domain of every product of the form $\Phi_s(f_1)\Phi_s(f_2)\cdots\Phi_s(f_n)$, and the set*

$$\left\{\Phi_s(f_1)\Phi_s(f_2)\cdots\Phi_s(f_n)\Omega_0 \,:\, n \geq 0,\ f_j \in F_0\right\}$$

is total in the symmetric Fock space.

(iii) *For all $\psi \in F_0$ and all $f, g \in \mathcal{H}$, we have the commutation relation*

$$\Phi_s(f)\Phi_s(g)\psi - \Phi_s(g)\Phi_s(f)\psi = i\,\mathrm{Im}\,\langle f, g\rangle\,\psi\,.$$

Moreover, the unitary operators $W(f) = e^{i\Phi_s(f)}$ satisfy

$$W(f+g) = e^{-i\,\mathrm{Im}\,\langle f,g\rangle/2}W(f)W(g)\,.$$

(iv) *If $f_n \to f$ in $\mathcal{H}$, then*

$$W(f_n)\psi \to W(f)\psi \quad \text{for all } \psi \in \mathcal{F}_s(\mathcal{H})$$

$$\Phi_s(f_n)\psi \to \Phi_s(f)\psi \quad \text{for all } \psi \in F_0\,.$$

(v) *For every unitary operator $U : \mathcal{H} \to \mathcal{H}$, the corresponding unitary operator $\Gamma(U)$ in Fock space maps the closure of $D(\Phi_s(f))$ into the closure of $D(\Phi_s(Uf))$, and for all $\psi \in \overline{D(\Phi_s(Uf))}$, we have*

$$\Gamma(U)\overline{\Phi_s(f)}\Gamma^{-1}\psi = \overline{\Phi_s(Uf)}\psi\,.$$

6.3.4 Free scalar bosonic QFT

Using this theorem, one can establish the quantization of the free scalar bosonic field of mass m. The end result is a free QFT theory in the sense of Wightman. The starting point of the construction is to consider the Hilbert space $\mathcal{H} = L^2(M_m,\, d\mu_m)$, where $M_m \subset \mathbb{R}^{1,3}$ is the hyperboloid of Example 4.22 and μ_m is the hyperbolic volume form on M_m (recall that M_m is isometric to hyperbolic 3-space). Next we let $E : \mathcal{S}(\mathbb{R}^4) \to \mathcal{H}$ be the map given by

$$Ef = \sqrt{2\pi}\,\hat{f}\big|_{M_n}\,,$$

where $\hat{f}$ denotes the Fourier transform in Minkowski space, namely

$$\hat{f}(p) = \frac{1}{(2\pi)^2}\int_{\mathbb{R}^4} e^{ip\cdot\tilde{x}}\,f(x)\,dx\,.$$

Here, $\tilde{x} = (x^0, -x^1, -x^2, -x^3)$ whenever $x = (x^0, x^1, x^2, x^3)$. Using the Segal quantization map Φ_s for $\mathcal{H}$, we take the bosonic fields to be given by $\phi_m(f) = \Phi_s(Ef)$ for all test functions $f \in \mathcal{S}(\mathbb{R}^4)$. Finally, we consider the unitary representation $U = U_m$ of the Poincaré group in $\mathcal{H}$ given by

$$(U_m(a;\Lambda)\psi)(p) = e^{ip\cdot\tilde{a}}\,\psi(\Lambda^{-1}p)\,.$$

The corresponding operators $\Gamma(U_m(a;\Lambda))$ yield a unitary representation of the Poincaré group in Fock space. We have, at last, the following theorem.

Theorem 6.12 *The symmetric Fock space $\mathcal{F}_s(L^2(M_m, d\mu_m))$, with its vacuum vector Ω_0 and its dense subspace F_0, the unitary representation $\Gamma(U_m(\cdot,\cdot))$ and the operator-valued distributions $\phi_m(\cdot)$ satisfy the Wightman axioms for a bosonic scalar QFT. Moreover, for every f in Schwartz space $\mathcal{S}(\mathbb{R}^4)$ we have*

$$\phi_m\left(\left(\Box + m^2\right) f\right) = 0\,.$$

Proof. The proof uses Theorem 6.11. See [RS2, p. 213]. □

Note that the last part of the statement of Theorem 6.12 expresses the fact that the field operator ϕ_m satisfies the Klein–Gordon equation in the distributional sense.

6.3.5 The physicist's viewpoint

Let us now discuss the quantization of free scalar fields from a different point of view, which is more familiar to physicists. In the above mathematically precise construction of scalar fields, we started with the appropriate definition of Fock space and ended with a field satisfying the Klein–Gordon equation in the distributional sense. The method explained here in some sense reverses this process. We start with a classical solution to the free Klein–Gordon equation and in the end, having the quantum field at hand, we reconstruct the Fock space. We do not mean to imply that the physicist's approach is less valid in any way: in fact, it is equivalent to what has already been done through Segal quantization.

Recall that the Klein–Gordon action for a classical free scalar field of mass m is given by

$$S = \frac{1}{2}\int \left(\partial_\mu\phi\partial^\mu\phi - m^2\phi\right)\, d^4x\,.$$

As we know, the Euler–Lagrange equation for this functional yields the free Klein–Gordon equation

$$\Box\phi + m^2\phi = 0\,. \tag{6.1}$$

A plane wave of the form $e^{\pm ip\cdot x}$ is a solution to this equation if and only if $p \in \mathbb{R}^4$ satisfies the mass-shell condition $(p^0)^2 - \boldsymbol{p}^2 = m^2$. These are the Fourier modes into which a general solution of (6.1) should decompose. Using the superposition of such plane waves (in other words, using the Fourier transform) and taking into account that we are looking for real scalar solutions to (6.1), we have the following formula for the general solution of the Klein–Gordon equation:

$$\phi(x) = \frac{1}{(2\pi)^3} \int_{R^3} \frac{1}{\sqrt{2\omega_p}} \left(a_p e^{-ip\cdot x} + a_p^* e^{ip\cdot x} \right) \Big|_{p^0=\omega_p} d^3 p \,, \tag{6.2}$$

where $\omega_p = \sqrt{p^2 + m^2}$ and $d^3 p = dp^1 dp^2 dp^3$. Note that each coefficient a_p appears alongside its complex conjugate a_p^*. Remember also that the conjugate field $\pi(x)$ to $\phi(x)$ is obtained from the latter by simply differentiating with respect to the time component x^0, so that

$$\pi(x) = \frac{1}{(2\pi)^3} \int_{R^3} \frac{(-i\omega_p)}{\sqrt{2\omega_p}} \left(a_p e^{-ip\cdot x} - a_p^* e^{ip\cdot x} \right) \Big|_{p^0=\omega_p} d^3 p \,. \tag{6.3}$$

The idea now is that, upon quantization, the coefficients a_p, a_p^* become operators that we represent respectively by $a(\boldsymbol{p})$, $a^\dagger(\boldsymbol{p})$ on some complex, separable Hilbert space, with $a^\dagger(\boldsymbol{p})$ being the Hermitian adjoint of $a(\boldsymbol{p})$, and the resulting quantized Klein–Gordon field operators $\phi(x)$, $\pi(x)$ (which we still denote by the same symbols) should satisfy the commutator relation

$$[\phi(x), \pi(x)] = i\delta^3(\boldsymbol{x} - \boldsymbol{y}) \,. \tag{6.4}$$

Proceeding formally, if we impose the commutator relation (6.4) on the field operators obtained from (6.2) and (6.3), replacing the Fourier coefficients by the corresponding operators, we see after a straightforward computation that it is necessary that

$$[a(\boldsymbol{p}), a^\dagger(\boldsymbol{q})] = (2\pi)^3 \delta^3(\boldsymbol{p} - \boldsymbol{q}) \,, \quad [a(\boldsymbol{p}), a(\boldsymbol{q})] = [a^\dagger(\boldsymbol{p}), a^\dagger(\boldsymbol{q})] = 0 \,.$$

The bosonic Fock space is formally constructed from a given vector, declared the vacuum state and usually denoted $|\Omega\rangle$, simply by applying to this vector all possible finite words on the creation operators, taking into account the above commutator relations. See [PS] for the details we are omitting here.

6.4 Quantization of fermionic fields

In this section, we describe the quantization of free fermionic fields. The idea behind the quantization of the free Dirac field is to transform the coefficients of the Fourier expansion of the solution to Dirac's equation (see Chapter 5) into (spinor) operators. Then plug the solution with these operator coefficients into the quantum Hamiltonian. The correct (anti-)commutation rules for these operators is imposed by the physical condition that the Hamiltonian be bounded from below. The construction of the appropriate Hilbert space uses Grassmann algebras, as we shall see in the following section.

6.4.1 Grassmann calculus in n variables

When we come to study the path integral approach to QFT, the appropriate way to deal with fermionic fields will be through *Grassmann algebras*. In studying fermionic fields, it is necessary to consider Dirac and Weyl spinors, and these lead us into the realms of real and complex Grassmann algebras, respectively.

6.4.1.1 Real Grassmann algebras We shall discuss the subject of Grassmann algebras from an abstract point of view. The exposition to follow is perhaps a lengthy digression, but we deem it necessary from a mathematical standpoint. Let us first consider the case of *real finite-dimensional* Grassmann algebras, i.e. real algebras generated by n anti-commuting variables. These can be formally defined as follows.

Let V be a finite-dimensional vector space defined over the reals, with $\dim_{\mathbb{R}}(V) = n$. Let $\wedge(V)$ denote the exterior algebra of V, namely, the subspace of the full tensor algebra of V,

$$\mathcal{T}(V) = \mathbb{R} \oplus V \oplus (V \otimes V) \oplus (V \otimes V \otimes V) \oplus \cdots ,$$

consisting of all completely anti-symmetric tensors. Given a basis for V, say $\theta^1, \theta^2, \ldots, \theta^n$, and any k-tuple $I = (i_1, i_2, \ldots, i_k) \in \{1, 2, \ldots, n\}^k$, we write

$$\theta^I = \frac{1}{k!} \sum_{\sigma} (-1)^\sigma \theta^{i_{\sigma(1)}} \otimes \cdots \otimes \theta^{i_{\sigma(k)}} , \tag{6.5}$$

where the sum is over all permutations of $\{1, 2, \ldots, k\}$ and $(-1)^\sigma$ denotes the sign of the permutation σ. These anti-symmetric elements generate $\wedge(V)$. Now, define the Grassmann product (or exterior product) as follows: if $I \in \{1, 2, \ldots, n\}^k$ and $J \in \{1, 2, \ldots, n\}^l$, let

$$\theta^I \theta^J = \theta^{IJ} ,$$

where IJ is the $(k+l)$-tuple $(i_1, \ldots, i_k; j_1, \ldots, j_l)$. The convention here is that $\theta^\varnothing = 1$. This product is extended by bilinearity and associativity to all of $\wedge(V)$. One easily verifies (exercise) that

$$\theta^I \theta^J = (-1)^{|I||J|} \theta^J \theta^I ,$$

where $|I|$ denotes the number of indices in I. The resulting algebra is the (real) Grassmann algebra in dimension n, denoted $\mathbb{G}_n$. For each $k \geq 0$ we let

$$\wedge^k(V) = \langle \theta^I \; |I| = k \rangle ,$$

the linear span of all θ^I with $|I| = k$. This is a vector subspace of $\mathbb{G}_n$ with dimension over $\mathbb{K}$ equal to

$$\binom{n}{k} = \frac{n!}{k!(n-k)!} .$$

Because we clearly have

$$\mathbb{G}_n = \bigoplus_{k=0}^{n} \wedge^k(V) ,$$

it follows that $\dim \mathbb{G}_n = 2^n$.

It should be clear from (6.5) that $\theta^I = 0$ whenever I has a repeated index. The notation has been set up in such a way that, if $I = (i_1, i_2, \ldots, i_k)$, then $\theta^I = \theta^{i_1}\theta^{i_2}\cdots\theta^{i_k}$.

Remark 6.13 Because what we really need are anti-commuting *variables*, not just *numbers*, it may be more natural to replace the vector space V by its dual V^*, regarding each θ^i as a 1-form over V.

Let us now consider polynomial maps $P : \mathbb{G}_n \to \mathbb{G}_n$, given by an expression of the form

$$P(\theta^1, \ldots, \theta^n) = a_0 + \sum_i a_i\theta^i + \sum_{i,j} a_{ij}\theta^i\theta^j + \cdots$$
$$+ \sum_{i_1,i_2,\cdots,i_n} a_{i_1i_2\cdots i_n}\theta^{i_1}\theta^{i_2}\cdots\theta^{i_n} .$$

Such a polynomial map has degree $\leq n$: monomials with $\geq n+1$ terms will necessarily repeat an index, and therefore will be equal to zero. Note that this fact also means that the polynomial algebra $\mathbb{P}(\mathbb{G}_n)$, defined as the algebra of all polynomial maps of this form, is in fact identical to the algebra of all *formal power series* in the non-commuting variables $\theta^1, \ldots, \theta^n$.

In particular, we may consider the following case, of special interest in the path integral approach to fermionic fields. Let $A = (a_{ij})$ be a complex $n \times n$ matrix, and let

$$\Theta = \begin{pmatrix} \theta^1 \\ \theta^2 \\ \vdots \\ \theta^n \end{pmatrix} .$$

Then we define

$$\exp\left(-\frac{1}{2}\Theta^t A\Theta\right) = \sum_{k=0}^{\infty} \frac{(-1)^k}{2^k k!}(\Theta^t A\Theta)^k .$$

This is the exponential of a quadratic form. Note that, because

$$\Theta^t A \Theta = \sum_{i,j} a_{ij} \theta^i \theta^j ,$$

we have $(\Theta^t A \Theta)^k = 0$ for all $k > n/2$. Therefore

$$\exp\left(-\frac{1}{2}\Theta^t A \Theta\right) = \sum_{k=0}^{\lfloor n/2 \rfloor} \frac{(-1)^k}{2^k k!} (\Theta^t A \Theta)^k ,$$

which is, of course, a polynomial in the Grassmann variables.

Now, the the polynomial algebra $\mathbb{P}(\mathbb{G}_n)$ is, of course, a differential algebra. In fact, it is a differential *graded* algebra, graded over $\mathbb{Z}/2$. This motivates us to define *integration* with respect to the Grassmann variables to be *the same as differentiation* with respect to such variables! More precisely, the integral can be taken to be a collection of maps

$$\mathcal{I}^I : \mathbb{P}(\mathbb{G}_n) \to \mathbb{P}(\mathbb{G}_n), \quad I \in \{1, 2, \ldots, n\}^k$$

defined as follows: if $f \in \mathbb{P}(\mathbb{G}_n)$ and $I = (i_1, i_2, \ldots, i_k)$, then

$$\mathcal{I}^I(f) = \frac{\partial}{\partial \theta^{i_1}} \frac{\partial}{\partial \theta^{i_2}} \cdots \frac{\partial}{\partial \theta^{i_k}} f(\theta^1, \theta^2, \ldots, \theta^n) .$$

The notation we shall use is

$$\mathcal{I}^I(f) = \int f \, d\theta^{i_1} d\theta^{i_2} \cdots d\theta^{i_k} .$$

This *Grassmannian* or *fermionic* integral has the following easy properties:

(i) $\displaystyle\int d\theta^i = 0$ for all i.

(ii) $\displaystyle\int \theta^i \, d\theta^j = \delta_{ij}$ for all i, j.

(iii) $\displaystyle\int d\theta^i d\theta^j = 0$ for all i, j.

(iv) $\displaystyle\int f(\theta^i) g(\theta^j) \, d\theta^i d\theta^j = \int f(\theta^i) \, d\theta^i \int g(\theta^j) \, d\theta^j$ for all i, j.

Let us go back to the example of the exponential of a quadratic form $\Theta \mapsto \frac{1}{2}\Theta^t A \Theta$ examined above. We are interested in calculating the fermionic integral of such a generalized Gaussian, namely

$$\int \exp\left(-\frac{1}{2}\Theta^t A \Theta\right) d\theta^1 d\theta^2 \cdots d\theta^n = \sum_{k=0}^{\lfloor n/2 \rfloor} \frac{(-1)^k}{2^k k!} \int (\Theta^t A \Theta)^k \, d\theta^1 d\theta^2 \cdots d\theta^n . \tag{6.6}$$

It is easy to verify that each integral on the right-hand side of (6.7) with $k < n/2$ vanishes. In particular, the Gaussian integral on the left-hand side will vanish whenever n is odd. Let us then assume that n is even, say $n = 2m$, so that in this case

$$\int \exp\left(-\frac{1}{2}\Theta^t A\Theta\right) d\theta^1 d\theta^2 \cdots d\theta^n = \frac{(-1)^m}{2^m m!} \int (\Theta^t A\Theta)^m \, d\theta^1 d\theta^2 \cdots d\theta^{2m} . \tag{6.7}$$

Now, if $A = (a_{ij})$ happens to be a *symmetric* matrix, then this last integral also vanishes (this is an exercise). Thus, we will suppose from now on that A is a *skew-symmetric* matrix. Here it pays off to interpret our quadratic form as a 2-form,

$$\omega_A = \sum_{i,j=1}^{2m} a_{ij}\, \theta^i \wedge \theta^j .$$

The iterated exterior product of this 2-form with itself m times yields a volume form in V, and we can write

$$\omega_A \wedge \omega_A \wedge \cdots \wedge \omega_A = 2^m m! \mathrm{Pf}(A)\, \theta^1 \wedge \theta^2 \wedge \cdots \wedge \theta^{2m} ,$$

where $\mathrm{Pf}(A)$ is called the *Pfaffian* of A. Expanding out the m-fold exterior product on the left-hand side, the reader will have no trouble in verifying that

$$\mathrm{Pf}(A) = \frac{1}{m!} \sum_{\sigma \in S_{2m}} (-1)^{|\sigma|} a_{\sigma(1)\sigma(2)} a_{\sigma(3)\sigma(4)} \cdots a_{\sigma(2m-1)\sigma(2m)} ,$$

where S_{2m} denotes the symmetric group of order $2m$.

Lemma 6.14 *We have* $\mathrm{Pf}(A)^2 = \det(A)$.

Proof. First note that the Pfaffian is invariant under orthogonal changes of variables. In other words, if Λ is an orthogonal $2m \times 2m$ matrix, then $\mathrm{Pf}(\Lambda^t A \Lambda) = \mathrm{Pf}(A)$. This follows easily from the definition (work out the effect of Λ on the 2-form ω_A). Then recall from linear algebra that, because A is skew-symmetric, there exists an orthogonal matrix Λ such that $B = \Lambda^t A \Lambda$ has the form

$$B = \begin{pmatrix} 0 & b_1 & & & & & & \\ -b_1 & 0 & & & & & & \\ & & 0 & b_2 & & & & \\ & & -b_2 & 0 & & & & \\ & & & & \ddots & & & \\ & & & & & & 0 & b_m \\ & & & & & & -b_m & 0 \end{pmatrix} .$$

Now, the Pfaffian of B can be evaluated explicitly from

$$\omega_B = \sum_{i,j=1}^{2m} b_{ij}\,\theta^i \wedge \theta^j = 2\sum_{k=1}^{m} b_k\,\theta^{2k-1} \wedge \theta^{2k} \,,$$

and the result is $\mathrm{Pf}(B) = b_1 b_2 \cdots b_k$. But then $\det(B) = b_1^2 b_2^2 \cdots b_k^2 = \mathrm{Pf}(B)^2$. Because $\det(A) = \det(B)$ and $\mathrm{Pf}(A) = \mathrm{Pf}(B)$, the lemma is proved. □

Theorem 6.15 *Let A be a skew-symmetric $2m \times 2m$ real matrix. Then we have*

$$\int \exp\left(-\frac{1}{2}\Theta^t A \Theta\right) d\theta^1 d\theta^2 \cdots d\theta^n = \det(A)^{1/2} \,. \tag{6.8}$$

Proof. It follows from (6.7) that the integral on the left-hand side of (6.8) is equal to $\mathrm{Pf}(A)$, so the desired result is an immediate consequence of Lemma 6.14. □

6.4.1.2 Complex Grassmann algebras and Hilbert spaces Everything that we did above for real algebras could be repeated here, replacing the real numbers with the complex numbers. But because we already know what a real Grassmann algebra is, it is easier to define complex Grassmann algebras as complexifications of real Grassmann algebras. For instance, in the case of one degree of freedom, we have the following.

Definition 6.16 *A* complex Grassmann algebra $\mathcal{A}$ with one degree of freedom *is an associative algebra over the complex numbers having two generators, θ and θ^*, subject to the following relations:*

$$\theta\theta^* + \theta^*\theta = 0\,, \quad \theta^2 = 0\,, \quad (\theta^*)^2 = 0\,.$$

The *existence* (and uniqueness up to isomorphism) of such an algebra is established using the real Grassmann algebra $\mathbb{G}_2$ with generators θ_1 and θ_2. We define $\mathcal{A} = \mathbb{G}_n^{\mathbb{C}}$, taking as its generators

$$\theta = \frac{\theta_1 + i\theta_2}{\sqrt{2}}\,, \quad \theta^* = \frac{\theta_1 - i\theta_2}{\sqrt{2}}\,.$$

A general element of $\mathcal{A}$ can be written as

$$f(\theta^*, \theta) = f_{00} + f_{10}\theta^* + f_{01}\theta + f_{11}\theta^*\theta\,.$$

As in the real case, we can make $\mathcal{A}$ into a differential algebra, introducing the differential operators

$$\frac{d}{d\theta} f(\theta^*, \theta) = f_{01} - f_{11}\theta^*$$
$$\frac{d}{d\theta^*} f(\theta^*, \theta) = f_{10} + f_{11}\theta^* .$$

These operators satisfy the relation

$$\frac{d}{d\theta}\frac{d}{d\theta^*} = -\frac{d}{d\theta^*}\frac{d}{d\theta} .$$

Again, there is a suitable integral calculus to go along with these differential operators. The algebraic rules for this integral calculus are the following:

$$\int f\, d\theta = \frac{d}{d\theta} f , \quad \int f\, d\theta^* = \frac{d}{d\theta^*} f$$
$$\int d\theta\, d\theta^* = \int d\theta \int d\theta^* = -\int d\theta^*\, d\theta .$$

In every such Grassmann algebra $\mathcal{A}$, there is an involution $\star : \mathcal{A} \to \mathcal{A}$ given by $\star\theta = \theta^*$, $\star\theta^* = \theta$, $\star(\theta^*\theta) = \theta^*\theta$, and extended by linearity to all of $\mathcal{A}$. For each $f \in \mathcal{A}$, we let $f^* = \star f$.

Let us take the time to define a Hilbert space associated to a complex Grassmann algebra $\mathcal{A}$ with one degree of freedom. There is a sub-algebra of $\mathcal{A}$ consisting of all holomorphic functions of θ^*, namely

$$\mathcal{H} = \{f(\theta^*) = f_0 + f_1\theta^* \mid f_0, f_1 \in \mathbb{C}\} .$$

Now, this sub-algebra can be made into a Hilbert space, in perfect analogy with the way we made the holomorphic functions in one complex variable into a Hilbert space in Chapter 2. We define an inner product in $\mathcal{H}$ as follows: if $f, g \in \mathcal{H}$ then

$$\langle f, g \rangle = \int f(\theta^*)^* g(\theta^*) e^{-\theta^*\theta}\, d\theta^* d\theta .$$

Here, $e^{-\theta^*\theta} = 1 - \theta^*\theta$ (all subsequent terms of the power series expansion vanish!). With this inner product, $\mathcal{H}$ is a two-dimensional Hilbert space over the complex numbers, and $\{1, \theta^*\}$ is an orthonormal basis.

Next, we define two operators on $\mathcal{H}$ that will play the role of the creation and annihilation operators. Let $a : \mathcal{H} \to \mathcal{H}$ be given by

$$(af)(\theta^*) = \frac{d}{d\theta^*} f(\theta^*) ,$$

and let $a^* : \mathcal{H} \to \mathcal{H}$ be given by

$$(a^* f)(\theta^*) = \theta^* f(\theta^*) .$$

In the natural identification of $\mathcal{H}$ with $\mathbb{C}^2$ as vector spaces, these operators become

$$a = \begin{pmatrix} 0 & 1 \\ 0 & 0 \end{pmatrix} , \quad a^* = \begin{pmatrix} 0 & 0 \\ 1 & 0 \end{pmatrix} .$$

Note that every linear operator $A : \mathcal{H} \to \mathcal{H}$ can be written uniquely in the form

$$A = k_{00} + k_{10}a^* + k_{01}a + k_{11}a^*a ,$$

where $k_{ij} \in \mathbb{C}$. In fact, we can also write

$$(Af)(\theta^*) = \int A(\theta^*, \alpha) f(\alpha^*) e^{-\alpha^*\alpha} \, d\alpha^* d\alpha ,$$

where $A(\theta^*, \theta)$ is the kernel of A. This kernel is clearly given by

$$A(\theta^*, \theta) = e^{\theta^*\theta} K(\theta^*, \theta) ,$$

where $K(\theta^*, \theta) = k_{00} + k_{10}\theta^* + k_{01}\theta + k_{11}\theta^*\theta$ is the so-called normal symbol of A.

Later (next chapter), when we perform the Lagrangian quantization of fermionic systems (in perturbative QFT), we shall need to know how to calculate fermionic Gaussian integrals over complex fermionic fields. For one degree of freedom, all we need is given by the following result.

Lemma 6.17 *Let $A : \mathcal{H} \to \mathcal{H}$ be a linear operator, and let $b \in \mathcal{H}$. Then*

$$\int \exp\{\theta^* A\theta + \theta^* b + b^*\theta\} \, d\theta^* d\theta = -A \exp\{-b^* Ab\} .$$

Let us now generalize this discussion to arbitrarily many (but a finite number of) degrees of freedom.

Definition 6.18 *A* complex Grassmann algebra $\mathcal{A}_n$ with n degrees of freedom *is an associative algebra over the complex numbers with $2n$ generators $\theta_1, \ldots, \theta_n, \theta_1^*, \ldots, \theta_n^*$ subject to the following relations (valid for all $i, j = 1, \ldots, n$):*

$$\theta_i^*\theta_j = -\theta_j\theta_i^* , \quad \theta_i\theta_j = -\theta_j\theta_i , \quad \theta_i^*\theta_j^* = -\theta_j^*\theta_i^* .$$

As in the case of one degree of freedom, the square of every generator is equal to zero. And, as in that case, the existence (and uniqueness up to isomorphism) of such a complex algebra is resolved by letting $\mathcal{A}_n = \mathbb{G}_{2n}^{\mathbb{C}}$, the complexification of the real Grassmann algebra with $2n$ degrees of freedom, taking as generators

$$\theta_j = \frac{\varphi_j + i\varphi_{j+n}}{\sqrt{2}} , \quad \theta_j^* = \frac{\varphi_j - i\varphi_{j+n}}{\sqrt{2}} , \quad j = 1, 2, \ldots, n ,$$

where φ_j, $j = 1, 2, \ldots, 2n$ are the generators of $\mathbb{G}_{2n}$. Also, as in that case, we can consider the differential operators

$$\frac{\partial}{\partial \theta_i} , \quad \frac{\partial}{\partial \theta_i^*} .$$

Integration with respect to these Grassmann variables enjoys the same properties as before, namely

$$\int f(\theta^*, \theta)\, d\theta_j = \frac{\partial}{\partial \theta_j} f(\theta^*, \theta)$$

(and similarly for θ_j^*), as well as

$$\int d\theta_i^* d\theta_j = \int d\theta_i^* \int d\theta_j = -\int d\theta_j \int d\theta_i^* = 0 .$$

Let us define $\mathcal{H}^{(n)}$ to be the sub-algebra of holomorphic functions of $\theta^* = (\theta_1^*, \ldots, \theta_n^*)$. As a vector space over the complex numbers, $\mathcal{H}^{(n)}$ has dimension 2^n, and a basis is given by the monomials $\theta_{i_1}^* \cdots \theta_{i_r}^*$, where $\{i_1, \ldots, i_r\}$ is an arbitrary subset of $\{1, 2, \ldots, n\}$. We make $\mathcal{H}^{(n)}$ into a Hilbert space, introducing the inner product

$$\langle f, g \rangle = \int f(\theta^*)^* g(\theta^*) e^{-\sum_{j=1}^n \theta_j^* \theta_j} \prod_{j=1}^n d\theta_j^* \theta_j .$$

We have also two families of special creation and annihilation operators $a_i^*, a_i : \mathcal{H}^{(n)} \to \mathcal{H}^{(n)}$, given by

$$a_i^* f(\theta^*) = \theta_i^* f(\theta^*) \quad \text{and} \quad a_i f(\theta^*) = \frac{\partial}{\partial \theta_i^*} f(\theta^*) .$$

These operators satisfy the anti-commuting relations

$$a_i a_j + a_j a_i = 0 , \quad a_i^* a_j + a_j a_i^* = \delta_{ik} , \quad a_i^* a_j^* + a_j^* a_i^* = 0 .$$

Now, if $A : \mathcal{H}^{(n)} \to \mathcal{H}^{(n)}$ is any linear operator, we can ask whether A can be expressed in terms of these creation and annihilation operators. This is indeed the case, as the following theorem shows.

Theorem 6.19 *If $A : \mathcal{H}^{(n)} \to \mathcal{H}^{(n)}$ is a linear operator, then there exist unique complex constants $k_{i_1,\ldots,i_r;j_1,\ldots,j_s}$ such that*

$$A = \sum_{r,s=1}^{n} \sum_{i_1<\cdots<i_r;j_1<\cdots<j_s} k_{i_1,\ldots,i_r;j_1,\ldots,j_s} a_{i_1}^* \cdots a_{i_r}^* a_{j_1} \cdots a_{j_s} .$$

Moreover, there exist unique constants $A_{i_1,\ldots,i_r;j_1,\ldots,j_s}$ such that, if

$$\hat{A}(\theta^*, \theta) = \sum A_{i_1,\ldots,i_r;j_1,\ldots,j_s} \theta_{i_1}^* \cdots \theta_{i_r}^* \theta_{j_1} \cdots \theta_{j_s} ,$$

then

$$Af(\theta^*) = \int \hat{A}(\theta^*, \alpha) f(\alpha^*) e^{-\sum \alpha_j^* \alpha_j} \prod_{j=1}^{n} d\alpha_j^* d\alpha_j .$$

Again, when we deal with the perturbative theory, we will need to know how to evaluate Gaussian integrals in several Grassmann variables. All we need is contained in Theorem 6.22 below. First we need an auxiliary result.

Lemma 6.20 *Let U be a unitary $n \times n$ complex matrix, and let us consider the n independent Grassmann variables $\theta_1, \theta_2, \ldots, \theta_n$ of a Grassmann algebra with n degrees of freedom. If $\vartheta_i = \sum_j U_{ij}\theta_j$ for each $i = 1, 2, \ldots, n$, then*

$$\prod_{i=1}^{n} \vartheta_i = (\det U) \prod_{i=1}^{n} \theta_i .$$

Moreover,

$$\prod_{i=1}^{n} \vartheta_i^* \vartheta_i = (\det U^*)(\det U) \prod_{i=1}^{n} \theta_i^* \theta_i = \prod_{i=1}^{n} \theta_i^* \theta_i .$$

Proof. An exercise for the reader. □

Note however that, under a general linear change of variables T, we have

$$\prod_{i=1}^{n} d\vartheta_i^* d\vartheta_i = (\det T)^{-1} \prod_{i=1}^{n} d\theta_i^* d\theta_i ,$$

which is the opposite of the usual Jacobian formula in ordinary n-variables integration. As a consequence of Lemma 6.20, we have the following fact.

Lemma 6.21 *The integral*

$$\int f(\theta, \theta^*) \prod_{j=1}^{n} d\theta_j^* d\theta_j$$

is invariant under unitary changes of coordinates.

Proof. Again, the proof is left as an exercise. □

Theorem 6.22 *Let B be an $n \times n$ Hermitian matrix with eigenvalues $b_1, b_2, \dots, b_n$. Then the Gaussian integral with covariance matrix B in the Grassmann algebra of n variables $\theta_1, \dots, \theta_n$ is given by*

$$\int e^{-\sum \theta_j^* B_{ij} \theta_j} \prod_{j=1}^{n} d\theta_j^* d\theta_j = \prod_{i=1}^{n} b_i = \det B \, .$$

Proof. Because B is Hermitian, there exists a unitary matrix U such that $UBU^* = D$ is a diagonal matrix, in fact $D = \mathrm{diag}\,(b_1, b_2, \dots, b_n)$. Consider the change of variables $\vartheta = U\theta$. Then, by Lemma 6.21, we have

$$\int e^{-\sum \theta_i^* B_{ij} \theta_j} \prod_{j=1}^{n} d\theta_j^* d\theta_j = \int e^{-\sum \vartheta_i^* D_{ij} \vartheta_j} \prod_{j=1}^{n} d\vartheta_j^* d\vartheta_j \, .$$

But because $D_{ij} = \delta_{ij} b_i$, this last integral is equal to

$$\int e^{-\sum \vartheta_i^* b_i \vartheta_i} \prod_{j=1}^{n} d\vartheta_i^* d\vartheta_i = \prod_{i=1} \int e^{-b_i \vartheta_i^* \vartheta_i} \, d\vartheta_i^* d\vartheta_i = \prod_{i=1}^{n} b_i = \det B \, ,$$

as was to be proved. □

6.4.2 The physicist's viewpoint

Just as we did for bosonic fields, let us briefly describe the physicist's *ad hoc* approach to the quantization of free fermionic fields. We start with the observation that a general solution to Dirac's equation in Minkowski space-time (see the exercises at the end of Chapter 5) can be Fourier-expanded in terms of basic plane-wave solutions, according to the formula

$$\psi(x) = \frac{1}{(2\pi)^3} \int_{\mathbb{R}^3} \frac{1}{\sqrt{2\omega_{\boldsymbol{p}}}} \sum_{s=-,+} \left(a_{\boldsymbol{p},s} u_s(p) e^{-ip\cdot x} + b_{\boldsymbol{p},s}^{\dagger} v_s(p) e^{ip\cdot x} \right) d^3 p \, . \tag{6.9}$$

Here, s denotes the field helicity state, and $u_s(p)$ and $v_s(p)$ are four-component spinors whose general expressions in terms of p need not concern us. Upon quantization, the coefficients $a_{\boldsymbol{p},s}$ and $b_{\boldsymbol{p},s}$ are promoted to operators $a_s(\boldsymbol{p})$ and $b_s(\boldsymbol{p})$, respectively. These operators, written in terms of their spinor components $a_s^j(\boldsymbol{p})$, $b_s^j(\boldsymbol{p})$, $1 \le j \le 4$, must obey the anti-commutation relations

$$\{a_s^j(\boldsymbol{p}), a_s^{k\,\dagger}(\boldsymbol{q})\} = \{b_s^j(\boldsymbol{p}), b_s^{k\,\dagger}(\boldsymbol{q})\} = (2\pi)^3 \delta^{(3)}(\boldsymbol{p} - \boldsymbol{q}) \delta_{jk} \, ,$$

all other possible anti-commutators being equal to zero. These relations imply that the spinor components of the quantized field Ψ must satisfy the relations

$$\{\Psi^j(t, \boldsymbol{x}), \Psi^{k\dagger}(t, \boldsymbol{y})\} = \delta^{(3)}(\boldsymbol{x} - \boldsymbol{y})\,\delta_{jk} \,.$$

The operators $a_s(\boldsymbol{p})$ and $b_s(\boldsymbol{p})$ are called *destruction operators* (of particles and antiparticles, respectively), and their formal adjoints $a_s^\dagger(\boldsymbol{p})$ and $b_s^\dagger(\boldsymbol{p})$ are called *creation operators*.

Now, the fermionic Fock space is constructed following the same procedure as in Section 6.3.5. One starts by defining a vacuum state $|\Omega\rangle$ annihilated by all destruction operators:

$$a_s(\boldsymbol{p})|\Omega\rangle = b_s(\boldsymbol{p})|\Omega\rangle = 0 \,.$$

Then the vectors representing multi-particle states are obtained by letting any finite *word* on the creation operators act on this vacuum state. The fermionic Fock Hilbert space will be the space generated by these vectors (which are declared to be unit vectors after normalization by a suitable scalar factor). Because the creation operators anti-commute, exchanging any two of them in a given word (applied to $|\Omega\rangle$) will change the sign of the resulting vector. This expresses the fact that a quantized Dirac field obeys the so-called Fermi–Dirac statistics. See [PS] and [SW] for more details.

6.4.3 Wightman axioms for vector and spinor fields

The axioms for a fermionic field are similar to the ones for a scalar bosonic field, with a few changes that we indicate below. Let ρ be an irreducible representation of $SL(2, \mathbb{C})$ on a finite-dimensional space of dimension d (recall that all the unitary representations of $SL(2, \mathbb{C})$ are infinite-dimensional, so ρ is certainly not unitary). The first three axioms are the same. In Axiom WA4, we have now a d-tuple of fields $(\phi_1(f), \ldots, \phi_d(f))$. The new Axiom WA5 is similar to the old one, whereas Axiom WA6 becomes the statement that

$$U(a, \Lambda)\phi_i(f)U(a, \Lambda)^{-1} = \sum_{j=1}^{d} \rho(A^{-1})_{ij}\phi_j(\{a, \Lambda\}f) \,,$$

where Λ is the Lorentz transformation that corresponds to $A \in SL(2, \mathbb{C})$ under the two-to-one covering homomorphism of the Lorentz group. To state Axiom WA7, we have two cases to consider. If ρ has integer spin, i.e. $\rho(-id) = id$ (bosons), then, for f and g spacelike separated,

$$\phi_i(f)\phi_j(g) - \phi_j(g)\phi_i(f) = 0 \quad \text{and} \quad \phi_i^*(f)\phi_j(g) - \phi_j(g)\phi_i^*(f) = 0 \,.$$

If ρ has half-integer spin then

$$\phi_i(f)\phi_j(g)+\phi_j(g)\phi_i(f)=0 \quad \text{and} \quad \phi_i^*(f)\phi_j(g)+\phi_j(g)\phi_i^*(f)=0\,.$$

Finally, the vacuum is only required to be cyclic for

$$\{\phi_1(f),\ldots,\phi_d(f),\phi_1^*(f),\ldots,\phi_d^*(f)\}\,.$$

The standard reference for this part is [SW, pp. 146–150].

6.5 Quantization of the free electromagnetic field

The Hamiltonian quantization of the free electromagnetic field follows a route similar to the ones used for the other free-field theories described above, with one important difference: here we have to take into account the gauge invariance of the action.

Recall from Chapter 5 that the free electromagnetic Lagrangian density is given by

$$\mathcal{L}=-\frac{1}{4}F_{\mu\nu}F^{\mu\nu}=\frac{1}{2}\left(\|\boldsymbol{E}\|^2-\|\boldsymbol{B}\|^2\right)\,.$$

Here, as before, $F=F_{\mu\nu}dx^\mu dx^\nu$ is the field strength, which we know is an exact 2-form; in other words $F=dA$, where $A=A_i dx^i$, $A_0=\phi$ is the scalar potential, and $\boldsymbol{A}=(A_1,A_2,A_3)$ is the vector potential. Thus, F is a closed 2-form, which yields the first pair of Maxwell's equations, whereas the Euler–Lagrange equations for the action functional built from the above Lagrangian are precisely the other pair of Maxwell's equations, namely $\nabla\cdot\boldsymbol{E}=0$ and $\nabla\wedge\boldsymbol{B}=\partial_t\boldsymbol{E}$. One can easily verify (exercise) that the Hamiltonian in the present case is given by

$$H=\frac{1}{2}\int_{\mathbb{R}^3}\left(\|\boldsymbol{E}\|^2+\|\boldsymbol{B}\|^2\right)d^3x\,.$$

Now, if we add an exact 1-form $\omega=d\Omega$ to A, the field strength F does not change, and neither does the Lagrangian density. In other words, we have invariance under gauge transformations of the form

$$A_\mu\mapsto A_\mu+\partial_\mu\Omega\,.$$

Here, Ω is an arbitrary function on spacetime M. Hence, we can choose the gauge in such a way that two things happen: (i) $A_0=0$; (ii) $\nabla\cdot\boldsymbol{A}=0$. That we can, indeed, make such a choice without conflict between (i) and (ii) is left as an exercise to the reader.

After fixing the gauge in this fashion, we can proceed with quantization as in the case of the Klein–Gordon field. We apply the Fourier transform method and get an expansion for the vector potential of the form

$$A(x) = \frac{1}{\sqrt{2k_0(2\pi)^3}} \int_{\mathbb{R}^3} \sum_{j=1}^{2} \epsilon_j(k) \left(a_j(k) e^{-ik\cdot x} + a_j^*(k) e^{ik\cdot x} \right) d^3k \, .$$

The $\epsilon_j(k)$ are called polarization vectors. The second gauge-fixing condition (ii) above implies that each polarization vector is Fourier orthogonal to the corresponding momentum k.

The quantization procedure now is to regard $a_j(k)$ and $a_j^*(k)$ simply as operators (on a suitable Fock space). These are the operators that, respectively, create and annihilate a quantum of the electromagnetic field with momentum k and polarization $\epsilon_j(k)$. The construction of Fock space, and of these creation and annihilation operators, is completely analogous to that performed for the Klein–Gordon field, so we omit the details.

6.6 Wick rotations and axioms for Euclidean QFT

We have seen earlier in this chapter that the Wightman functions

$$W(x_1, \ldots, x_n) = \langle \Omega, \phi(x_1) \cdots \phi(x_n) \Omega \rangle$$

can be used to completely reconstruct a scalar bosonic field theory. These functions, as it turns out, can be analytically continued to the so-called Schwinger points $z_j = (ix_j^0, x_j^1, x_j^2, x_j^3)$, provided the points $x_j \in \mathcal{R}^4$ are such that $x_j \neq x_j$ if $j \neq k$. The map $(\mathbb{R}^4)^n \to (i\mathbb{R} \times \mathbb{R}^3)^n$ given by $(x_1, x_2 \ldots, x_n) \mapsto (z_1, z_2, \ldots, z_n)$ is called a *Wick rotation*. The functions

$$S_n(x_1, x_2, \ldots, x_n) = W(z_1, z_2, \ldots, z_n)$$

are called *Schwinger functions*.

A set of axioms for the Schwinger functions was devised by Osterwalder and Schrader in [OS]. These axioms can be used to construct (via the inverse of Wick rotations) the Wightman correlation functions, and therefore the whole scalar bosonic theory, by Wightman's reconstruction theorem. The Osterwalder–Schrader axioms can be informally stated as follows (cf. [Ri, pp. 21–2]):

OSA-1 (regularity). The Schwinger functions, S_n do not grow too quickly with n.

OSA-2 (Euclidean covariance). The Schwinger functions are invariant under a global Euclidean transformation (for scalar bosonic fields – in the case of a fermionic theory, appropriate changes have to be made here).

OSA-3 (O–S positivity). The expectation value of a function F (defined via the Schwinger functions) of the fields multiplied by the function F^* obtained from F by reflection across a hyperplane and complex conjugation is a positive number. This axiom yields the positive-definite metric of the Hilbert space for the Wightman fields.

OSA-4 (symmetry). The Schwinger functions – in a theory for bosons – are symmetric under permutations of its (external) arguments. For fermions, symmetry should of course be changed to anti-symmetry.

OSA-5 (cluster property). A Schwinger functions can be written asymptotically as a product when two sets of its arguments are taken far apart.

Now, the point is that one has a reconstruction theorem akin to Wightman's.

Theorem 6.23 (O–S reconstruction theorem) *Suppose we are given a set of functions satisfying the axioms OSA-1 to OSA-5 above. Then these functions are the Schwinger functions, a unique field theory in the sense of Wightman.*

For a more precise (and much more formal) treatment of the Ostwalder–Schrader axioms, see Kazhdan's lectures [Kh, Chapter 2]. See also [Sch, pp. 97–101] for a precise formulation of a version of these axioms, and a proof of the reconstruction theorem, in the context of conformal field theory.

6.7 The CPT theorem

We would like to add a few informal remarks on the so-called CPT theorem. There are three very important finite-order symmetries on fields that stand out. These translate, in QFT, to certain unitary or anti-unitary operators acting on quantum fields. To keep the discussion short, we only talk about the effect of these symmetries on *fermionic* fields. We refer to Section 6.4.2 for notation. The symmetries under scrutiny are the following.

(i) *Charge conjugation operator.* As we have seen in Remark 5.3 (Section 5.4.2), at the classical level we can consider the operator $C : \psi \mapsto \psi^C = -i\gamma^2\psi^*$ on Dirac fields. This charge conjugation operator maps solutions of Dirac's equation with charge q to solutions of Dirac's equation with charge $-q$. In other words, it changes matter to antimatter and vice versa. At the quantum level this operator becomes a unitary operator acting on fermionic Fock space. In terms of the generators $a_s(\boldsymbol{p})$ and $b_s(\boldsymbol{p})$ of fermionic Fock space, C acts as follows:

$$C^{-1}a_s(\boldsymbol{p})C = \eta_a b_s(\boldsymbol{p}) ,$$
$$C^{-1}b_s(\boldsymbol{p})C = \eta_b a_s(\boldsymbol{p}) .$$

Here, η_a, η_b are phase factors (recall that a unit vector representing a quantum-mechanical state is determined only up to a phase factor). It can be shown in the present context that $\eta_a^2 = \eta_b^2 = 1$, and that $\eta_a = -\eta_b$. We obviously expect $C^2 = C \circ C$ to be the identity, so in fact $C^{-1} = C$. With these relations, one can extend C to the space of all fermionic fields by linearity, using the Fourier decomposition given in 6.9. The resulting operator on fermionic fields is unitary.

(ii) *Parity operator*. At the classical level, the parity operator corresponds to the reflection $(t, \boldsymbol{x}) \mapsto (t, -\boldsymbol{x})$ on spacetime. Thus, under this parity transformation, the momentum $\boldsymbol{p}$ of a particle is changed to $-\boldsymbol{p}$, whereas its *spin* remains unchanged (classically, the spatial coordinate involution $\boldsymbol{x} \mapsto -\boldsymbol{x}$ does *not* change angular momentum). Upon quantization, we should therefore expect the parity operator P to be a unitary operator such that

$$P^{-1} a_s(\boldsymbol{p}) P = a_s(-\boldsymbol{p}) ,$$
$$P^{-1} b_s(\boldsymbol{p}) P = b_s(-\boldsymbol{p}) .$$

Again, with these relations one can extend P to the space of all fermionic fields by linearity. And again, the resulting P is an involution as well: $P = P^{-1}$.

(iii) *Time-reversal operator*. At the classical level, the operation of switching t to $-t$ corresponds to the transformation on fermionic fields given by $\psi(t, \boldsymbol{x}) \mapsto -\gamma^1\gamma^3\psi(-t, \boldsymbol{x})$. Here, γ^1 and γ^3 are the usual Dirac matrices. Time reversal changes a particle's momentum $\boldsymbol{p}$ to $-\boldsymbol{p}$, and it also reverses spin, $s \mapsto -s$. Hence, upon quantization, the time-reversal operator T changes the signs of both momentum and spin of every destruction operator, namely:

$$T^{-1} a_s(\boldsymbol{p}) T = a_{-s}(-\boldsymbol{p}) ,$$
$$T^{-1} b_s(\boldsymbol{p}) T = b_{-s}(-\boldsymbol{p}) .$$

It can be verified that T, after being extended by linearity to the entire space of fermionic fields, cannot be unitary. Rather, it must be an anti-unitary operator (recall Wigner's theorem in Chapter 2).

One should not expect that a quantum theory of fields might be invariant under any of these symmetries taken in isolation. As it turns out, it is an experimentally observed fact in particle physics that *parity* is violated in weak interactions (see Chapter 9). However, the interesting thing is that any such reasonable theory will remain invariant under the combined action of C, P, and

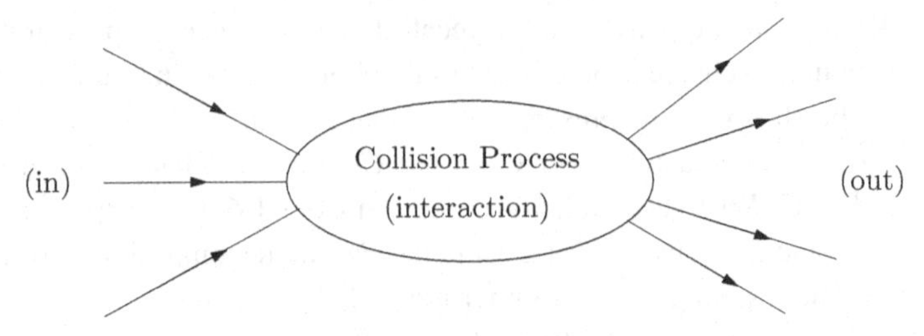

Figure 6.1 Scattering.

T. More precisely, the *CPT theorem* states that a Lorentz-invariant field theory (whose corresponding Hamiltonian is a Hermitian operator) is necessarily invariant under the composite symmetry $C \circ P \circ T$. One of the great merits of the axiomatic approach to QFT is to provide a mathematically rigorous proof of this theorem. The reader may consult [SW] for a complete exposition of this result and its close relationship to the so-called *spin-statistics theorem* of Pauli. See also [B, Chapter 5] for an interesting conceptual discussion of these topics.

6.8 Scattering processes and LSZ reduction

The main point of contact between quantum field theory and experiment is provided by scattering phenomena. What physicists actually measure in the laboratory is scattering cross-sections of particle collisions. Typically, one has an incoming beam of particles (henceforth called *in-particles*) that get together and interact in some manner, producing an outgoing beam of particles (the *out-particles*) that scatter away (see Figure 6.1). For each given state of the in-particles and each desirable state of the out-particles, one would like to compute the corresponding probability amplitude.

To simplify the discussion to follow, we consider only spinless particles, such as scalar bosons. We think of the in-particles as having momenta $\boldsymbol{q}_1, \boldsymbol{q}_2, \ldots, \boldsymbol{q}_r$, and of the out-particles as having momenta $\boldsymbol{p}_1, \boldsymbol{p}_2, \ldots, \boldsymbol{p}_s$. We shall write $P = \{\boldsymbol{p}_i\}$ and $Q = \{\boldsymbol{q}_j\}$. In general, we have $r \neq s$: we can have creation and annihilation of particles when the interaction occurs. We should bear in mind, however, that the in-particles represent a common eigenstate for a commuting set of field operators (corresponding to the incoming momenta), and similarly for the out-particles. Using Dirac's notation, we represent the incoming state by $|Q, \text{in}\rangle$ and the outgoing state by $|P, \text{out}\rangle$. These states should be thought of as asymptotic states – intuitively, if we recede to the past, the in-particles do not interact with each other, and similarly for the out-particles in the future. Hence, they are elements of a Hilbert (Fock) space $\mathcal{H}$, and correspond to free fields. The scattering process can therefore be thought of as an operator $S : \mathcal{H} \to \mathcal{H}$, the so-called scattering operator. What one measures about this process is the so-called scattering amplitudes, namely the inner products

$$S_{P,Q} = \langle P, \mathrm{out} | Q, \mathrm{in} \rangle \ .$$

As we know from quantum mechanics, these complex numbers – known collectively as the scattering matrix, or S-matrix – represent the probability amplitudes that the in-particles, initially measured to be in the state $|Q, \mathrm{in}\rangle$, will scatter away after the interaction into a set of out-particles in the state $|P, \mathrm{out}\rangle$.

How can we calculate these scattering amplitudes? The answer was given in 1955 by Lehmann, Symanzik, and Zimmermann, through what is now known as the LSZ reduction formula. The intuitive idea behind LSZ reduction is that, because the in-particles before collision can be thought of as created from the vacuum and the out-particles after collision can be thought of as annihilating into the vacuum, the values of the S-matrix entries should be expressible in terms of *vacuum-to-vacuum* correlations. In this section, we shall present a fairly detailed mathematical treatment of this formula for scalar bosons. In the exposition to follow, when dealing with quantum fields – operator-valued distributions – we shall entirely omit their evaluation against test functions (in Schwartz space), thereby glossing over some very technical details. Although this may not be considered satisfactory from a strictly mathematical point of view, we advise the reader that a completely rigorous treatment is available. See [Kh, pp. 405–12] for what is missing here.

6.8.1 The in and out free fields

Let us consider the case of a bosonic field ϕ of mass m. This field will be, in a suitable sense to be made precise below, asymptotic in the past to a given incoming free field ϕ_{in}, and in the future to a given outgoing free field ϕ_{out}. We assume that the field ϕ is described by a Lagrangian density that is a perturbation of a Klein–Gordon Lagrangian, possibly with a different mass m_0, say

$$\mathcal{L} = \frac{1}{2}\partial_\mu \phi \partial^\mu \phi - \frac{1}{2} m_0^2 \phi^2 + \mathcal{L}_{\mathrm{int}} \ .$$

The interaction part of the Lagrangian, $\mathcal{L}_{\mathrm{int}}$, may contain self-interaction terms such as $\phi^3/3!$, or even interaction terms with other fields (e.g. fermionic fields). We do not want, however, interaction terms involving the partial derivatives of ϕ. Therefore, if we consider the Euler–Lagrange equations for the above

Lagrangian, we get the equation

$$\left(\square + m_0^2\right)\phi(x) = j_0(x)\,,$$

where $j_0 = \partial\mathcal{L}_{\text{int}}/\partial\phi$. In terms of the Klein–Gordon operator for the boson of mass m, this can be recast as

$$\left(\square + m^2\right)\phi(x) = j(x)\,, \tag{6.10}$$

where $j(x) = j_0(x) + (m^2 - m_0^2)\phi(x)$. We may solve equation (6.10) using the method of Green's function – in this case, a retarded Green's function given by

$$G^-(x) = \frac{i}{(2\pi)^3}\theta\left(x^0\right)\int \frac{1}{2\omega_k}\left(e^{-ik\cdot x} - e^{ik\cdot x}\right)_{k^0=\omega_k} d^3k\,,$$

where $\omega_k = \sqrt{\boldsymbol{k}^2 + m^2}$. The function θ is the Heaviside step function ($\theta(s) = 0$ when $s < 0$ and $\theta(s) = 1$ when $s \geq 0$). Using the incoming field ϕ_{in} as an asymptotic boundary condition and the above Green's function, we arrive at the following result.

Lemma 6.24 *The field given by*

$$\phi(x) = \sqrt{Z}\phi_{\text{in}}(x) + \int G^-(x-y)j(y)\,d^4y\,, \tag{6.11}$$

where Z is a constant, is a solution to (6.10).

Proof. An exercise for the reader. □

The number Z is a normalizing constant about which we will have more to say later. The important thing to remember here is that ϕ_{in} is a free scalar field of mass m, i.e. a solution to the Klein–Gordon equation

$$\left(\square + m^2\right)\phi_{\text{in}}(x) = 0\,.$$

Therefore, we can express ϕ_{in} in terms of its Fourier modes (flat waves) as follows. First let

$$e_k(x) = \left.\frac{e^{-ik\cdot x}}{(2\pi)^{3/2}(2\omega_k)^{1/2}}\right|_{k^0=\omega_k}.$$

Then write, as in Section 6.3.4,

$$\phi_{\text{in}}(x) = \int \left\{e_k(x)a_{\text{in}}(\boldsymbol{k}) + e_k^*(x)a_{\text{in}}^\dagger(\boldsymbol{k})\right\} d^3k\,.$$

Using the inverse Fourier transform, we can express each coefficient in terms of the in-field ϕ_{in} itself. For the annihilation coefficients, we get

$$a_{\text{in}}(\boldsymbol{k}) = i \int e_k^*(x) \overleftrightarrow{\partial_0} \phi_{\text{in}}(x)\, d^3x \,,$$

where we use the notation $f \overleftrightarrow{\partial_0} g = f \cdot (\partial_0 g) - (\partial_0 f) \cdot g$. A similar formula holds for the creation coefficients $a_{\text{in}}^\dagger(\boldsymbol{k})$.

Upon quantization, as we saw in Section 6.3.4, these coefficients are promoted to operators in Fock space. Recall that these creation and annihilation operators satisfy the commutation relations

$$[a_{\text{in}}(\boldsymbol{p}), a_{\text{in}}(\boldsymbol{q})] = 0; \quad [a_{\text{in}}(\boldsymbol{p}), a_{\text{in}}^\dagger(\boldsymbol{q})] = \delta^3(\boldsymbol{p} - \boldsymbol{q}) \,.$$

Starting from the vacuum vector Ω, one builds any given in-particle state $|\boldsymbol{q}_1, \boldsymbol{q}_2, \ldots, \boldsymbol{q}_r \text{ in}\rangle$ using these operators in the usual way, namely

$$|\boldsymbol{q}_1, \boldsymbol{q}_2, \ldots, \boldsymbol{q}_r \text{ in}\rangle = 2^{r/2} \left(\prod_{i=1}^{r} \omega_{k_i} \right)^{1/2} a_{\text{in}}^\dagger(\boldsymbol{q}_1) \ldots a_{\text{in}}^\dagger(\boldsymbol{q}_r) |\Omega\rangle \,.$$

We remark also that everything we did above using the in-field ϕ_{in} can be done also for the out-field ϕ_{out}. In particular, we have an integral representation

$$\phi(x) = \sqrt{Z}\phi_{\text{out}}(x) + \int G^+(x-y) j(y)\, d^4y \,, \tag{6.12}$$

where G^+ is the *advanced* Green's function for the Klein–Gordon operator, whose explicit computation we leave as an exercise for the reader.

Finally, we need to clarify in which sense the in-field ϕ_{in} and the out-field ϕ_{out} represent the asymptotic behavior of our interacting field ϕ. We make the *hypothesis* that the following weak asymptotic relations hold. Given any pair of states $|P, \text{ out}\rangle$ and $|Q, \text{ in}\rangle$, and given any solution f to the Klein–Gordon equation, we have

$$\begin{aligned} &\lim_{x^0 \to -\infty} \int \langle P, \text{out}| f(x) \overleftrightarrow{\partial_0} \phi(x) |Q, \text{in}\rangle\, d^3x \\ &\quad = \sqrt{Z} \int \langle P, \text{out}| f(x) \overleftrightarrow{\partial_0} \phi_{\text{in}}(x) |Q, \text{in}\rangle\, d^3x \,. \end{aligned} \tag{6.13}$$

Likewise, for the out-field we have

$$\begin{aligned} &\lim_{x^0 \to +\infty} \int \langle P, \text{out}| f(x) \overleftrightarrow{\partial_0} \phi(x) |Q, \text{in}\rangle\, d^3x \\ &\quad = \sqrt{Z} \int \langle P, \text{out}| f(x) \overleftrightarrow{\partial_0} \phi_{\text{out}}(x) |Q, \text{in}\rangle\, d^3x \,. \end{aligned} \tag{6.14}$$

Using the Klein–Gordon equation for both ϕ_{in} and f, it is easy to show (exercise) that the right-hand side of (6.13) is indeed time-independent, as one should expect from glancing at the left-hand side. Similarly for (6.14).

Given these considerations, we are now ready to formulate a *definition* of scattering (or interacting) fields.

Definition 6.25 *Given two free scalar fields ϕ_{in} and ϕ_{out}, an* interacting scalar field *of mass m, with in-field ϕ_{in} and out-field ϕ_{out}, is a quantum field ϕ (i.e. an operator-valued distribution) having the following properties:*

(i) It satisfies the equation (6.10) *in the distributional sense;*
(ii) It admits the integral representations (6.11) *and* (6.12)*, also in the distributional sense;*
(iii) It satisfies the weak asymptotic relations (6.13) *and* (6.14)*, again in the sense of distributions.*

It is for quantum interacting fields defined in this manner (and whose existence will be taken for granted) that we shall prove the LSZ formula.

6.8.2 The LSZ formula for bosons

We are now almost ready to state and prove the LSZ reduction formula for bosons. It will be convenient at this point to introduce the notion of the *time-ordered product* of a finite number of field operators. Given x, y in Lorentzian spacetime, and the corresponding field operators $\phi(x)$, $\phi(y)$, we define their time-ordered product to be

$$T\{\phi(x)\phi(y)\} = \theta(x^0 - y^0)\phi(x)\phi(y) + \theta(y^0 - x^0)\phi(y)\phi(x)\,,$$

where θ is the step function introduced earlier. More generally, if we have n field operators $\phi(x_1), \ldots, \phi(x_n)$, we define

$$T\{\phi(x_1)\cdots\phi(x_n)\} = \phi(x_{\sigma(1)})\cdots\phi(x_{\sigma(n)})\,,$$

where $\{\sigma(1), \ldots, \sigma(n)\}$ is a permutation of $\{1, \ldots, n\}$ with the property that the corresponding time coordinates decrease from left to right: $x^0_{\sigma(1)} \geq \cdots \geq x^0_{\sigma(n)}$. Time ordering is a way of incorporating causality into quantum field theory.

The main result of this section can now be stated as follows.

Theorem 6.26 (LSZ reduction formula) *The scattering matrix elements of an interacting scalar field ϕ having mass m are functions of the Wightman vacuum-to-vacuum correlations of the field. More precisely, if $Q = \{\boldsymbol{q}_j\}_{1\leq j\leq r}$ and $P = \{\boldsymbol{p}_k\}_{1\leq k\leq s}$ represent the in-particles and out-particles, respectively, and if no in-particle momentum $\boldsymbol{q}_j$ is equal to an out-particle momentum $\boldsymbol{p}_k$,* *then*

$$
\begin{aligned}
&\langle P \text{ out} | Q \text{ in} \rangle \\
&= \frac{i^{r+s}}{((2\pi)^{3/2} Z^{1/2})^{r+s}} \int \prod_{j=1}^{r} \{ e^{-iq_j \cdot x_j} (\Box_{x_j} + m^2) \} \\
&\quad \times \prod_{k=1}^{s} \{ e^{+ip_k \cdot y_k} (\Box_{y_k} + m^2) \} \langle \Omega | T\, \phi(x_1) \cdots \phi(x_r) \phi(y_1) \cdots \phi(y_s) | \Omega \rangle \\
&\quad \times dx_1 \cdots dx_r \, dy_1 \cdots dy_s \, .
\end{aligned} \tag{6.15}
$$

Proof. In the proof, we will write $|Q\rangle$ and $|P\rangle$ instead of $|Q \text{ in}\rangle$ and $|P \text{ out}\rangle$, respectively. Let us calculate something more general than the S-matrix element appearing on the left-hand side of the above formula, namely

$$\mathcal{M} = \langle P | T\{\phi(z_1) \cdots \phi(z_n)\} | Q \rangle \, .$$

This is the expected value of the time-ordered product of the field operators $\phi(z_1), \ldots, \phi(z_n)$ with respect to the in and out states. If there are no field operators in the time-ordered product, then what we have is precisely the S-matrix element that interests us. We can extract from the in-state Q the particle with momentum $\boldsymbol{q} = \boldsymbol{q}_1$ using a creation operator,

$$|Q\rangle = \sqrt{2\omega_q} a_{\text{in}}^{\dagger}(\boldsymbol{q}) |Q'\rangle \, ,$$

where $|Q'\rangle$ indicates the state of in-particles $\boldsymbol{q}_j$ with $j \geq 2$ (i.e. with $\boldsymbol{q}_1$ removed). Hence we can write

$$\mathcal{M} = \sqrt{2\omega_q} \, \langle P | T\{\phi(z_1) \cdots \phi(z_n)\} a_{\text{in}}^{\dagger}(\boldsymbol{q}) | Q' \rangle \, .$$

Because we are assuming that there are no out-particles having momentum $\boldsymbol{q}$, this last equality can be rewritten as

$$\mathcal{M} = \sqrt{2\omega_q} \, \langle P | T\{\phi(z_1) \cdots \phi(z_n)\} a_{\text{in}}^{\dagger}(\boldsymbol{q}) - a_{\text{out}}^{\dagger}(\boldsymbol{q}) T\{\phi(z_1) \cdots \phi(z_n)\} | Q' \rangle \, . \tag{6.16}$$

The reason is the following. In the extra term that has just been introduced on the right-hand side of (6.16), the creation operator $a_{\text{out}}^{\dagger}(\boldsymbol{q})$ can be replaced by its adjoint, the annihilation operator $a_{\text{out}}(\boldsymbol{q})$, acting on the *left* side of the Dirac bracket, i.e. on the out-state vector $|P\rangle$. This vector, however, is built from the vacuum state $|\Omega\rangle$ by successive application of the creation operators $a_{\text{out}}^{\dagger}(\boldsymbol{p}_k)$. Our hypothesis says that no $\boldsymbol{p}_k$ is equal to $\boldsymbol{q}$. Therefore, the annihilation operator $a_{\text{out}}(\boldsymbol{q})$ commutes with each creation operator $a_{\text{out}}^{\dagger}(\boldsymbol{p}_k)$; so it can be made to act directly on the vacuum state, producing 0 as a result. This shows that the extra term introduced on the right-hand side of (6.16) is indeed equal to zero. Replacing the creation operators in (6.16) with their expressions in terms of the inverse Fourier transforms of the in-field and out-field, we see that the right-hand side is equal to

$$-i\sqrt{2\omega_q}\int e_q(x)$$
$$\times\overleftrightarrow{\partial_0}\langle P|T\{\phi(z_1)\cdots\phi(z_n)\}\phi_{\text{in}}(x)-\phi_{\text{out}}(x)T\{\phi(z_1)\cdots\phi(z_n)\}|Q'\rangle d^3x\,.$$

Using the weak asymptotic relations (6.13) and (6.14) with $f = e_q$ (clearly a solution of the KG equation), we deduce that

$$\mathcal{M} = -i\sqrt{\frac{2\omega_q}{Z}}\left\{\lim_{x^0\to-\infty}\int e_q(x)\overleftrightarrow{\partial_0}\langle P|T\{\phi(z_1)\cdots\phi(z_n)\}\phi(x)|Q'\rangle\, d^3x\right.$$
$$\left.-\lim_{x^0\to+\infty}\int e_q(x)\overleftrightarrow{\partial_0}\langle P|\phi(x)T\{\phi(z_1)\cdots\phi(z_n)\}|Q'\rangle d^3x\right\}. \tag{6.17}$$

At this point, we note that as $x^0 \to -\infty$, the time coordinate x^0 becomes smaller than the time coordinate of each z_i, and therefore the field $\phi(x)$ in the first of the two integrals above can be absorbed into the time-ordered product; in other words, we have

$$T\{\phi(z_1)\cdots\phi(z_n)\}\phi(x) = T\{\phi(z_1)\cdots\phi(z_n)\phi(x)\}\quad \text{as } x^0\to-\infty\,.$$

Likewise, we have

$$\phi(x)T\{\phi(z_1)\cdots\phi(z_n)\} = T\{\phi(x)\phi(z_1)\cdots\phi(z_n)\}\quad \text{as } x^0\to+\infty\,.$$

Therefore, we can rewrite (6.17) as

$$\mathcal{M} = -i\sqrt{\frac{2\omega_q}{Z}}\left(\lim_{x^0\to-\infty}-\lim_{x^0\to+\infty}\right)\int e_q(x)\overleftrightarrow{\partial_0}\beta(x)\,d^3x\,,$$

where we have introduced the function

$$\beta(x) = \langle P|T\{\phi(x)\phi(z_1)\cdots\phi(z_n)\phi(x)\}|Q'\rangle\,.$$

Yet another way to write the last expression for $\mathcal{M}$ is to integrate and differentiate with respect to the time coordinate x^0, as follows:

$$\mathcal{M} = i\sqrt{\frac{2\omega_q}{Z}}\int\partial_0\left(\int \mathrm{d}^3x e_q(x)\overleftrightarrow{\partial_0}\beta(x)\,d^3x\right)dx^0\,.$$

Calculating the time derivative explicitly, we see that

$$\mathcal{M} = i\sqrt{\frac{2\omega_q}{Z}}\int x\left\{e_q(x)\partial_0^2\beta(x)-\beta(x)\partial_0^2 e_q(x)\right\}d^4x\,.$$

Using the fact that $e_q(x)$ is a solution of the Klein–Gordon equation – so that $\partial_0^2 e_q = (\Delta - m^2)e_q$ – and integrating by parts, we get

$$\mathcal{M} = i\sqrt{\frac{2\omega_q}{Z}} \int e_q(x) \left(\partial_0^2 - \Delta + m^2\right) \beta(x)\, d^4x \, .$$

Finally, if we take into account the explicit expression for $e_q(x)$ and $\beta(x)$, we arrive at

$$\mathcal{M} = \frac{i}{(2\pi)^{3/2} Z^{1/2}} \int e^{-iq\cdot x} \left(\Box_x + m^2\right)$$
$$\times \langle P|T\{\phi(x)\phi(z_1)\cdots\phi(z_n)\phi(x)\}|Q'\rangle \, d^4x \, .$$

Let us step back and look at what we have accomplished. Using a suitable creation operator for the in-field, we have been able to extract an in-particle from the initial state $|Q\rangle$, inserting a field factor $\phi(x)$ in the time-ordered product, whose correlation with the new initial and the final states is now acted upon by the Klein–Gordon operator with respect to x. One can now proceed inductively, extracting one particle at a time from the initial state, until we are left with the vacuum vector on the right-hand side of the inner product (accumulating field factors in the time-ordered product, and Klein–Gordon operators in the integrand). One can then follow an entirely similar inductive procedure to remove particles one at a time from the final state $|P\rangle$ (using the creation operators for the out-field), until we are left with the vacuum state on the left side of the inner product. The integrand in the final expression will be a vacuum-to-vacuum correlation of a time-ordered product of the original factors $\phi(z_i)$ with field factors corresponding to the extracted particles, acted upon by a product of $r + s$ Klein–Gordon operators. In particular, if initially there are no factors of the form $\phi(z_i)$, in the end we will have the expression on the right-hand side of (6.15), so the theorem is proved. □

There is an alternative way to write down the LSZ formula, which is in fact much more elegant. The idea is to use the Fourier transform of the time-ordered correlation functions, namely

$$\Gamma(\xi_1, \ldots, \xi_n) = \int \langle \Omega|T\{\phi(w_1)\ldots\varphi(w_n)\}|\Omega\rangle \prod_{i=1}^{n} e^{i\xi_k\cdot w_k} d^4w_1 \cdots d^4w_n \, .$$

When we use this together with the inverse Fourier transform, the LSZ reduction formula becomes

$$\langle P \text{ out}|Q \text{ in}\rangle = \frac{(-i)^{r+s}}{((2\pi)^{3/2} Z^{1/2})^{r+s}}$$
$$\times \prod_{k=1}^{s} \left(p_k^2 - m^2\right) \prod_{j=1}^{r} \left(q_j^2 - m^2\right) \Gamma(p_1, \ldots, p_s\, ; -q_1, \ldots, -q_r) \, .$$

Here, we are writing $p_k^2 = (p_k^0)^2 - (p_k^1)^2 - (p_k^2)^2 - (p_k^3)^2$ for the Minkowski norm of p_k, and similarly for q_j. This formula shows us that the values of the S-matrix entries are the *residues of poles* of the Fourier transforms of the corresponding correlation functions, up to a normalization factor (these poles appear when we put the 4-momenta p_k, q_j on-shell, i.e. when their Minkowski norms are equal to m^2).

We close with two remarks. First, as promised, we have shown that the elements of the scattering matrix of a scalar interacting field can be computed from the corresponding vacuum-to-vacuum Wightman correlations. These are, however, very difficult to compute. Perturbative methods for computing these Wightman correlations will be developed in the next chapter, and even then we shall see that there are serious divergences that can only be resolved with the help of renormalization methods (see Chapter 8). Second, a similar LSZ formula holds for self-interacting fermionic fields (in the absence of an external field); see for instance [PS] or [IZ]. However, LSZ reduction does not make sense for gauge fields, the primary reason being that in such cases the free part of the Lagrangian is a degenerate quadratic form, and hence non-invertible.

7
Perturbative quantum field theory

In this chapter, we present the basics of the perturbative approach to the quantization of interacting fields. This approach was pioneered by Schwinger, Tomonaga, Feynman, and Dyson, halfway through the last century. At center stage, here lies the so-called *Feynman path integral* (a simpler version of which we already encountered in Chapter 2).

7.1 Discretization of functional integrals

Heuristically, Feynman's path integral approach to the quantization of fields starts from a functional integral of the form

$$Z = \int e^{iS(\phi)}\,\mathcal{D}\phi\,,$$

called the *partition function*, taken over the infinite-dimensional space of all fields. Here,

$$S(\phi) = \int_M \mathcal{L}(\phi(x))\,d^4x$$

is the action, where $\mathcal{L}$ is the Lagrangian density of the system, and $\mathcal{D}\phi$ is a heuristic measure in the space of all fields. Starting out from such a partition function, one attempts to construct the quantization of the system by writing down other related path integrals giving the Wightman correlation functions (see Chapter 6). If $\mathcal{O}$ is a functional on the space of fields, its expectation is, again heuristically,

$$\langle \mathcal{O} \rangle = \frac{1}{Z}\int \mathcal{O}(\phi)e^{iS(\phi)}\,\mathcal{D}\phi\,.$$

In particular, the Wightman k-point correlation functions are

$$\langle\phi(x_1)\phi(x_2)\cdots\phi(x_k)\rangle = \frac{1}{Z}\int \phi(x_1)\phi(x_2)\cdots\phi(x_k)e^{iS(\phi)}\,\mathcal{D}\phi\ .$$

It is hard, if not downright impossible, to put the subject of Feynman path integrals on a sound mathematical basis, except in the case of free fields, when the Lagrangian of the system is a (non-degenerate, positive definite) quadratic form.

The strategy we will follow instead is to split the action as a sum $S = S_0 + \lambda S_{\rm int}$, where S_0 is the quadratic part and $S_{\rm int}$, the interactive action, is the integral over spacetime of a sum of higher order monomials in the field ϕ. The case $\lambda = 0$ corresponds to a free field and in the Euclidean case to a Gaussian probability measure supported in the Schwartz space of distributions. This measure can be obtained by a process of discretization as a limit of finite-dimensional Gaussian measures. The natural way to approach the general case perturbatively would be to expand the exponential of the action as a power series in the coupling constant λ. Unfortunately, the coefficients of this expansion involve powers of the field, and because the measure is supported in a space of *distributions* and not functions, we face the usual problem that such powers do not make sense (we cannot multiply distributions). To circumvent this problem, we will study the perturbative series of the discretized theory. Here, all the coefficients are given by sums of products of point-to-point correlations of Gaussian measures. These sums can be organized using the technique of Feynman diagrams. The idea would then be to take a (thermodynamic and continuum) limit, but a new problem arises: in many cases the limiting sums are finite-dimensional divergent integrals. This last problem will be treated by *renormalization* in Chapter 8.

7.2 Gaussian measures and Wick's theorem

7.2.1 Gaussian integrals

We now move to the study of Gaussian integrals in a finite-dimensional setting. Let $A : \mathbb{R}^n \to \mathbb{R}^n$ be a positive-definite, symmetric linear map, and let $J \in \mathbb{R}^n$. We are interested in the value of the Gaussian integral

$$Z(A, J) = \int_{\mathbb{R}^n} \exp\{-i(\langle Ax, x\rangle + \langle J, x\rangle)\}\,dx\ .$$

This value can be obtained by analytic continuation, as we shall see.

Let us first note that for each $z \in \mathbb{C}$ with positive real part the function

$$\mathbb{R}^n \ni x \mapsto \exp\{-z(\langle Ax, x\rangle + \langle J, x\rangle)\} \in \mathbb{C}$$

is integrable. Therefore the function

$$I(z) = \int_{\mathbb{R}^n} \exp\{-z(\langle Ax, x\rangle + \langle J, x\rangle)\}\, dx$$

is holomorphic in the half-plane $\{z |\, \mathrm{Re}\, z > 0\}$.

Next, let us compute $I(\rho)$, where $\rho > 0$. To do this, we change variables using an orthogonal transformation T that diagonalizes the quadratic form. In other words, we let T be an orthogonal transformation such that $T^t AT = D$, where $D = \mathrm{diag}\{\alpha_1, \alpha_2, \ldots, \alpha_n\}$, and each $\alpha_k > 0$. Writing $x = Ty$ and $\tilde{J} = T^t J = (\beta_1, \beta_2, \ldots, \beta_n)$, we have

$$\begin{aligned} I(\rho) &= \int_{\mathbb{R}^n} \exp\left\{-\rho\left(\sum_{k=1}^{n} \alpha_k y_k^2 + \beta_k y_k\right)\right\} dy_1 dy_2 \cdots dy_n \\ &= \prod_{k=1}^{n} \int_{\mathbb{R}} \exp\{-\rho\alpha_k y_k^2 - \rho\beta_k y_k\}\, dy_k\,. \end{aligned} \tag{7.1}$$

The integrals in this last product can be explicitly evaluated using the following well-known fact.

Lemma 7.1 *If $a > 0$, then*

$$\int_{\mathbb{R}} \exp\{-ax^2\}\, dx = \sqrt{\frac{\pi}{a}}\,.$$

More generally, if b is real, we have

$$\int_{\mathbb{R}} \exp\{-ax^2 + bx\}\, dx = \sqrt{\frac{\pi}{a}}\, \exp\left\{-\frac{b^2}{2a}\right\}.$$

Proof. Exercise. □

Going back to (7.1), we deduce from the above lemma that

$$\begin{aligned} I(\rho) &= \prod_{k=1}^{n} \exp\left\{-\frac{\rho\beta_k^2}{2\alpha_k}\right\} \sqrt{\frac{\pi}{\rho\alpha_k}} \\ &= \left(\frac{\pi}{\rho}\right)^{n/2} \frac{1}{\sqrt{\det A}} \exp\left\{-\frac{\rho}{2}\sum_{k=1}^{n} \frac{\beta_k^2}{\alpha_k}\right\}. \end{aligned}$$

But clearly,

$$\sum_{k=1}^{n} \frac{\beta_k^2}{\alpha_k} = \langle D^{-1}\tilde{J}, \tilde{J}\rangle = \langle T^t A^{-1} J, T^t J\rangle = \langle A^{-1} J, J\rangle\,.$$

Therefore,

$$I(\rho) = \left(\frac{\pi}{\rho}\right)^{n/2} \frac{1}{\sqrt{\det A}} \exp\left\{-\frac{\rho}{2}\left\langle A^{-1}J, J\right\rangle\right\} .$$

Now, the point is that this formula makes sense for complex ρ. More precisely, let $z \mapsto \sqrt{z}$ be the branch of the square root that maps $\mathbb{C}\backslash\mathbb{R}^-$ holomorphically onto the positive half-plane. Then, consider the holomorphic function

$$\hat{I}(z) = \frac{\pi^{n/2}}{(\sqrt{z})^n} \frac{1}{\sqrt{\det A}} \exp\left\{-\frac{z}{2}\left\langle A^{-1}J, J\right\rangle\right\} .$$

Because $\hat{I}(z) = I(z)$ for all $z \in \mathbb{R}^+$, it follows that $\hat{I}(z)$ is a holomorphic extension of $I(z)$. Hence we can define the value of the desired Gaussian integral $Z(A, J)$ to be $\hat{I}(i)$; that is

$$\int_{\mathbb{R}^n} e^{-i(\langle Ax,x\rangle+\langle J,x\rangle)}\, dx = \frac{\pi^{n/2}\left(e^{-\pi i/4}\right)^n}{\sqrt{\det A}}\, e^{-\frac{i}{2}\left\langle A^{-1}J,J\right\rangle} . \tag{7.2}$$

It is worth emphasizing that the main contribution to (7.2) on the right-hand side is provided by the critical point of the function in the exponent of the exponential on the left-hand side. Indeed, if $S(x) = \langle Ax, x\rangle + \langle J, x\rangle$, then

$$DS(x)v = \langle 2Ax + J, v\rangle \; .$$

This shows that S has a unique critical point $\bar{x}$ such that $2A\bar{x} + J = 0$, i.e. $\bar{x} = (-1/2)A^{-1}J$. This critical point is a global minimum of S, and $S(\bar{x}) = \left\langle A^{-1}J, J\right\rangle/4$.

These results can be generalized to the case where A is not necessarily positive-definite, but is still non-degenerate. The formula we obtain in the end is similar to (7.2) but involves, not surprisingly, the signature of A.

7.2.2 Gaussian measures

Using the above formulas, we can now define a Gaussian measure in $\mathbb{R}^n$ with given covariance matrix C as follows:

$$d\mu_C(x) = \frac{1}{\pi^{n/2}\sqrt{\det C}} \exp\{-\langle xC^{-1}x\rangle\}\, dx \; .$$

This is a probability measure:

$$\int_{\mathbb{R}^n} d\mu_C(x) = 1 \; .$$

The Fourier transform of this measure is

$$\int_{\mathbb{R}^n} e^{i\langle J,x\rangle}\, d\mu_C(x) = \exp\left\{-\frac{1}{2}\langle J, CJ\rangle\right\} .$$

The moment-generating function, on the other hand, is simply

$$\int_{\mathbb{R}^n} e^{\langle J,x\rangle}\, d\mu_C(x) = \exp\left\{\frac{1}{2}\, \langle J, CJ\rangle\right\} .$$

7.2.3 Wick's theorem

Let us now state and prove Wick's theorem. This result yields an explicit formula for the moments (or correlations) of a Gaussian measure.

Theorem 7.2 *Let $f_1, f_2, \ldots, f_n$ be linear polynomials. Then $\langle f_1 f_2 \cdots f_k\rangle = 0$ if n is odd, whereas if n is even*

$$\langle f_1 f_2 \cdots f_k\rangle = \sum C(f_{i_1}, f_{i_2})C(f_{i_3}, f_{i_4}) \cdots C(f_{i_{m-1}}, f_{i_m}) , \qquad (7.3)$$

where the sum extends over all pairings $(i_1, i_2), \ldots, (i_{m-1}, i_m)$ of the indices $1, 2, \ldots, m$, and where $C(f_i, f_j) = \langle f_i C f_j\rangle_\Gamma$.

To prove this theorem, we need the following auxiliary lemma in multilinear algebra.

Lemma 7.3 *Let V be a vector space over $\mathbb{K} = \mathbb{R}$ or $\mathbb{C}$, and let $B : V \times \cdots \times V \to \mathbb{K}$ be a symmetric n-linear map. If for all $x \in V$ we have $x^n = B(x, \ldots, x) = 0$, then $B = 0$.*

Proof. It suffices to show, by induction, that if $x^n = 0$ for all x then $x^{n-1}y = 0$ for all x, y. This because for each y the map $B_y : V \times \cdots \times V \to \mathbb{K}$ given by

$$B_y(x_1, x_2, \ldots, x_{n-1}) = B(x_1, x_2, \ldots, x_{n-1}, y)$$

is $(n-1)$-linear and vanishes at the diagonal. By hypothesis, for all $x, y \in V$ and each $z \in \mathbb{K}$, we have

$$0 = (x + zy)^n = x^n + z\binom{n}{1}x^{n-1}y + \cdots + z^{n-1}\binom{n}{n-1}xy^{n-1} + z^n y^n \quad (7.4a)$$

$$= z\binom{n}{1}x^{n-1}y + \cdots + z^{n-1}\binom{n}{n-1}xy^{n-1} . \qquad (7.4b)$$

Now, taking $z_1, z_2, \ldots, z_n$ $\mathbb{K}$ pairwise distinct non-zero scalars, we know that the Vandermonde determinant

$$\det\begin{pmatrix} z_1 & z_1^2 & \cdots & z_1^{n-1} \\ z_2 & z_2^2 & \cdots & z_2^{n-1} \\ \cdot & \cdot & \cdots & \cdot \\ z_{n-1} & z_{n-1}^2 & \cdots & z_{n-1}^{n-1} \end{pmatrix} = z_1 z_2 \cdots z_{n-1} \prod_{i<j}(z_i - z_j) \neq 0 .$$

Therefore, putting $z = z_1, \ldots, z_{n-1}$ in (7.4a), we get a homogeneous $(n-1) \times (n-1)$ linear system in the unknowns $x^{n-1}y, \ldots, xy^{n-1}$ whose only solution is the trivial solution $x^{n-1}y = \cdots xy^{n-1} = 0$. □

Given this lemma, Wick's theorem will be a consequence of the following result. Suppose $A : V \to V^*$ is a linear operator such that

$$A(x, y) = \langle x, Ay \rangle = \sum_{i,j=1}^{d} A_{ij} x_i y_j$$

is a positive-definite bilinear form. Let us denote by $C : V^* \to V$ the inverse of A, its associated positive-definite bilinear form being

$$C(\lambda, \mu) = \sum_{i,j=1}^{d} C_{ij} \lambda_i \mu_j \, .$$

Lemma 7.4 *If $f \in V^*$ is a linear functional, then for every $k \geq 1$ we have*

$$\int_V f(x)^{2k} \, e^{-\frac{1}{2}\langle x, Ax \rangle} \, dx = (2k-1) C(f, f) \int_V f(x)^{2(k-1)} \, e^{-\frac{1}{2}\langle x, Ax \rangle} \, dx \, . \quad (7.5)$$

Proof. Let us write $f(x) = \sum_{i=1}^{d} a_i x_i$, and let us denote the integral on the left-hand side of (7.5) by I. Then

$$I = \sum_{i=1}^{n} a_i \int_V x_i (f(x))^{2k-1} \, e^{-\frac{1}{2}\langle x, Ax \rangle} \, dx \, . \quad (7.6)$$

Note however that

$$x_i e^{-\frac{1}{2}\langle x, Ax \rangle} = \sum_{j=1}^{n} C_{ij} \frac{\partial}{\partial x_j} \, e^{-\frac{1}{2}\langle x, Ax \rangle} \, .$$

Hence, we can integrate by parts in (7.6) to get

$$\begin{aligned} I &= \sum_{i=1}^{d} a_i \sum_{j=1}^{d} C_{ij} \int_V (f(x))^{2k-1} \frac{\partial}{\partial x_j} \left(e^{-\frac{1}{2}\langle x, Ax \rangle} \right) dx \\ &= \sum_{i=1}^{d} a_i \sum_{j=1}^{d} C_{ij} \int_V (2k-1)(f(x))^{2k-2} a_j e^{-\frac{1}{2}\langle x, Ax \rangle} \, dx \\ &= (2k-1) \sum_{i,j=1}^{d} a_i C_{ij} a_j \int_V (f(x))^{2(k-1)} \, e^{-\frac{1}{2}\langle x, Ax \rangle} \, dx \, . \end{aligned}$$

□

Proof of Wick's theorem 7.2. Let us denote the left-hand side of (7.3) by $B_1(f_1, \ldots, f_n)$ and the right-hand side by $B_2(f_1, \ldots, f_n)$. Both B_1 and B_2 are symmetric n-linear forms; hence so is their difference $B = B_1 - B_2$. To show that this difference is zero, it suffices by Lemma 7.3 to show that

$$B_1(f, \ldots, f) = B_2(f, \ldots, f) \tag{7.7}$$

for all $f \in V^*$. Applying Lemma 7.4 and induction to the left-hand side of (7.7), we see that

$$B_1(f, \ldots, f) = \int_V f(x)^{2n}\, e^{-\frac{1}{2}\langle x, Ax\rangle}\, dx = (2n-1)\cdots 3 \cdot 1 C(f, f)^{2n}.$$

But, all terms in the summation defining the right-hand side of (7.7) are equal to $C(f, f)^{2n}$, and the number of such terms is the number of pairings of $\{1, 2, \ldots, 2n\}$, which is precisely $(2n-1)!! = (2n-1)\cdot(2n-3)\cdots 3 \cdot 1$. □

7.3 Discretization of Euclidean scalar fields

We consider fields on a lattice (in $\varepsilon \cdot \mathbb{Z}^d$). Looking in a finite volume (e.g. some cube in $\varepsilon \cdot \mathbb{Z}^d$), we get that the partition function and the field correlations are finite-dimensional Euclidean integrals. After we understand these integrals, we attempt to take limits, by letting ε go to zero (continuum limit) and also the volume go to infinity (thermodynamic limit). For free fields (quadratic Lagrangian), the thermodynamic and continuum limits exist, and we recover the Schwinger correlation functions.

To be specific, let us consider the case of the (Euclidean) Klein–Gordon field in d-dimensional space. Let $\Gamma = \Gamma_{\varepsilon,L} = \varepsilon\mathbb{Z}^d/L\mathbb{Z}^d$, where L is chosen so that $L/2\varepsilon$ is a positive integer. Then Γ is a cube in the lattice $\varepsilon\mathbb{Z}^d$; note that in fact

$$\Gamma = \left\{ x \in \varepsilon\mathbb{Z}^d \subset \mathbb{R}^d \;:\; -\frac{L}{2} \le x_j < \frac{L}{2},\; j = 1, 2, \ldots, d \right\}.$$

The volume of this cube is $|\Gamma| = (\varepsilon^{-1}L)^d$. If $F : \Gamma \to \mathbb{C}$, we write

$$\int_\Gamma F(x)\, dx = \varepsilon^d \sum_{x\in\Gamma} F(x).$$

A scalar field over the lattice Γ is simply a function $\varphi : \Gamma \to \mathbb{C}$. The space of all such fields is therefore $\mathbb{C}^\Gamma$. We define a Hermitian inner product on $\mathbb{C}^\Gamma$ by

$$\langle \varphi, \psi \rangle_\Gamma = \int_\Gamma \overline{\varphi(x)}\psi(x)\, dx.$$

Definition 7.5 *The* discrete Laplacian *on the finite lattice* Γ *is the difference operator* $-\Delta : \mathbb{C}^\Gamma \to \mathbb{C}^\Gamma$ *given by*

$$-\Delta\phi(x) = \frac{1}{\varepsilon^2}\sum_{k=1}^{d}(2\phi(x) - \phi(x+\varepsilon e_k) - \phi(x-\varepsilon e_k)) .$$

Here, $\{e_1, e_2, \ldots, e_d\}$ is the canonical basis of $\mathbb{R}^d$. It is a convention in the above expression for $-\Delta$ that ϕ is taken to be equal to zero at points outside Γ. Thus, up to a multiplicative constant, the value of the discrete Laplacian of ϕ at x is the difference between $\phi(x)$ and the average value of ϕ at the neighboring sites of x in Γ. The discrete Laplacian is a positive linear operator, in the sense that

$$\langle \phi, -\Delta\phi\rangle_\Gamma > 0 \text{ provided } \langle \phi, \phi\rangle_\Gamma > 0 .$$

It is also symmetric; in other words $\langle \phi, -\Delta\psi\rangle_\Gamma = \langle -\Delta\phi, \psi\rangle_\Gamma$, for all $\phi, \psi \in \mathbb{C}^\Gamma$.

Definition 7.6 *The* discrete Euclidian Klein–Gordon operator $A : \mathbb{C}^\Gamma \to \mathbb{C}^\Gamma$ *is the linear operator given by*

$$A\phi = -\Delta\phi + m^2\phi .$$

It is worth to summarize the basic properties of this operator in a simple lemma.

Lemma 7.7 *The discrete Klein–Gordon operator A is positive and self-adjoint. In particular, it has an inverse.*

Proof. Exercise. □

Because we are dealing with finite-dimensional spaces here, every linear operator $T : \mathbb{C}^\Gamma \to \mathbb{C}^\Gamma$ has a *kernel*, namely a map

$$\Gamma \times \Gamma \ni (x, y) \mapsto T(x, y) \in \mathbb{C} ,$$

such that

$$T\phi(x) = \int_\Gamma T(x, y)\phi(y)\, dy .$$

In particular, the discrete Laplacian has a kernel. To identify it, let us write $\delta_\Gamma(x, y) = \varepsilon^{-d}\delta_{xy}$, where δ_{xy} is 1 if $x = y$ and zero otherwise. This is precisely the kernel of the identity operator, because

$$\phi(x) = \int_\Gamma \delta_\Gamma(x, y)\phi(y)\, dy .$$

Now, we have the following fact.

Lemma 7.8 *The discrete Laplacian operator has kernel*

$$-\Delta(x, y) = \frac{1}{\varepsilon}\sum_{k=1}^{d}\left(2\delta_\Gamma(x, y) - \delta_\Gamma(x + \varepsilon e_k, y) - \delta_\Gamma(x - \varepsilon e_k, y)\right) .$$

Proof. Exercise. □

Functional derivatives can also be discretized. If $S : \mathbb{C}^\Gamma \to \mathbb{C}$ is a differentiable (typically non-linear) functional, then $\delta S/\delta\phi(x)$ is defined so that

$$DS(\phi)\delta\phi = \lim_{t\to 0}\frac{1}{t}\left(S(\phi + t\delta\phi) - S(\phi)\right) = \int_\Gamma \frac{\delta S}{\delta\phi(x)}\delta\phi(x)\, dx .$$

We are now ready for discrete path integrals. Let us introduce a measure in the space of all discretized fields; we simply take it to be Lebesgue measure in $\mathbb{C}^\Gamma$, written as

$$\mathcal{D}_\Gamma\phi = \prod_{x\in\Gamma} d\phi(x) .$$

Let $S_\Gamma : \mathbb{C}^\Gamma \to \mathbb{C}$ be an action functional in the space of fields, having the form

$$S_\Gamma(\phi) = \frac{1}{2}\langle\phi, A\phi\rangle + \lambda S_\Gamma^{\text{int}}(\phi) ,$$

where $S_\Gamma^{\text{int}}(\phi)$ represents the *interaction part of the action* and involves higher than quadratic terms in the field. The most important example is the so-called ϕ^4-action given by

$$S_\Gamma^{\text{int}}(\phi) = \int_\Gamma \phi(x)^4\, dx .$$

Let us also consider a linear functional $J : \Gamma \to \mathbb{R}$. The (generalized) discrete partition function $Z_\Gamma(J)$ is given by

$$Z_\Gamma(J) = \int_{\mathbb{C}^\Gamma} e^{-S_\Gamma(\phi)+\langle J,x\rangle_\Gamma}\, \mathcal{D}_\Gamma\phi .$$

Now, if we are given an observable $\mathcal{O}$ in the space of fields, its expectation is defined to be

$$\langle\mathcal{O}\rangle = \frac{1}{Z_\Gamma(0)}\int_{\mathbb{C}^\Gamma} \mathcal{O}e^{-S_\Gamma(\phi)}\, \mathcal{D}_\Gamma\phi .$$

Likewise, if we are given k such observables, their correlation is given by

$$\langle\mathcal{O}_1\mathcal{O}_2\cdots\mathcal{O}_k\rangle = \frac{1}{Z_\Gamma(0)}\int_{\mathbb{C}^\Gamma} \mathcal{O}_1(\phi)\mathcal{O}_2(\phi)\cdots\mathcal{O}_k(\phi)e^{-S_\Gamma(\phi)}\, \mathcal{D}_\Gamma\phi .$$

In particular, the k-point Schwinger correlation functions are given by

$$\langle \phi(x_1)\phi(x_2)\cdots\phi(x_k)\rangle = \frac{1}{Z_\Gamma(0)} \int_{\mathbb{C}^\Gamma} \prod_{j=1}^{k} \phi(x_j) e^{-S_\Gamma(\phi)}\, \mathcal{D}_\Gamma\phi \,.$$

Everything here, of course, is only true provided these integrals exist. The existence depends on imposing reasonable conditions on the action functional. When the action is a positive definite quadratic form (as is the case for the discrete Klein–Gordon action), the partition function is a Gaussian integral that can be explicitly evaluated (see Section).

Lemma 7.9 *The Schwinger k-point correlation functions satisfy the identity*

$$\langle \phi(x_1)\phi(x_2)\cdots\phi(x_k)\rangle = \left[\frac{1}{Z_\Gamma(J)} \frac{\delta}{\delta J(x_1)} \cdots \frac{\delta}{\delta J(x_k)} Z_\Gamma(J)\right]_{J=0} .$$

Proof. Differentiate under the integral sign. □

Let us now evaluate these integrals in the case of interest to us, namely the Klein–Gordon field. Let us denote by $C = C_\Gamma : \mathbb{C}^\Gamma \to \mathbb{C}^\Gamma$ the positive, self-adjoint operator that is the inverse of the Klein–Gordon operator A. The fields are now real, and so the partition function is

$$Z_\Gamma(J) = \int_{\mathbb{R}^\Gamma} e^{-\frac{1}{2}\langle\phi, A\phi\rangle_\Gamma + \langle J,\phi\rangle_\Gamma}\, \mathcal{D}_\Gamma\phi \,.$$

Using the results on Gaussian integrals of Section, we see that

$$Z_0 = Z_\Gamma(0) = (2\pi)^{|\Gamma|/2}\sqrt{\det C}\,,$$

and also that

$$Z_\Gamma(J) = Z_0 e^{\frac{1}{2}\langle J, CJ\rangle_\Gamma} \,.$$

Therefore, by Lemma 7.9, the k-point correlation functions are given by

$$\langle \phi(x_1)\phi(x_2)\cdots\phi(x_k)\rangle = \frac{1}{Z_0}\left(\frac{\delta}{\delta J(x_1)} \cdots \frac{\delta}{\delta J(x_k)}\right) e^{\frac{1}{2}\langle J, CJ\rangle_\Gamma}\Big|_{J=0} .$$

More generally, we have

Lemma 7.10 *If $f_1, f_2, \ldots, f_k$ are arbitrary polynomials, then*

$$\langle f_1 f_2 \cdots f_k\rangle = f_1\left(\frac{\delta}{\delta J}\right)\cdots f_k\left(\frac{\delta}{\delta J}\right) e^{\frac{1}{2}\langle J, CJ\rangle_\Gamma}\Big|_{J=0} .$$

Proof. Exercise. □

Oftentimes, when studying field theory, we need to pass from coordinate space to momentum space and vice versa. What allows us to do this is the Fourier transform, for fields on our finite lattice $\Gamma = \Gamma_{\varepsilon,L} = \varepsilon\mathbb{Z}^d/L\mathbb{Z}^d$ (where as before $L/2\varepsilon \in \mathbb{N}$). In order to define the Fourier transform in this finite setting, let us first consider the *dual lattice* to Γ, namely

$$\Gamma^* = \frac{2\pi}{L}\mathbb{Z}^d \Big/ \frac{2\pi}{\varepsilon}\mathbb{Z}^d .$$

Given $p \in \Gamma^*$ and $x \in \Gamma$, we write

$$p \cdot x = p_1 x_1 + p_2 x_2 + \cdots + p_d x_d .$$

Definition 7.11 *The* Fourier transform *of $F : \Gamma \to \mathbb{C}$ is given by*

$$\hat{F}(p) = \int_\Gamma F(x) e^{-ip\cdot x}\, dx .$$

Just as with the standard theory of the Fourier transform, here too we have an inversion formula,

$$F(x) = \int_{\Gamma^*} \hat{F}(p) e^{ip\cdot x}\, dp .$$

The integral over the dual lattice Γ^* is defined exactly as before. So is the Hermitian product of dual fields in $\mathbb{C}^{\Gamma^*}$.

Now, suppose we have a linear operator $A : \mathbb{C}^\Gamma \to \mathbb{C}^\Gamma$ on the space of fields, given say by

$$(A\phi)(x) = \int_\Gamma A(x, y)\phi(y)\, dy .$$

Then we define its Fourier transform to be the linear operator $\hat{A} : \mathbb{C}^{\Gamma^*} \to \mathbb{C}^{\Gamma^*}$ given by $\hat{A}(\hat{\phi}) = \widehat{A(\phi)}$. If we express $\hat{A}$ through its kernel,

$$(\hat{A}\hat{\phi})(p) = \int_{\Gamma^*} \hat{A}(p, q)\hat{\phi}(q)\, dq ,$$

then it is clear that this kernel is given by

$$\hat{A}(p, q) = \int_\Gamma e^{-ip\cdot x} \left[\int_\Gamma A(x, y) e^{ip\cdot y}\, dy \right] dx .$$

The main example is provided by the discrete Laplacian, whose kernel, as we saw earlier, is

$$-\Delta(x, y) = \varepsilon^{-2} \sum_{k=1}^{d} \left(2\delta_\Gamma(x, y) - \delta_\Gamma(x + \varepsilon e_k, y) - \delta_\Gamma(x - \varepsilon e_k, y) \right) .$$

One can prove (exercise) that in this case the kernel of the Fourier transform of the Laplacian is given by

$$-\hat{\Delta}(p, q) = \delta_{\Gamma^*}(p, q)\, D_\varepsilon(p) ,$$

where

$$D_\varepsilon(p) = \frac{2}{\varepsilon^2}\sum_{k=1}^{d}(1-\cos(\varepsilon p_k)) .$$

We can use this fact to study the discrete Klein–Gordon operator as well. In this situation, we find that the *free covariance* or *propagator* that we met earlier (which is in fact a discrete version of a fundamental solution to the Klein–Gordon equation) is given by

$$C_\Gamma(x,y) = \int_{\Gamma^*} \frac{e^{ip\cdot(x-y)}}{m^2+D_\varepsilon(p)}\, dp . \tag{7.8}$$

From this and the expression obtained earlier for the partition function $Z_\Gamma(0)$, we see that

$$Z_\Gamma(0) = \det(2\pi C_\Gamma) = \prod_{p\in\Gamma^*}\left(\frac{2\pi L^d}{\varepsilon^d}\,\frac{1}{m^2+D_\varepsilon(p)}\right)^{1/2} .$$

Going back to (7.8), we analyze the behavior of the propagator when we take the thermodynamic and continuum limits. This will be very important when we investigate the perturbative expansion of the scalar field; see the next section. First of all, when we let $L\to\infty$ (thermodynamic limit), we get

$$C^\varepsilon(x,y) = \lim_{L\to\infty} C_\Gamma(x,y) = \frac{1}{(2\pi)^d}\int_{Q(\varepsilon)} \frac{e^{ip\cdot(x-y)}}{m^2+D_\varepsilon(p)}\, d^d p ,$$

where $Q(\varepsilon) = \mathbb{R}^d/\left(\frac{2\pi}{\varepsilon}\right)\mathbb{Z}^d$. Second, when we now let $\varepsilon\to 0$ (continuum limit), then *as long as* $x\neq y$, we do get a finite limit, given by the integral

$$C(x,y) = \lim_{\varepsilon\to 0} C^\varepsilon(x,y) = \frac{1}{(2\pi)^d}\int_{\mathbb{R}^d} \frac{e^{ip\cdot(x-y)}}{m^2+p^2}\, d^d p .$$

This integral (for $x\neq y$!) converges, as one can easily check using the residue theorem. When $x=y$ the integral obviously diverges, but we have the following asymptotic behavior. As $|x-y|\to 0$, we have

$$C(x,y) \sim \begin{cases} |x-y|^{-(d-2)} & \text{for } d\geq 3 \\ \log|x-y| & \text{for } d=2 . \end{cases}$$

7.4 Perturbative quantum field theory

Here we consider perturbative theory, in which the Lagrangian is a perturbation of a quadratic Lagrangian, which once again involves

(i) discretization (to a lattice in $\mathbb{Z}^d$);
(ii) finite volume approximation;

(iii) perturbative expansion in terms of the coupling constants;
(iv) Feynman diagrams;
(v) continuum and thermodynamical limits of Feynman diagrams;
(vi) Euclidean field theory and statistical mechanics.

There are problems of two kinds here. There is a divergence problem, and there is also the problem that the Lagrangian is in general non-invertible.

7.4.1 Perturbative Euclidean ϕ^4-theory

We shall study here the so-called ϕ^4-theory, in its Euclidean form. In this scalar (bosonic) theory, the interaction part of the Lagrangian consists of a single quartic monomial of the form $\lambda\phi^4/4!$, where $\lambda > 0$ is the coupling constant.

7.4.1.1 Asymptotic expansion: heuristics Ideally, we know from Wightman's reconstruction theorem (Chapter 6) that we can recover a quantum field theory from its Wightman correlation functions. In Euclidean QFT (i.e. after a suitable Wick rotation), the relevant correlation functions are the Schwinger n-point functions. In the present situation (ϕ^4-theory), these Schwinger functions are given by

$$S_n(x_1, x_2, \ldots, x_n) = \frac{1}{Z}\int \phi(x_1)\phi(x_2)\cdots\phi(x_n)e^{-\frac{\lambda}{4!}\int \phi(x)^4\,dx}\,d\mu_C(\phi)\,,$$

where $d\mu_C$ is the Gaussian measure with correlation operator C given by the inverse of the Klein–Gordon operator, and Z is the quantum partition function

$$Z = \int e^{-\frac{\lambda}{4!}\int \phi(x)^4\,dx}\,d\mu_C(\phi)\,.$$

The naive standard approach to the study of these integrals is to write down an *asymptotic expansion* of the form

$$S_n(x_1, x_2, \ldots, x_n) \sim \sum_{k=0}^{\infty} S_{n,k}(x_1, x_2, \ldots, x_n)\,\lambda^k$$

for the n-point Schwinger function, where

$$S_{n,k}(x_1, x_2, \ldots, x_n) = \frac{(-1)}{Z}\,\frac{1}{k!(4!)^k}\int \phi(x_1)\cdots\phi(x_n) \times\left(\int_{\mathbb{R}^d}\phi(y)\,dy\right)^k d\mu_C(\phi)\,,$$

and a similar asymptotic expansion for Z.

This is all purely formal. Indeed, as we noted at the end of Section 7.2, although the measure $d\mu_C(\phi)$ exists in the infinite-dimensional setting, the typical field ϕ in its support is not a function but merely a *distribution* and therefore, the above integral expressions do not even make sense. Hence, the strategy to follow is to work with the discretization process described earlier, where everything makes sense.

7.4.1.2 Discrete Schwinger functions We shall work with the discretized field in a finite lattice $\Lambda \subset \mathbb{R}^d$, as in the previous section. We would like to calculate the Schwinger correlation functions given by

$$S_n^{\Lambda}(x_1, x_2, \ldots, x_n) = \int_{\mathbb{R}^{\Lambda}} \phi(x_1)\phi(x_2)\cdots\phi(x_n)\, e^{-\frac{\lambda}{4!}\int_{\Lambda}\phi(x)^4\,dx}\, d\mu_C(\phi)\,,$$

where, as before, $d\mu_C$ is the Gaussian measure with correlation operator $C = A^{-1}$ (the inverse of the discretized Klein–Gordon operator) and

$$\int_{\Lambda} \phi(x)^4\,dx = \varepsilon^d \sum_{x\in\Lambda} \phi(x)^4\,.$$

Here, everything is in principle computable, because we are in a finite lattice.

We can expand the exponential as a (formal) power series in λ, obtaining

$$\begin{aligned} &S_n^{\Lambda}(x_1, x_2, \ldots, x_n) \\ &\quad = \int_{\mathbb{R}^{\Lambda}} \phi(x_1)\phi(x_2)\cdots\phi(x_n) \sum_{k=0}^{\infty} \frac{(-1)^k}{k!}\frac{\lambda^k}{(4!)^k}\left(\int_{\Lambda}\phi(x)^4\,dx\right)^k d\mu_C(\phi) \\ &\quad = \sum_{k=0}^{\infty} \frac{(-1)^k}{k!(4!)^k}\lambda^k \int_{\mathbb{R}^{\Lambda}} \phi(x_1)\phi(x_2)\cdots\phi(x_n) \prod_{j=1}^{k}\left(\int_{\Lambda}\phi(y_j)^4\,dy_j\right)^k d\mu_C(\phi)\,. \end{aligned}$$

This shows that we can write each Schwinger function as a perturbative series

$$S_n^{\Lambda}(x_1, x_2, \ldots, x_n) = \sum_{k=0}^{\infty} \lambda^k S_{n,k}^{\Lambda}(x_1, x_2, \ldots, x_n)\,,$$

where

$$\begin{aligned} S_{n,k}^{\Lambda}(x_1, x_2, \ldots, x_n) &= \frac{(-1)^k}{k!(4!)^k}\lambda^k \int_{\Lambda} dy_1 \cdots \int_{\Lambda} dy_k \\ &\quad \times \left[\int_{\mathbb{R}^{\Lambda}} \phi(x_1)\cdots\phi(x_n)\,\phi(y_1)^4\cdots\phi(y_k)^4\, d\mu_C(\phi)\right]\,. \end{aligned} \tag{7.9}$$

Note that we have interchanged the order of integration (Gaussian measure first, then the product measure $dy_1 \cdots dy_k$). This interchange is legitimate because the integrals over Λ are just finite sums.

Now, the point is that the integral appearing on the right-hand side of (7.9) can be evaluated explicitly with the help of Wick's theorem. For n odd, the integral is equal to zero. When n is even, the final result will involve a sum of terms corresponding to all possible pairings of the $n + 4k$ terms appearing in the integrand, namely

$$\phi(x_1) \cdots \phi(x_n)\, \phi(y_1)\phi(y_1)\phi(y_1)\phi(y_1) \cdots \phi(y_k)\phi(y_k)\phi(y_k)\phi(y_k)\,.$$

The number of all such pairings is $(n + 4k - 1)!!$, a large number even for modest values of n and k. The task of computing the Wick sum can be simplified considerably by collecting together large groups of identical terms, each group being characterized by a combinatorial device known as a *Feynman diagram*, as we shall explain.

Before jumping to a formal definition of Feynman diagrams, let us examine a simple example in detail.

Example 7.12 *Let us consider the case when* $n = 2$ *(that is to say, we are interested in the* 2*-point correlation function* $S_2(x_1, x_2)$*). The first term in the sum defining* $S_2(x_1, x_2)$ *has* $k = 0$*. In this case, there is only one pairing* $\phi(x_1)\phi(x_2)$*, and we easily get*

$$S_{2,0}(x_1, x_2) = C(x_1, x_2)\,.$$

The second term in the sum has $k = 1$*. To apply Wick's theorem, we need to write down all possible pairings of the six terms in the product*

$$\phi(x_1)\phi(x_2)\phi(y_1)\phi(y_1)\phi(y_1)\phi(y_1)\,.$$

Here, there are two cases to consider.

(*i*) *The term* $\phi(x_1)$ *pairs up with* $\phi(x_2)$*, and the four copies of* $\phi(y_1)$ *make up two more pairs. This can be done in three ways, each contributing*

$$C(x_1, x_2)C(y_1, y_1)^2$$

to Wick's sum. This case can be represented graphically as in Figure 7.1 (a). The diagram displays an edge connecting the two white vertices labeled x_1 *and* x_2*, and two loops forming a figure eight connecting a single black vertex to itself (not labeled in the figure, but corresponding to* y_1*).*

(*ii*) *The term* $\phi(x_1)$ *pairs up with one of the four copies of* $\phi(y_1)$*, the term* $\phi(x_2)$ *pairs up with a second such copy, and the two remaining terms pair with each other. Here there are altogether* 12 *possibilities, each contributing*

$$C(x_1, y_1)C(x_2, y_1)C(y_1, y_1)$$

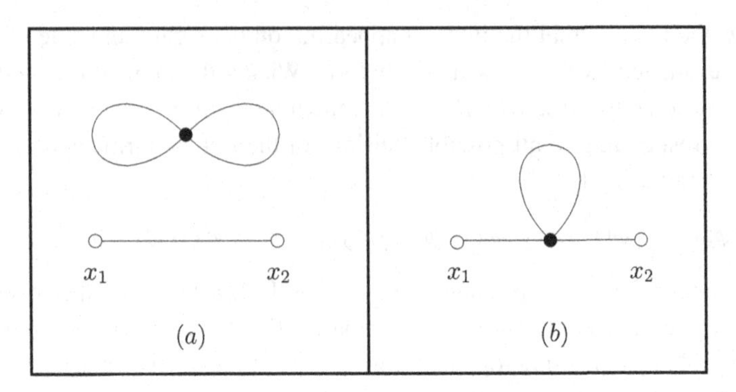

Figure 7.1 The two Feynman diagrams in $\mathcal{F}^e_{2,1}$.

to Wick's sum. Again, this information can be encoded graphically; see the diagram in Figure 7.1(b). Note that now the diagram is connected; there are two edges connecting the white vertices labeled x_1 and x_2 to the (unlabeled) black vertex, which is also connected to itself by a loop.

Note that adding up the total number of possibilities in both cases we get $3+12=15$, which is precisely the number of possible pairings of six objects ($(6-1)!! = 5\cdot 3\cdot 1 = 15$). Summarizing, we have shown that

$$S_{2,1}(x_1,x_2) = -\frac{1}{24}\int_\Lambda \left[3C(x_1,x_2)C(y_1,y_1)^2 + 12C(x_1,y_1)C(x_2,y_1)C(y_1,y_1)\right] dy_1 .$$

This can be rewritten as

$$S_{2,1}(x_1,x_2) = -\frac{1}{8}\int_\Lambda C(x_1,x_2)C(y_1,y_1)^2\,dy_1 - \frac{1}{2} \times \int_\Lambda C(x_1,y_1)C(x_2,y_1)C(y_1,y_1)dy_1 . \tag{7.10}$$

How are we to interpret the numbers appearing in front of these integrals? The answer pops up if we go back to Figure 7.1 and examine the symmetries of both Feynman diagrams. We are not allowed to change the positions of the labeled vertices x_1 and x_2. In Figure 7.1(a), this forces us to leave the edge connecting these two labeled vertices untouched; the remaining figure eight has a symmetry group generated by three elements of order two: for each loop in the figure, there are a reflection about the symmetry axis of that loop (switching its ends) – call these reflections a and b – and also an involution c that switches both loops. These generators satisfy the relations $a^2 = b^2 = c^2 = 1$ and $cac = b$.

The reader can easily check that the resulting group of symmetries has order 8. This is precisely the number that appears in the denominator of the fraction multiplying the first integral in (7.10). *The situation in Figure 7.1(b) is even simpler. Here the only possible symmetry besides the identity consists of a flip of the single loop in the picture about its (vertical) axis, keeping the labeled vertices in their places. Hence, the symmetry group has order 2, which is once again the denominator of the fraction that multiplies the second integral in* (7.10). *These remarkable facts are valid in general, as we shall see in the sequel.*

7.4.2 Feynman diagrams

Now that we have seen some examples of Feynman diagrams and how they appear in the perturbative expansion of a scalar field theory, let us proceed to a formal definition of these combinatorial objects.

Definition 7.13 *A* Feynman diagram *with n external vertices and k internal vertices consists of three finite sets V, E, and F, having the following properties.*

- *(i) The set V is the disjoint union $V_e \sqcup V_i$ where V_e has n elements called external vertices and V_i has k elements called internal vertices.*
- *(ii) The set F has an even number of elements, called ends of edges, and the set E – whose elements are called edges – is a pairing of the set F; i.e. there exists a two-to-one map of F onto E.*
- *(iii) There exist a map $\partial : E \to V \times V/\sim$, where $\sim$ is the equivalence relation $(a, b) \sim (b, a)$, called the incidence map, and a surjective map $\gamma : F \to V$ such that if $e \in E$, if $\partial(e) = \{a, b\}$ and $e = \{\alpha, \beta\} \subseteq F$, then $\{\gamma(\alpha), \gamma(\beta)\} = \{a, b\}$.*
- *(iv) If $v \in V_e$ then $\gamma^{-1}(v)$ has only one element, called an external end; if $v \in V_i$, then $\gamma^{-1}(v)$ has at least two elements.*

An edge having an external end is called, not surprisingly, an *external* edge. We also remark that in the ϕ^4 theory under study in this chapter, all graphs appearing in the perturbative expansion have the property that $\gamma^{-1}(v)$ has exactly four elements for every $v \in V_i$.

The reader can check that the formal definition given above indeed corresponds to the intuitive notion of Feynman diagram given before through examples.

Definition 7.14 *In a Feynman diagram, the* valency *of a vertex v is the number of elements in $\gamma^{-1}(v) \subseteq F$.*

In other words, the valency of a vertex v is the number of ends-of-edges that meet at v. Thus, every external vertex has valency equal to 1. For the Feynman diagrams appearing in the ϕ^4-theory considered in this chapter, every internal

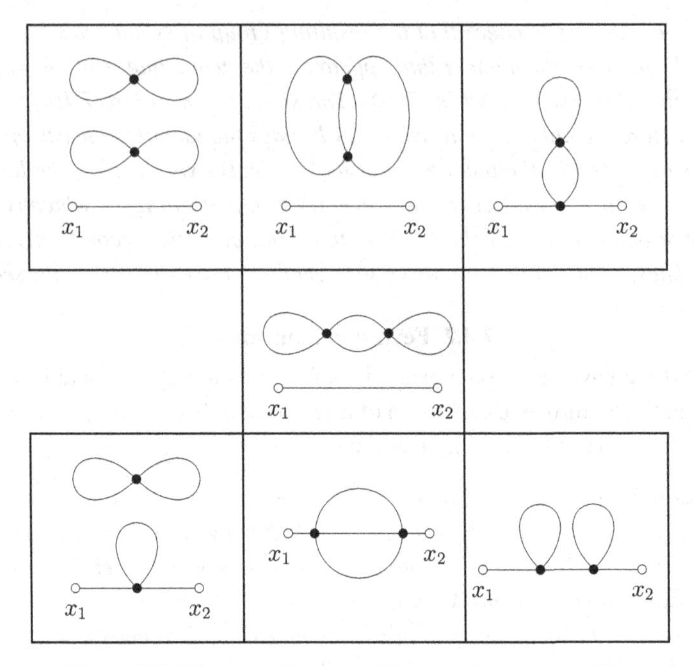

Figure 7.2 The seven Feynman diagrams that make up $\mathcal{F}^e_{2,2}$.

vertex has valency 4. This is the case, of course, of the diagrams in Figures 7.1 and 7.2.

Now we need a notion of equivalence between two diagrams. This is akin to the notion an isomorphism of graphs, the main difference here being that we have to take into account the *ends of edges*, in addition to vertices and edges.

Definition 7.15 *An* equivalence *between two diagrams* $G_j = (V_j, E_j, F_j; \partial_j, \gamma_j)$ $(j = 1, 2)$ *is a triple of bijections* $V_1 \to V_2$, $E_1 \to E_2$, *and* $F_1 \to F_2$ *that are compatible with the incidence and pairing relations given by* ∂_j, γ_j.

We also need the notion of an *automorphism* of a Feynman diagram. Roughly speaking, an automorphism of a diagram is a bit more special than an equivalence of that diagram to itself: we do not want an automorphism to move the external vertices around. Here is the formal definition.

Definition 7.16 *An* automorphism $\theta : (V, E, F; \partial, \gamma) \hookleftarrow$ *of a Feynman diagram consists of the following three maps:*

(i) *a bijection* $\theta_V : V \to V$ *such that* $\theta_V|_{V_e} = \mathrm{id}_{V_e}$;

(ii) *a bijection* $\theta_E : E \to E$ *that is compatible with* θ_V *with respect to the incidence relation;*
(iii) *a bijection* $\theta_F : F \to F$ *that is compatible with both* θ_V *and* θ_E.

It is worth remarking that even when $\theta_V = \mathrm{id}_V$ and $\theta_E = \mathrm{id}_E$, it may still happen that $\theta_F \neq \mathrm{id}_F$. It is clear how to compose two automorphisms of a Feynman graph G, and it is clear also that, under this composition operation, the set of all automorphisms of G is a group, denoted $\mathrm{Aut}(G)$.

Let us denote by $\mathcal{F}_{n,k}$ the set of all Feynman diagrams with n external vertices and k internal vertices up to equivalence. We have the following basic combinatorial result.

Lemma 7.17 *Let* $G \in \mathcal{F}_{n,k}$ *be a Feynman diagram whose internal vertices have valencies* $n_1, n_2, \ldots, n_k$. *Then the automorphism group of* G *has order given by*

$$|\mathrm{Aut}(G)| = \frac{k!}{P_G}\,(n_1!)(n_2!)\cdots(n_k!)\,,$$

where P_G *is the number of possible pairings of adjacent internal vertices of* G.

Proof. An exercise in combinatorics for the reader. □

For the purposes we have in mind (namely, perturbative expansions), it will be convenient to *label* our Feynman diagrams. We will consider two possible types of labeling. Given $G \in \mathcal{F}_{n,k}$, we can label its external vertices as $x_1, x_2, \ldots, x_n$, keeping the internal vertices unlabeled. The class of all such externally labeled diagrams (up to equivalence) will be denoted by $\mathcal{F}^e_{n,k}$. The second type of labeling is that in which, in addition to labeling the external vertices of G as before, we label its internal vertices by $y_1, y_2, \ldots, y_k$. The set of all such (externally and internally) labeled diagrams will be denoted by $\mathcal{G}_{n,k}$. Clearly, every automorphism of an externally labeled diagram preserves the external labels.

We need also the notion of a vacuum diagram.

Definition 7.18 *A vacuum diagram is a Feynman diagram without external vertices.*

Some examples of connected vacuum diagrams are shown in Figure 7.3. We shall assume that the countable collection of all connected vacuum diagrams (up to equivalence) is enumerated $V_1, V_2, \ldots, V_j, \ldots$.

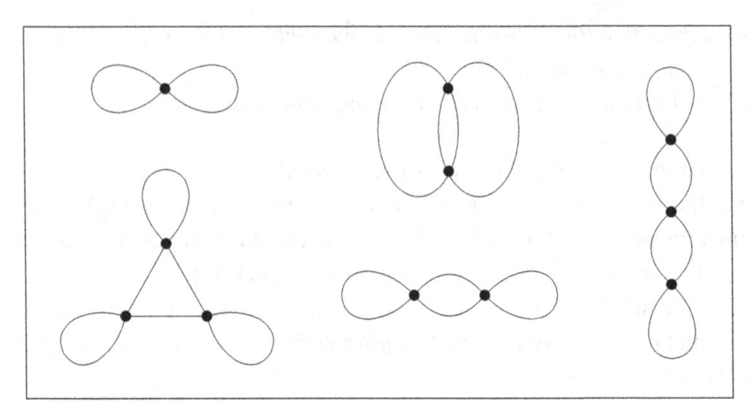

Figure 7.3 Some examples of (connected) vacuum diagrams.

7.4.3 Algebraic structure of Feynman diagrams

Let $\mathcal{F} = \cup \mathcal{F}^e_{n,k}$ be the collection of all (equivalence classes) of externally labeled Feynman diagrams. Then $\mathcal{F}$ has the structure of an abelian semi-group under the operation of disjoint union:

$$G_1 \in \mathcal{F}^e_{n_1,k_1}, G_2 \in \mathcal{F}^e_{n_2,k_2} \Longrightarrow G_1 \circledast G_2 = G_1 \sqcup G_2 \in \mathcal{F}^e_{n_1+n_2,k_1+k_2} .$$

The neutral element is, of course, the empty diagram: $G \circledast \varnothing = \varnothing \circledast G = G$.

Now let $\mathcal{A}(\mathcal{F})$ be the algebra of all complex-valued functions defined on $\mathcal{F}$. An element of $\mathcal{A}(\mathcal{F})$ is a formal, and possibly infinite, linear combination of diagrams, such as

$$\sum_j \alpha_j G_j , \quad \text{where } G_j \in \mathcal{F} \quad \text{and} \quad \alpha_j \in \mathbb{C} \quad \text{for all } j .$$

Note that $\mathcal{A}(\mathcal{F})$ contains all formal power series in any number of variables, e.g.

$$\sum_{(j_1, j_2, \cdots, j_k)} \alpha_{j_1, j_2, \cdots, j_k} G_1^{j_1} \circledast G_2^{j_2} \circledast \cdots \circledast G_k^{j_k} .$$

Here, naturally, G^n means the product $G \circledast G \circledast \cdots \circledast G$ (n times). In particular, we can exponentiate any Feynman diagram, namely

$$e^G = \varnothing + G + \frac{1}{2!} G^2 + \cdots + \frac{1}{n!} G^n + \cdots .$$

This has the familiar properties of the exponential, such as $e^{G_1+G_2} = e^{G_1} \cdot e^{G_2}$.

We can assign a scalar value, or *weight*, to each element of our algebra, in a way that will be most convenient in our study of perturbative expansions. This value will depend also on the correlation operator C and on the finite lattice Λ

introduced earlier. Let $\mathbb{C}(\lambda)$ be the algebra of formal power series in λ. We define the *valuation map* $\nu : \mathcal{A}(\mathcal{F}) \to \mathbb{C}(\lambda)$ as follows. Given $G \in \mathcal{F}$ we let

$$\nu(G) = (-\lambda)^{|V_i(G)|} \int_{\Lambda^k} \prod_{e \in E(G)} C(e)\, dy_1 dy_2 \cdots dy_k \,, \tag{7.11}$$

where $V_i(G)$ is the set of internal vertices of G, $E(G)$ is the set of edges, and $C(e) = C(v, w)$ is the value of the propagator when the vertices of e are v, w and when the $k = |V_i(G)|$ internal vertices of G are labeled $y_1, y_2, \ldots, y_k$. We extend the definition of ν to the whole algebra $\mathcal{A}(\mathcal{F})$ in the obvious fashion. However, to make sure the definition is consistent, we need the following simple lemma.

Lemma 7.19 *The right-hand side of* (7.11) *is independent of the internal labeling of G.*

Proof. If $\sigma : \{y_1, y_2, \ldots, y_k\} \hookleftarrow$ is a permutation, then its induced change of variables in (7.11) has Jacobian determinant equal to 1. □

We have also the following natural result.

Lemma 7.20 *The valuation map* $\nu : \mathcal{A}(\mathcal{F}) \to \mathbb{C}(\lambda)$ *is a homomorphism of algebras.*

Proof. The map ν is $\mathbb{C}$-linear by construction. Hence, it suffices to show that it is multiplicative (in $\mathcal{F}$). If $G = G' \circledast G''$ then obviously $V_i(G) = V_i(G') \sqcup V_i(G'')$ and $E(G) = E(G') \sqcup E(G'')$. Let us write $k' = |V_i(G')|, k'' = |V_i(G'')|$, and let us label the vertices of G so that $y_1, \ldots, y_{k'}$ are the vertices of G' and $y_{k'+1}, \ldots, y_{k'+k''}$ are the vertices of G''. Because there are no edges joining vertices of G' to the vertices of G'', the product in the integrand of the expression defining $\nu(G)$ splits. Therefore, applying Fubini's theorem, we get

$$\begin{aligned}
\nu(G) &= (-\lambda)^{k'+k''} \int_{\Lambda^{k'+k''}} \prod_{e \in E(G)} C(e)\, dy_1 \cdots dy_{k'} dy_{k'+1} \cdots dy_{k'+k''} \\
&= (-\lambda)^{k'+k''} \int_{\Lambda^{k'}} \prod_{e' \in E(G')} C(e')\, dy_1 \cdots dy_{k'} \\
&\quad \times \int_{\Lambda^{k''}} \prod_{e'' \in E(G'')} C(e'')\, dy_{k'+1} \cdots dy_{k'+k''} \\
&= \nu(G') \cdot \nu(G'') \,.
\end{aligned}$$

□

Before we go back to the perturbative expansion of ϕ^4-theory, we need one more result concerning the relationship between the algebraic structure of a Feynman diagram and its group of automorphisms. Recall that we have enumerated all connected vacuum diagrams as $V_1, V_2, \ldots, V_j, \ldots$.

Lemma 7.21 *Every externally labeled Feynman diagram G admits a unique representation of the form*

$$G = V_1^{k_1} \circledast V_2^{k_2} \circledast \cdots \circledast V_m^{k_m} \circledast G^\sharp , \tag{7.12}$$

where each $k_j \geq 0$ and m is the smallest possible, and $G^\sharp$ is the maximal subdiagram of G all of whose connected components contain external vertices of G. In particular, we have

$$|\mathrm{Aut}(G)| = k_1! \cdots k_m! \, |\mathrm{Aut}(V_1)|^{k_1} \cdots |\mathrm{Aut}(V_m)|^{k_m} \, |\mathrm{Aut}(G^\sharp)| . \tag{7.13}$$

Proof. We present a brief sketch of the proof, leaving the details as a combinatorial exercise to the reader. A connected component of G either contains external vertices, or does not. In the latter case, it is necessarily a (connected) vacuum diagram, i.e. it is equal to some V_j. We can factor out from G all vacuum diagrams, putting equivalent diagrams together (writing their disjoint union as powers of the same vacuum diagram). The subdiagram $G^\sharp$ of G that is left after this finite process contains all the external vertices of G. In this way, we obtain the representation in (7.12). Now, every automorphism of G maps connected components onto connected components, and it also fixes the set of external vertices pointwise. Hence, it must map $G^\sharp$ onto itself, and it must also permute the vacuum components of the same type. There are $k_j!$ possible permutations of the k_j components equivalent to V_j. This accounts for the product of factorial terms in the right-hand side of (7.13). The rest of the formula is clear. □

7.4.4 Back to ϕ^4-theory

Now we have all the tools at hand to calculate the perturbative expansion of the discrete Schwinger functions in the case of scalar ϕ^4-theory.

Theorem 7.22 *For all $n \geq 0$ and each $k \geq 0$ we have*

$$Z\lambda^k S_{n,k}^{\Lambda}(x_1, x_2, \ldots, x_n) = \sum_{G \in \mathcal{F}_{n,k}} \frac{\nu(G)}{|\mathrm{Aut}(G)|} .$$

Proof. The proof combines Wick's theorem with the combinatorial lemmas of the previous section, and is left as a (challenging) exercise for the reader. □

Corollary 7.23 *For all $n \geq 0$, the discrete n-point Schwinger correlation function is given by*

$$S_n^{\Lambda}(x_1, x_2, \ldots, x_n) = \sum_{G^\sharp} \frac{\nu(G^\sharp)}{|\operatorname{Aut}(G^\sharp)|} .$$

Proof. Apply Theorem 7.22 and Lemma 7.21. □

7.4.5 The Feynman rules for ϕ^4 theory

We summarize the above discussion by stating the Feynman rules for computing the amplitudes of graphs in the scalar ϕ^4 theory. First we state these rules in coordinate space. It is necessary to undo the Wick rotation we performed in the beginning, going back to Minkowski space. Then, the Euclidean propagator (for $d = 4$) given by

$$C(x, y) = \frac{1}{(2\pi)^4} \int_{\mathbb{R}^4} \frac{e^{ip\cdot(x-y)}}{p^2 + m^2} \, d^4 p ,$$

becomes, after the change $p^0 \mapsto ip^0$, the Feynman propagator

$$D_F(x, y) = \frac{1}{(2\pi)^4} \int_{\mathbb{R}^4} \frac{-i e^{ip\cdot(x-y)}}{p^2 - m^2 + i\varepsilon} \, d^4 p ,$$

where now it is implicit that the dot product represents the Minkowski inner product (in particular, $p^2 = (p^0)^2 - (p^1)^2 - (p^2)^2 - (p^3)^2$). We are now ready to state the Feynman rules in coordinate space:

(i) To each edge with endpoints labeled x and y, we associate a propagator $D_F(x - y)$.
(ii) To each internal vertex labeled by y, we associate the weighted integral $(-i\lambda) \int d^4 y$.
(iii) To each external vertex labeled by x, we simply associate the factor 1.
(iv) We divide the product of all these factors by the diagram's symmetry factor.

These rules can be Fourier transformed into corresponding rules in momentum space. In this situation, each edge of our Feynman diagram is labeled by a momentum variable and is given an arbitrary orientation. The rules become the following:

(i) To each edge labeled by p, we associate a propagator given by

$$\frac{i}{p^2 - m^2 + i\varepsilon} .$$

(ii) To each internal vertex, we associate the weight $(-i\lambda)$ times a delta function that imposes momentum conservation at that vertex.

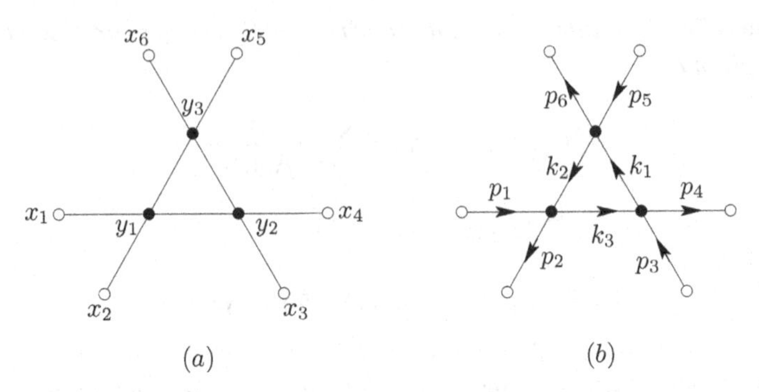

Figure 7.4 Feynman diagram for Example 7.24: here, momentum conservation tells us that $p_1 + p_3 + p_5 = p_2 + p_4 + p_6$.

(iii) To each external vertex labeled x, we associate the factor $e^{-ip\cdot x}$, where p is the momentum (pointing away from x) labeling its external edge.

(iv) We integrate over each momentum variable, normalizing each integral by the factor $1/(2\pi)^4$.

(v) We divide everything by the diagram's symmetry factor.

The computation of the Feynman amplitudes can therefore be performed in two ways (coordinate or momentum space), and what takes us from one to the other is the Fourier transform. Let us illustrate this equivalence by means of two examples. In the first example, the Feynman amplitude will be a convergent integral, whereas in the second the amplitude will diverge (and renormalization will be necessary; see Chapter 8).

Example 7.24 *Let us consider the Feynman diagram G of Figure 7.4. This diagram has six external vertices labeled $x_1, \ldots, x_6$ and three internal vertices labeled y_1, y_2, y_3, with each y_j $(j = 1, 2, 3)$ adjacent to x_{2j} and x_{2j-1}. Note that the symmetry group for this diagram is trivial. Indeed, the external vertices x_i remain fixed; hence each y_j is fixed as well, which fixes the three edges connecting these three vertices. Therefore, the symmetry factor is equal to 1. Applying the Feynman rules in coordinate space, we see that the amplitude for this graph is*

$$A_G(x_1, \ldots, x_6) = (-i\lambda)^3 \iiint \prod_{j=1}^{3} D_F(y_j - x_{2j-1}) D_F(x_{2j} - y_j)$$

$$\times D_F(y_2 - y_1) D_F(y_3 - y_2) D_F(y_1 - y_3)\, d^4y_1 d^4y_2 d^4y_3 \,.$$

On the other hand, in the dual diagram in momentum space, there are six external momenta $p_1, p_2, \ldots, p_6$, and three internal momenta k_1, k_2, k_3 (integration *variables). Hence, if we apply the Feynman rules in momentum space, we get*

$$A_G(x_1,\ldots,x_6) = \frac{(-i\lambda)^3}{(2\pi)^{36}} \int \cdots \int e^{-i(p_1x_1 - p_2x_2 + p_3x_3 - p_4x_4 + p_5x_5 - p_6x_6)}$$

$$\times\, \delta(p_1 - p_2 + k_2 - k_3)\delta(p_3 - p_4 + k_3 - k_1)\delta(p_5 - p_6 + k_1 - k_2)$$

$$\times \prod_{j=1}^{6} \frac{i}{p_j^2 - m^2 + i\varepsilon} \prod_{\ell=1}^{3} \frac{i}{k_\ell^2 - m^2 + i\varepsilon}\, d^4k_1 d^4k_2 d^4k_3\, d^4p_1 \cdots d^4p_6\,.$$

This second value for the amplitude is supposed to be equal to the first. Showing that they are equal seems a rather daunting task. But it could be done, if only we could make sense of the product of delta functions appearing in the integrand on the right-hand side of this last formula. This will be done below, after this example and the next. There is something more we can say here, without having to do the integrations explicitly. Using the conservation of momenta at each vertex, we deduce easily that the external momenta satisfy the relation $p_6 = p_1 - p_2 + p_3 - p_4 + p_5$. *This means that the Fourier transform of the Feynman amplitude in momentum space is a function of only five of the six external momenta. When this information is used in the last formula above, we get that* A_G *is a function of the five differences* $x_1 - x_6, \ldots, x_5 - x_6$, *a fact that may not be apparent if we look at the first expression for the amplitude. But it is consistent with the fact that the Feynman amplitudes (or the correlation functions) are translation-invariant (this comes from the Lorentz invariance of the field). We emphasize that the amplitude in this example is given by a convergent integral.*

Example 7.25 *Now, let us consider the Feynman diagram* G *of Figure 7.5, with two external vertices labeled* x_1 *and* x_2 *and two internal vertices labeled* y_1 *and* y_2. *The symmetry group for this diagram has order* $3! = 6$, *corresponding to the permutations of the three internal edges, which are the only possible symmetries. Following the Feynman rules in coordinate space, the amplitude of this diagram is given by*

$$A_G(x_1,x_2) = \frac{(-i\lambda)^2}{6} \iint D_F(y_1 - x_1)D_F(y_2 - y_1)^3 D_F(x_2 - y_2)\, dy_1 dy_2\,.$$

On the other hand, if we apply the Feynman rules in momentum space, we arrive at

$$A_G(x_1,x_2) = \frac{i^5(-i\lambda)^2}{6(2\pi)^{20}} \int \cdots \int \frac{e^{-i(p_1x_1 - p_2x_2)}}{\prod_{j=1}^{2}(p_j^2 - m^2 + i\varepsilon)\prod_{\ell=1}^{3}(k_\ell^2 - m^2 + i\varepsilon)}$$

$$\times\, \delta(p_1 - k_1 - k_2 - k_3)\delta(k_1 + k_2 + k_3 - p_2) d^4k_1 d^4k_2 d^4k_3 d^4p_1 d^4p_2.$$

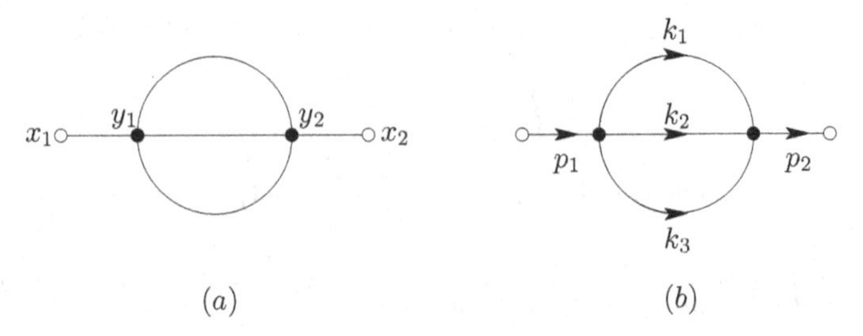

Figure 7.5 Feynman diagram for Example 7.25: here, momentum conservation implies that $p_1 = p_2$.

If the rules are consistent, these two values for the amplitude must be equal. They are, and once again this involves understanding the meaning of the product of delta distributions in the integrand of this last formula. See the discussion below. As in the previous example, we note that momentum conservation tells us that $p_1 = p_2$, and this in turn implies that the amplitude $A_G(x_1, x_2)$ is in fact a function of the difference $x_1 - x_2$. Unlike the previous example, however, this amplitude is divergent (and in order to extract from it a physically meaningful number, renormalization must be applied; see Chapter 8).

How are we to make sense of the products of delta functions that appear in these examples? What do physicists mean by them? Although in general we cannot multiply distributions, here the situation is much simpler because the expressions of momentum conservation at the vertices, taken in as arguments of the delta functions, are *linear expressions* in the momenta. Thus, suppose G is a Feynman graph with m external momenta $p_1, \ldots, p_m$ and n internal momenta $k_1, \ldots, k_n$, collectively denoted by p and k respectively. If s is the number of internal vertices of G, the momentum conservation at each vertex is a linear form on p and k, say $\ell_j(p, k)$, $1 \le j \le s$. Hence, for each given p, we have a surjective linear map $L_p : (\mathbb{R}^4)^n \to (\mathbb{R}^4)^s$, whose s image components are precisely these linear forms. For instance, in Example 7.25 we have

$$L_p(k_1, k_2, k_3) = (p_1 - k_1 - k_2 - k_3\,,\; k_1 + k_2 + k_3 - p_2)\,.$$

We *define* the product of delta distributions as follows:

$$\prod_{j=1}^{s} \delta(\ell_j(p, k)) = \delta(L_p(k))\,.$$

Thus, when computing the Feynman amplitude A_G in momentum space, if we first integrate over the internal momentum variables k (ignoring the exponential factors of the form $e^{ip\cdot x}$), we get the Fourier transform $\hat{A}_G$, which is given by an expression of the form

$$\hat{A}_G(p) = \int f(p,k)\,\delta(L_p(k))\,dk\ ,$$

where $f(p,k)$ is a (rational) function of the momenta. Now, for each p, we can certainly find an invertible linear map $T_p : (\mathbb{R}^4)^n \to (\mathbb{R}^4)^n$ such that $\pi \circ T = L_p$, where $\pi : (\mathbb{R}^4)^n \to (\mathbb{R}^4)^s$ is the linear projection onto the first s components. Using T_p as a change of variables, i.e. letting $q = T_p(k)$, we can rewrite the Fourier transform of the amplitude as

$$\begin{aligned}\hat{A}_G(p) &= \int f(p, T_p^{-1}(q))\,\delta(\pi(q))\det T_p^{-1}\,dk\\ &= \int f(p, T_p^{-1}(q_1,\ldots,q_{n-s},0,0,\ldots,0))\det T_p^{-1}\,dq_1\cdots dq_{n-s}\ .\end{aligned}$$

The role of the product of delta functions is now clear: it amounts to a reduction in the number of integrations over internal momenta. The reader can verify that there is a relationship between the number of integrations performed and the number of *loops* in the Feynman diagram.

7.4.6 Perturbative theory for fermions

The perturbative theory for bosons developed so far can be adapted for the Dirac field, with some significant changes that we briefly indicate. Recall from Chapter 6 that the Dirac field can be thought of as a Grassmann field, whose values in spacetime are given by anti-commuting variables. Just as in the bosonic case, we can write an action functional. This functional, with a source term, can be written as follows:

$$Z[\eta,\bar{\eta}] = \int \exp\left\{ i\int \left(\bar{\psi}(i\partial\!\!\!/ - m)\psi + \bar{\eta}\psi + \bar{\psi}\eta\right)\,d^4x\right\}\,\mathcal{D}\bar{\psi}\mathcal{D}\psi\ ,$$

where the source field $\eta(x)$ takes values in a Grassmann algebra. Using the standard trick of completing the square, we can rewrite this formula as

$$Z[\eta,\bar{\eta}] = Z_0\,\exp\left\{-\int \bar{\eta}(x)S_F(x-y)\eta(y)\,d^4x d^4y\right\}\ ,$$

where Z_0 is the value of the generating functional when the source field is equal to zero, and where S_F is the propagator

$$S_F(x-y) = \frac{1}{(2\pi)^4}\int \frac{ie^{-ip\cdot(x-y)}}{p\!\!\!/ - m + i\varepsilon}\,d^4p\ .$$

With this propagator at hand, we can try to imitate the procedure used for scalar fields in order to calculate the correlations of this fermionic theory. The idea is the same: we differentiate the generating functional with respect to the source components, which are Grassmann variables. We learned about Grassmann differentiation in Chapter 6. In any case, we find that everything works provided we have a version of Wick's theorem in the present context. This amounts to knowing how to evaluate Gaussian integrals involving Grassmann variables. We know how to do that already: this is precisely the content of Theorem 6.22 of Chapter 6.

Hence, everything we have done so far in this chapter can be adapted to fermions. We have perturbative expansions of correlation functions in terms of Feynman diagrams and, just as before, only the connected diagrams will matter. We also have Feynman rules for computing the amplitudes associated with such diagrams. We shall not elaborate on these rules beyond merely stating them. For details, the reader is invited to look in Chapter 4 of [PS]. We will in fact be a bit more general here and consider the rules for a theory involving a Dirac fermionic field ψ coupled with a bosonic field ϕ through the so-called *Yukawa potential* $V = g\bar{\psi}\psi\phi$ (g being the coupling constant). In this theory, we will have two types of edges: the scalar particles (bosons) are represented by dashed lines, and the fermions are represented by solid lines. The Feynman rules in momentum space are stated in Figure 7.6.

7.4.7 Feynman rules for QED

We can turn to the case of quantum electrodynamics (QED), where we have a fermionic field (whose particles and antiparticles are electrons and positrons, respectively) interacting with a background electromagnetic field. The full Lagrangian density for QED is

$$\mathcal{L}_{QED} = \bar{\psi}(i\partial\!\!\!/ - m)\psi - \frac{1}{4}F_{\mu\nu}F^{\mu\nu} - e\bar{\psi}\gamma^{\mu}\psi A_{\mu}\,.$$

Here, as we saw before, A_{μ} represents the connection (vector potential), and $F_{\mu\nu}$ is its curvature (the field strength tensor). We interpret the constant e as a coupling constant; one can therefore write down a perturbative expansion as a power series in powers of e. As before, we have sums of amplitudes over connected Feynman diagrams, and corresponding Feynman rules for computing such amplitudes. The Feynman rules for QED are given in Figure 7.7.

7.4.8 Power counting

Let us consider a perturbative field theory whose Lagrangian involves both bosons and fermions. We assume that we have already performed a perturbative expansion, and we examine a given Feynman amplitude of a (connected) Feynman diagram G in this expansion. Let us evaluate the contribution to the degree of superficial divergence of G, denoted ω_G, of a term in the Lagrangian

1. Propagators:

2. Vertices:

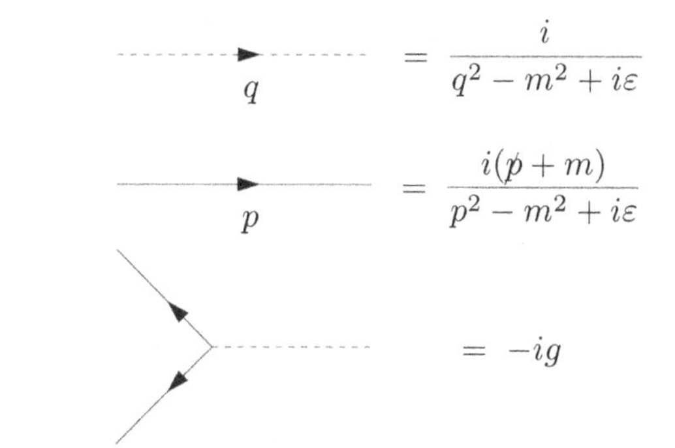

q
$= \frac{i}{q^2 - m^2 + i\varepsilon}$
p
$= \frac{i(\not{p} + m)}{p^2 - m^2 + i\varepsilon}$
$= -ig$

3. External legs:

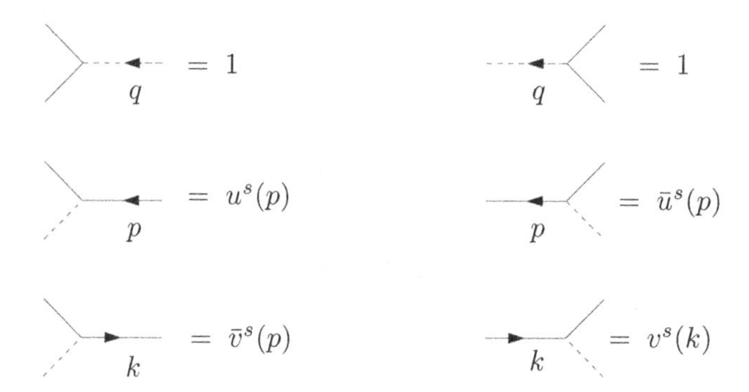

Figure 7.6 Feynman rules for a Yukawa theory.

that involves r derivatives acting on the product of b bosons and f fermions. The propagators for bosons contribute $\frac{1}{2}(-2b)$, and the propagators for fermions contribute $\frac{1}{2}(-f)$; here, it is necessary to divide by 2 because each propagator corresponds to an edge of G having two vertices. The integrations over internal momenta contribute $\frac{1}{2}(b + f) \cdot d$ (where d is the dimension). The delta function imposing momentum conservation at the vertex reduces the number of integrations by 1, and therefore contributes $-d$ to ω_G. Therefore, because each derivative contributes with 1, the net contribution of this vertex to ω_G is

$$\frac{1}{2}(-2b - f + bd + fd) - d + r .$$

This can be written as $\omega_G = \delta - d$, where

$$\delta = \frac{b(d-2)}{2} + \frac{f(d-1)}{2} + r .$$

Photon propagator: $\mu \sim\sim\sim \nu$ (q) $= \dfrac{-ig_{\mu\nu}}{q^2 + i\varepsilon}$

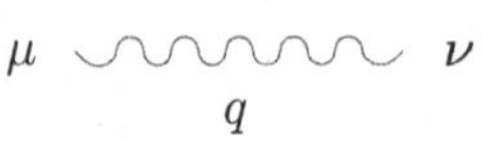

New vertex: 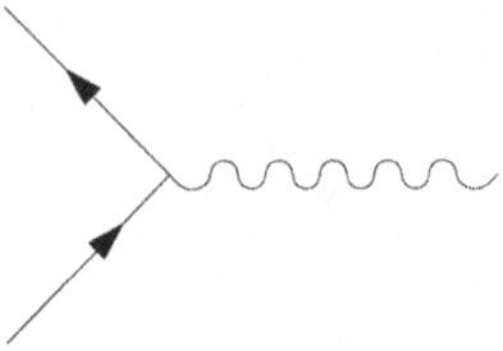 $= -ie\gamma^{\mu}$

External photon lines:

$A_\mu|\boldsymbol{p}\rangle =$ (μ, p) $= \epsilon_\mu(p)$ $\langle \boldsymbol{p}|A_\mu =$ (μ, p) $= \epsilon^*_\mu(p)$

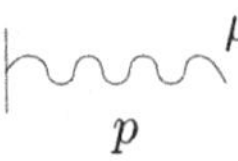

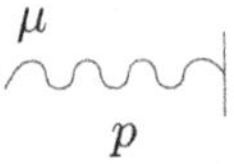

Figure 7.7 Additional Feynman rules for QED.

7.5 Perturbative Yang–Mills theory

In Yang–Mills theory, the gauge invariance implies that the quadratic form appearing in the Lagrangian is non-invertible.

7.5.1 Ghosts and gauge-fixing

This problem was solved by Faddeev and Popov in 1967 with their theory of ghosts and gauge fixing (Landau gauge) and by Fradkyn and Tyutin in 1969 (Feynman gauge). In 1971, t'Hoof gave a formal proof of the validity of these theories.

The idea behind the Faddeev–Popov construction is to proceed through the following steps:

(i) *Gauge fixing*: Once the gauge invariance is understood one performs a *gauge fixing*. This means that one finds a submanifold of the space of connections, $\{A; F(A) = 0\}$, that is intersected by each orbit of the gauge group at exactly one point. This yields the notion of a determinant (the Faddeev–Popov determinant) of an operator M such that the partition function of the theory can be written as

$$Z = \int e^{iS(A)} \det(M)\, \delta(F(A)) \mathcal{D}A \, .$$

(ii) *Ghosts*: One writes the determinant of M as the exponential of a trace by introducing Grassmann fields $\eta, \overline{\eta}$ such that

$$\det(M) = \exp\left\{\int (\cdots)\, d\eta d\overline{\eta}\right\} .$$

(iii) One incorporates the determinant as well as the δ-function into the exponential, writing

$$Z = \int e^{iS(A,\eta,\overline{\eta})}\, dA d\eta d\overline{\eta} .$$

(iv) *Perturbative expansion*: Now one writes

$$S(A, \eta, \overline{\eta}) = S_Q + g \cdot S_{NQ} ,$$

where S_Q is the quadratic part of the action, S_{NQ} is the non-quadratic part, and g is a coupling constant, and then one performs a perturbative series expansion as in the case of invertible Lagrangians.

The main difficulty in extending perturbation theory to the case of gauge theories is that the quadratic part of the Lagrangian is no longer invertible, due to the gauge invariance. One possible strategy for overcoming this problem is to perform the path integral in the quotient space by the gauge group action. To construct a formal measure on the quotient space, we start by gauge fixing: we represent the quotient space by a submanifold Σ of the space of connections $\mathcal{C}$ that is transversal to the orbits of the gauge group $\mathcal{G}$, intersecting each orbit at a unique point. Such a submanifold is usually given by the zero level set of the function F on the space of connections. The idea is to construct an induced "measure" on Σ, restrict the gauge-invariant functional we want to integrate to Σ, and integrate it with respect to the induced "measure." Of course the result must be independent of the choice of Σ, which requires that the induced "density of the measure" should contain a Jacobian determinant related to the derivative of F, the so-called Faddeev–Popov determinant. To make this picture clearer, let us examine an analogous finite-dimensional problem.

7.5.2 A finite-dimensional analogue

Let G be a compact Lie group of dimension l acting on the right on $\mathbb{R}^n$ by isometries and such that the projection of $\mathbb{R}^n$ onto the orbit space $\mathbb{R}^n/G$ is a trivial principal bundle. Let $S\colon \mathbb{R}^n \to \mathbb{R}$ be a G-invariant function. We want to express the integral

$$Z = \int_{\mathbb{R}^n} \exp[iS(x)]dx$$

in terms of an integral in the orbit space. To do that, we will represent the orbit space by a submanifold $\Sigma \subset \mathbb{R}^n$ that intersects each orbit at a unique point. We may assume that $\Sigma = \{x \in \mathbb{R}^n; F(x) = 0\}$, where F is a smooth fibration onto $\mathbb{R}^l$. Let $d\lambda_\Sigma = \delta(F(x))dx$ be the restriction of the Lebesgue measure on Σ and take a volume form $d\gamma$ on G, invariant by right translation. On $\Sigma \times G$ we will consider the product volume form $d\lambda_\Sigma \times d\gamma$. By invariance, the Jacobian of the diffeomorphism $\Phi\colon \Sigma \times G \to \mathbb{R}^n$ given by $\Phi(y, g) = R_g(y)$ is a function $\det(\Delta)$ that depends only on the first variable. Because $S(\Phi(y, g)) = S(y)$, we have

$$Z = \int_{\Sigma \times G} \exp[iS(y)] \det(\Delta(y)) d\lambda_\Sigma \times d\gamma$$
$$= \int_G d\gamma \int_\Sigma \exp[iS(y)] \det(\Delta(y))\, d\lambda_\Sigma \, .$$

We may also write this equation as

$$Z = \int_{\mathbb{R}^n} \exp[iS(x)] \det(\Delta_\Sigma(x)) \delta(F(x))\, dx \, ,$$

where $\det(\Delta_\Sigma(x)) = \det(\Delta(y))$ if x is in the orbit of $y \in \Sigma$. The next step in the Faddeev–Popov procedure is to write the delta function and the determinant as an exponential. For the determinant, the trick is to use anti-commuting variables and write

$$\det(\Delta_\Sigma(x)) = \int \exp[i(\overline{c}\, \Delta_\Sigma\, c)]\, d\overline{c}\, dc \, .$$

If the gauge-fixing map F can be translated to give another gauge fixing $\Sigma_z = \{x \in \mathbb{R}^n; F(x) - z = 0\}$ in such a way that $\det(\Delta_{\Sigma_z}(x)) = \det(\Delta_\Sigma(x))$, we again have

$$Z = \int_{\mathbb{R}^n} \exp[iS(x)] \det(\Delta_\Sigma(x)) \delta(F(x) - z)\, dx$$

for all z, and we can integrate over z with a Gaussian density to get

$$Z = N(\xi) \int_{\mathbb{R}^l} dz \exp\left[\frac{-\|z\|^2}{2\xi}\right] \int_{\mathbb{R}^n} \exp[iS(x)] \det(\Delta_\Sigma(x)) \delta(F(x) - z)\, dx \, .$$

Interchanging the integrals and using that

$$\int_{\mathbb{R}^l} \exp\left[\frac{-i\|z\|^2}{2\xi}\right] \delta(F(x) - z)\, dz = \exp\left[\frac{-i\|F(x)\|^2}{2\xi}\right] ,$$

we get the following expression for the integral:

$$Z = N(\xi) \int dx \int d\overline{\eta} \int d\eta \exp[i(S(x) - S_{FP}(x, \overline{\eta}, \eta) - S_{GF}(x))] \, ,$$

where $S_{FP}(x, \overline{\eta}, \eta) = \overline{\eta}\Delta_\Sigma \eta$ is the Faddeev–Popov action and

$$S_{GF}(x) = \frac{-i||F(x)||^2}{2\xi}$$

is the gauge-fixing action. The new action is a function of more variables. The gauge invariance of the initial action implies that its quadratic part cannot be inverted, and this problem is eliminated by the introduction of the new variables.

7.5.3 The Faddeev–Popov and gauge-fixing actions

We now consider the pure Yang–Mills integral

$$Z = \int \mathcal{D}A \exp\left[i \int \left(-\frac{1}{4}(F^a_{\mu\nu})^2\right)\right] d^4x\ .$$

As gauge fixing, we can take a Lorentz gauge: $F(A) = \partial^\mu A^a_\mu(x)$ or some translation of it $F(A) - \omega^a(x)$. As in the finite-dimensional example, we can write this integral as

$$Z = \int \mathcal{D}\alpha \int \mathcal{D}A e^{iS(A)} \delta F(A)) \det\left(\frac{\delta F(A^\alpha)}{\delta \alpha}\right) ,$$

where $(A^\alpha)^a_\mu = A^a_\mu + \frac{1}{g}\partial_\mu \alpha^a + f^{abc} A^b_\mu$. For an arbitrary function ω^a, the Faddeev–Popov operator associated to the gauge fixing $F(A) = \partial^\mu A^a_\mu(x) - \omega(x)$ is

$$\frac{\delta F(A^\alpha)}{\delta \alpha} = \frac{1}{g}\partial^\mu D_\mu\ ,$$

where $D_\mu(\alpha^a) = \partial_\mu \alpha^a + g f^{abc} A^b_\mu \alpha_c$ is the covariant derivative. To lift the determinant of this operator to an exponential, we use anti-commuting fields, called Faddeev–Popov ghosts:

$$\det\left(\frac{1}{g}\partial^\mu D_\mu\right) = \int \mathcal{D}c\mathcal{D}\overline{c} \exp\left[i \int d^4x\, \overline{c}\,(-\partial^\mu D_\mu)\, c\right] .$$

As in the finite-dimensional example, we can integrate over ω^a with a Gaussian weight to lift the delta function also to a exponential, thereby getting the gauge-fixing action

$$S_{GF}(A) = \int d^4x\, \frac{1}{2\xi}(\partial^\mu A^a_\mu)^2$$

and the ghost action

$$S_{Gh} = i \int d^4x\, \overline{c}\,(-\partial^\mu D_\mu)\, c\ .$$

We remark that in the calculation of the correlation functions, the infinite factor $\int \mathcal{D}A$ appears both in the numerator and in the denominator, and so it cancels out. In the case of QED, the Faddeev–Popov determinant does not depend on the connection A and so it can be moved outside of the integral, and therefore it also cancels when correlation functions are computed. So in this case the ghosts do not appear. We can also incorporate fermions into the theory as before. We can see that the quadratic part of the total Lagrangian can be used to define the propagators, which are the following:

(i) Ghost propagator:

$$-\frac{i\delta_{ab}}{k^2 + i\epsilon}$$

(ii) Vector boson propagator:

$$\left(-g_{muv} + (1-\xi)\frac{k_\mu k_\nu}{k^2}\right)\frac{i\delta_{ab}}{k^2 + i\epsilon} .$$

To complete the Feynman rules we need also to describe the vertices:

(i) Boson–ghost vertex:

$$g f^{abc} k^a_\mu$$

(ii) Three-bosons vertex:

$$-igf^{abc}(k_1 - k_2)_\lambda g_{\mu\nu} + (k_2 - k_3)_\mu g_{\nu\lambda} + (k_3 - k_1)_\nu g_{\lambda\mu}$$

(iii) Four-bosons vertex:

$$-ig^2\left[f^{abc} f^{ade}(g_{\mu\rho} g_{\nu\sigma} - g_{\mu\sigma} g_{\nu\rho}) + \text{ permutations}\right]$$

(iv) Integrals over propagators and delta functions at the vertices, as before.

7.5.4 The BRST cohomological methods

As we have seen, the Faddeev–Popov method consists of breaking the gauge invariance by a gauge fixing and extending the phase space by introducing new fields, the ghost fields and anti-ghost fields. This approach was substantially improved by Becchi, Rout, Stora, and Tyutin who introduced some cohomological methods inspired by the work of Koszul in the 1950s into the theory. They discovered a global symmetry of the Faddeev–Popov action, the BRST differential, which is a nilpotent derivation acting on the extended space containing the fields, anti-fields, ghosts, and anti-ghosts. The zero-dimensional cohomology of the associated complex is the space of gauge-invariant physical fields.

To introduce these new ideas, we will analyze a very simple finite-dimensional situation where the space of fields is a finite-dimensional vector space V and the group acts as translations by vectors of a subspace $W \subset V$. The orbit space is therefore the quotient space V/W. We consider the algebra $A = Pol^+(V) \otimes Pol^-(W)$, where $Pol^+(V)$ is the symmetric tensor algebra of V and $Pol^-(W)$ is the anti-symmetric tensor algebra of W. The elements of a basis $\{v^1, \ldots, v^n\}$ of V generate $Pol^+(V)$ with the relations $v_i v_j = v_i v_j$, whereas a basis $\{w^1, \ldots, w^m\}$, $w^\alpha = C^\alpha_j v^j$, generates $Pol^-(W)$, with $w^\alpha w^\beta = -w^\beta w^\alpha$. This is a graded algebra, $A = A_0 \oplus A_1 \cdots \oplus A_m$, where an element of A_k is of the form $P(v^1, \cdots, v^n) w^{\alpha_1} \ldots w^{\alpha_k}$, where P is a polynomial. We can also write A as a super-algebra $A = A_0 \oplus A_1$, where the elements of A_0 have even grade and the elements of A_1 have odd grade. The parity of an element $a \in A_0 \cup A_1$ is defined as $|a| = 0$ if $a \in A_0$ and $|a| = 1$ if $a \in A_1$. Note that $A_k \cdot A_l \subset A_{k+l}$, $A_0 \cdot A_0 \subset A_0$, $A_0 \cdot A_1 \subset A_1$, $A_1 \cdot A_1 \subset A_0$, the elements of A_0 commute with everybody, and two elements of A_1 anticommute. An odd derivation of a graded algebra A is a linear map $s\colon A \to A$ that maps A_0 into A_1 and A_1 into A_0 and satisfies the graded Leibnitz rule

$$s(a \cdot b) = s(a) \cdot b + (-1)^{|a|} a \cdot s(b)$$

for homogeneous elements. A differential is a nilpotent odd derivation of order two: $s \circ s = 0$. In our example we can define a differential by

$$s(v^i) = C^i_\alpha w^\alpha$$

$$s(w^\alpha) = 0\,,$$

extending it to the whole algebra by the Leibnitz rule. The grade of s is one because it maps A_k into A_{k+1}, so it defines a cohomology $H^k(A) = \ker(s|A_k)/\mathrm{Im}(s|A_{k-1})$. It is easier to compute this cohomology if we make a better choice of generators, adapting the basis of V to W: $v^j = w^j$; $j = 1 \ldots m$. Then, we can define a homotopy h as a graded derivation by

$$h(w^\alpha) = v^\alpha$$

$$h(v^j) = 0\,, \quad j = 1, \ldots, m.$$

If $N\colon A \to N$ is the linear operator such that $N(a) = ka$ for $a \in A_k$, then we have

$$hs + sh = N, \quad [h, N] = [s, N] = 0\,.$$

Thus, for $a \in A_k$, $k > 0$ with $s(a) = 0$, we have

$$a = \frac{1}{k}(hs + sh)(a) = \frac{1}{k}s(hp)\,,$$

which implies that $H^k(A) = 0$ for $k > 0$. Furthermore, $H^0(A)$ is clearly isomorphic to the algebra $(Pol^+(V/W))$ of invariant polynomials. Thus, by introducing the ghost variables, we have described the gauge-invariant polynomials cohomologically.

We can also describe cohomologically the gauge-fixing procedure. Let us consider, in the finite-dimensional space V, a submanifold Σ transversal to the gauge orbits and given by a set of polynomial equations $G_a = 0$, which we assume are independent (they define a submersion on another vector space whose dimension is the codimension of Σ in V). We introduce the anti-ghost variables ω_a and define

$$s(\omega_a) = G_a \, ,$$

$$s(v_j) = 0 \, .$$

Now the algebra is negatively graded. Again, the cohomology H^{-k} vanishes for $k > 0$ and H^0 is isomorphic to $Pol(V)/(I_\Sigma)$, where (I_Σ) is the ideal of polynomials that vanishes on Σ.

Let us use these ideas to construct the infinitesimal BRST symmetry of the Faddeev–Popov action for pure Yang–Mills.

We may write the Yang–Mills Lagrangian as

$$-\frac{1}{4} G^a_{\mu\nu} G^{a\mu\nu} \, ,$$

where $G^a_{\mu\nu} = \partial_\mu A^a_\mu - \partial_\nu A^a_\mu + g f^a_{bc} A^b_\mu A^c_\nu$ is the field strength and f^a_{bc} are the structure constants of the Lie algebra ($A_\mu = A^a_\mu T_a$, where T_a, T_b are the generators of the Lie algebra with commutators $[T_a, T_b] = f^c_{a,b} T_c$). We have incorporated a coupling constant g into the self-interaction of the gauge field for the perturbation expansion.

The Faddeev–Popov action is

$$L_{FP}(A, \eta, \overline{\eta}) = -\frac{1}{4} G^a_{\mu\nu} G^{a\mu\nu} + \overline{\eta}^a \partial^\mu D^{ac}_\mu \eta^c - \frac{1}{2\xi} (\partial_\mu A^{a\mu})^2 \, ,$$

where $D^{ac}_\mu = \delta_{ac} \partial_\mu + g f^{abc} A^b_\mu$ is the covariant derivative. Let us construct an infinitesimal symmetry of the Faddeev–Popov action as an odd derivation s. So it is enough to define s in the generators, $A_\mu, \eta, \overline{\eta}$. We start by defining

$$s A_\mu = -D_\mu \eta \, .$$

Because the Lagrangian of pure Yang–Mills is invariant by an infinitesimal action of any element X of the Lie algebra, and the infinitesimal action of X has the same form as s, it follows that $sL_{YM} = 0$. Now, let us consider the gauge-fixing part of the Lagrangian:

$$s\left(-\frac{1}{2\alpha}(\partial_\mu A^\mu)^2\right) = \frac{1}{\alpha}(\partial_\mu A^\mu)\partial_\nu D^\nu \eta\,.$$

On the other hand, for the ghost part of the Lagrangian we would have

$$s(\overline{\eta}\partial^\mu D_\mu \eta) = s(\overline{\eta})\cdot(\partial^\mu D_\mu \eta)) - \overline{\eta}\cdot s(D_\mu \eta)$$

and

$$\begin{aligned} s(D_\mu \eta) &= D_\mu(s\eta) + (sD_\mu)\eta = D_\mu(s\eta) + g[sA_\mu, \eta] \\ &= D_\mu(s\eta) - g[D_\mu \eta, \eta] = D_\mu(s\eta) - D_\mu\left(\frac{1}{2}g[\eta,\eta]\right)\,. \end{aligned}$$

Now, it is clear that the whole Lagrangian will be invariant if we define

$$s(\eta) = \frac{1}{2}g[\eta,\eta]$$

and

$$s(\overline{\eta}) = \frac{1}{\alpha}(\partial_\mu A^\mu)\,.$$

It is easy to see that the derivation s is nilpotent of order two. From the definition of s it is clear how to define the grades, which are called ghost numbers: A_μ has ghost number 0, η has ghost number 1, and $\overline{\eta}$ has ghost number -1.

The Lagrangian BRST formalism uses the following steps:

(i) Start with a gauge-invariant Lagrangian $L_0 = L_0(\phi)$

(ii) For each gauge degree of freedom introduce a ghost field η^a whose ghost number is defined to be $+1$.

(iii) Construct a BRST transformation s in the extended space of fields such that the initial action is invariant and $s^2 = 0$.

(iv) Enlarge the space again by introducing an anti-ghost field β of ghost number -1 and an auxiliary field α of ghost number 0. Extend s by $s(\beta) = \alpha$ and $s(\alpha) = 0$.

(v) Define the effective action by adding a BRST invariant term to the original action of the form $s(\psi)$ for some fermion in the algebra of fields with ghost number -1.

(vi) The zero cohomology should give the space of physical observables (gauge-invariant fields).

In [BV], Batalin and Vilkovisky extended the above cohomological methods to cover constrained quantum systems much more general than Yang–Mills. Their method is closer to a Hamiltonian formulation.

They extended the phase space as follows. For each commuting (anti-commuting) gauge parameter they introduced a ghost fermionic (bosonic) field. For more complicated constraints they also introduced ghost variables. Then they doubled the space by introducing an anti-field corresponding to each field (including the anti-ghosts). The fields and anti-fields form a graded algebra. Representing all fields by the same symbol Φ^A and all anti-fields by Φ^*_A, they defined the anti-bracket between two functionals $F(\Phi^A, \Phi^*_A)$ by

$$(F, G) = \frac{\delta^R F}{\delta \Phi^A} \frac{\delta^L G}{\delta \Phi^*_A} - \frac{\delta^R F}{\delta \Phi^*_A} \frac{\delta^L G}{\delta \Phi^A} ,$$

where δ^R and δ^L denote left and right functional derivatives, which are equal to each other modulo the signs of the fields involved. This anti-bracket has ghost number $+1$ and is odd. The BRST transformation is represented in terms of this antibracket by a functional S:

$$sF = (S, F) .$$

The nilpotency of the BRST transformation is then equivalent to the so-called classical *master equation*

$$(S, S) = 0 .$$

This equation is solved inductively as $S = S_0 + S_1 + \cdots$, starting with S_0, the original classical action. The solution of the classical master equation gives the action that is used in the path integral quantization. By construction this action is BRST invariant, but the formal measure of the path integral may not be. The invariance of the formal measure under the BRST transformation is given by the equation

$$\Delta S = 0 ,$$

where Δ is the second-order differential operator

$$\Delta = \pm \frac{\delta^L}{\delta \Phi^A} \frac{\delta^L}{\delta \Phi^*_A} ,$$

where the sign depends on the parity of the field Φ^A. The second-order differential operator Δ is nilpotent of order two, is odd, and has ghost number -1. It is not a derivation of the product and the failure to be a derivation is given by the anti-bracket:

$$\Delta(AB) - \Delta(A) \cdot B - (-1)^{|A|} A \cdot \Delta(B) = (-1)^{|A|} (A, B) .$$

If the formal measure is also BRST invariant, then the path integral is reduced to the zero set of the BRST transformation (fixed point of symmetry), which in several cases is finite-dimensional. If this is not the case, we have to construct a quantum action by adding quantum corrections to the classical action,

$$W = S + \hbar W_1 + \hbar^2 W_2 + \cdots ,$$

in order to solve the *quantum master equation*

$$\frac{1}{2}(W, W) = i\hbar \Delta W .$$

Notice that the quantum master equation reduces to the classical master equation when $\hbar \mapsto 0$. Although the classical master equation has a solution for general gauge theories, there may exist an obstruction for the existence of the quantum master equation, which we call an *anomaly* that prevents quantization. In theories, such as Yang–Mills, in which there exists a solution of the quantum master equation, there exists a quantum BRST symmetry σ such that

$$\sigma A = (W, A) ,$$

which is again nilpotent of order two. For more details on classical and quantum BRST the reader should consult [HT]; see also [FHM] and [Sth].

8
Renormalization

We saw in Chapter 3 that, when attempting to construct a relativistic quantum theory of particles, Dirac and others realized that the number of particles in the system could not be taken to be constant, because new particles are constantly created and also destroyed. The solution to this problem was to develop instead a relativistic quantum theory of fields, because a field corresponds already to infinitely many particles. However, for many years there were serious doubts in the physical community concerning the foundations of this theory because, as we have seen, many of the calculations done in the perturbative theory give rise to infinities. The situation changed completely in the late 1940s when Feynman, Schwinger, Tomonaga, Dyson, and others developed a procedure to remove these infinities from the calculations, and the finite results obtained had an impressive agreement with experiment. The idea of *renormalization* was born. For this discovery, the first three shared the Nobel prize in 1965. In the early 1970s a new breakthrough was achieved when t'Hooft and Veltman proved the renormalizability of gauge theories (Nobel Prize 1999). Finally Wilson developed the concept of a *renormalization group*, related to critical phenomena connected with phase transitions. His ideas were later used in perturbative quantum field theory, as we will describe below, and also in *constructive* field theory. For these developments, Wilson received the Nobel prize in 1982.

8.1 Renormalization in perturbative QFT

As we have seen, the coefficients of the perturbative expansion of the correlation functions of field theory in terms of powers of the Planck constant and of the coupling constants can be described by a finite sum of amplitudes that are finite-dimensional integrals of rational functions, each integral being associated to a Feynman diagram. This formula, which is a consequence of the Wick theorem on Gaussian integration, is problematic because many Feynman diagrams give rise to divergent integrals. This is to be expected, because a typical Gaussian field is not a smooth function (or a smooth section of a fiber bundle) but is a Schwartz distribution, and the perturbative expansion of the correlation function involves integration of products of distributions, which in general are ill-defined mathematical objects. The formula for each amplitude of a Feynman integral is given by the Feynman rules of the theory. Theses rules are based on the Lagrangian of the theory, and they start by describing all possible Feynman diagrams of each order, and then associating propagators to each line of the diagram and coupling constants multiplying an integration on the internal vertices. The formula for the amplitude of a diagram D is therefore a product of distributions $\prod_{l\in L(D)} \Delta_l^F(x_{i_l} - x_{f_l})$ over all lines l of the diagram with endpoints labeled by four vectors x_{i_l}, x_{f_l} of Minkowski space. The Fourier transform of each of these distributions is given by

$$\widetilde{\Delta}_l^F(p) = \lim_{\epsilon\downarrow 0} i\, P_l(p)\big(p^2 - m_l^2 + i\epsilon\big)^{-1} ,$$

where $m_l > 0$ and $P_l(p)$ is a polynomial (of degree zero in the case of a scalar field). The divergences of the Feynman amplitudes are due to the slow decay of the propagators in momentum space (ultraviolet divergences) and also to singularities of propagators at zero momentum, which generally occur in the presence of massless fields (infrared divergences). In order to get finite results for the correlation functions and S-matrix, we have to renormalize the theory. This procedure involves two steps. The first one is to regularize the theory by introducing a cut-off. One possible cut-off is the lattice discretization, where the cut-off is the lattice width and the box volume. This is very natural, because it is a finite-dimensional approximation to an infinite-dimensional problem. However, it has the disadvantage of a severe loss of symmetry.

A choice of cut-off that preserves Lorentz symmetry is the following. First we write

$$\widetilde{\Delta}_l^F(p) = \lim_{\epsilon\downarrow 0} P_l(p) \int_0^\infty d\alpha \;\exp[i\alpha(p^2 - m_l^2 + i\epsilon)] .$$

The corresponding regularized expression is

$$\widetilde{\Delta}_l^{r,\epsilon}(p) = \lim_{\epsilon\downarrow 0} P_l(p) \int_r^\infty d\alpha \exp[i\alpha(p^2 - m_l^2 + i\epsilon)] .$$

The corresponding distribution in coordinate space,

$$\Delta_l^{r,\epsilon}(x) = -\frac{i}{4} P_l\left(i\frac{\partial}{\partial x}\right) \int_r^\infty \frac{d\alpha}{\alpha^2} \exp\left[-i\alpha(m_l^2 - i\epsilon) - i\frac{x^2}{4\alpha}\right] ,$$

is now a smooth function with moderate growth. Hence the product

$$\prod_{l \in L(D)} \Delta_l^{r,\epsilon}(x_{i_l} - x_{f_l})$$

is a well-defined smooth function as well as a distribution on $\mathbb{R}^{4n}$. Its limit as $r \to 0$ and $\epsilon \to 0$ has singularities at coincident points (the generalized diagonal). However, the limit defines a continuous functional in the subspace $\mathcal{S}_N(\mathbb{R}^{4n})$ of test functions $\phi \in \mathcal{S}(\mathbb{R}^{4n})$ that vanish on the diagonal to sufficiently high order $N = N(D)$. The renormalization procedure consists of subtracting cut-off-dependent counter-terms from the Lagrangian, by splitting the bare parameters of the Lagrangian (masses and coupling constants) into physical parameters and cut-off-dependent divergent parameters, so that the amplitude of each diagram computed with the modified Lagrangian will be equal to the original amplitude where counter-terms that vanish on $\mathcal{S}_N(\mathbb{R}^{4n})$ are subtracted, so that the remainder extends to a functional on $\mathcal{S}(\mathbb{R}^{4n})$. As Hepp remarked in [He], this renormalization procedure may be interpreted as a constructive form of the Hahn–Banach theorem. That this construction can be made in a consistent way in all orders was first established by Dyson in [Dy]. However, the proof was incomplete, because it is necessary to develop a combinatorial way to organize the construction of the counter-terms, which is rather complicated, due to the presence of sub-diagrams with overlapping divergences. This combinatorial construction was performed by N. Bogoliubov and O. Parasiuk in [BP]. This proof was rather complicated and also had some mathematical gaps that were fixed by Hepp in [He]. Zimmermann in [Zi] gave a much cleaner proof working directly in momentum space, using very clever and elegant combinatorial formulas that we will describe below.

In the case of gauge fields, the above regularization does not preserve gauge symmetry. A new regularization, called dimension regularization, was developed by t'Hooft and Veltman in [tHV] to extend this procedure to gauge-field-preserving gauge invariance in the limit where the cut-off is removed. As we have seen before, to formulate the perturbative expansion of the correlation function of gauge fields, we have to fix the gauge and introduce ghost fields. In this setting, the gauge invariance is expressed by a set of identities among the correlation functions, called the Ward identities in the case of QED or Slavnov–Taylor identities in the case of non-abelian gauge theories. The regularization and counter-term subtraction destroy these identities, and a major effort of the theory is to prove that these identities are restored in the limit when the cut-off is removed, given finite and gauge-invariant renormalized correlation functions that can be used for physical predictions.

Let us discuss in more detail the approach of Zimmermann. The Feynman rules in momentum space associate a Feynman propagator to each line of a diagram. This propagator may be written in the form

$$\frac{1}{l_0^2 - \boldsymbol{l}^2 - m_l^2 + i\epsilon(\boldsymbol{l}^2 + m_l^2)} .$$

This way of presenting the Feynman propagator breaks Lorentz invariance, and we have to prove that it will be restored in the limit as $\epsilon \downarrow 0$. The amplitude of a diagram is a function of the external momenta $q = (q_1, \ldots, q_r)$, the masses $m = (m_1, \ldots, m_l)$, and ϵ and is given by the integral

$$I(q, m, \epsilon) = \int dk \, \frac{P(k, q)}{\prod_{j=1}^n f_j(k, q, m, \epsilon)} ,$$

where $k = (k_1, \ldots, k_m)$ are the loop momenta, $f_j(k, q, m, \epsilon) = l_0^2 - \boldsymbol{l}^2 - m_l^2 + i(\boldsymbol{l}^2 + m_l^2)$, and the 4-vectors l_j are linear combinations of the external momenta and loop momenta. We consider also the associated Euclidean integral

$$I_E(q, m) = \int dk \, \frac{P(k, q)}{\prod_{j=1}^n e_j(k, q, m)} ,$$

with $e_j(k, q, m) = l_{j,o}^2 + \boldsymbol{l}_j^2 + m_j^2$. To compare the two integrals when $\epsilon > 0$ we just need the following simple estimates:

$$\frac{\boldsymbol{l}^2 + m_l^2}{|l_0^2 - \boldsymbol{l}^2 - m_l^2 + i\epsilon(\boldsymbol{l}^2 + m_l^2)|} \leq \frac{1}{\epsilon}$$

and

$$\frac{l_0^2}{|l_0^2 - \boldsymbol{l}^2 - m_l^2 + i\epsilon(\boldsymbol{l}^2 + m_l^2)|} \leq \sqrt{1 + \frac{1}{\epsilon^2}} .$$

These estimates imply that

$$\frac{1}{\sqrt{1+\epsilon^2}} \leq \frac{\boldsymbol{l}^2 + m_l^2}{l_0^2} + |l_0^2 - \boldsymbol{l}^2 - m_l^2 + i\epsilon(\boldsymbol{l}^2 + m_l^2)| \leq \frac{1}{\epsilon} + \sqrt{1 + \frac{1}{\epsilon^2}} .$$

Therefore, the Minkowski integral is absolutely convergent if and only if the Euclidean integral is. With this observation, Zimmermann extends the power counting theorem that Weinberg [We] proved for Euclidean integrals to the following.

Theorem 8.1 *If all masses are positive, the Minkowski integral is absolutely convergent if and only if it has negative dimension as well as all sub-integrals* *of the form*

$$I(q, m, \epsilon, H) = \int_H dV \frac{P(k, q)}{\prod_{j=1}^n f_j(k, q, m, \epsilon)},$$

where H is any affine subspace of $\mathbb{R}^{4m}$. Here, the dimension of a rational integral is $d = d_1 + d_2$, where d_1 is the number of integration variables (dimension of H) and d_2 is the degree of the integrand with respect to the integration variables.

If the criteria of the above theorem are satisfied, we deduce that the Feynman integral defines, for $\epsilon > 0$, a smooth function of the external momenta. As we mentioned before, this function is not Lorentz invariant. So the next step is to prove that as $\epsilon \downarrow 0$ it converges to a Lorentz invariant distribution in $\mathcal{S}'(\mathbb{R}^{4r})$. In this proof one uses the so-called *Feynman trick*,

$$I(q, m, \epsilon) = (n-1)! \int dk \int_\Delta d\alpha \frac{P(k, q)}{(\sum_{j=1}^n \alpha_j f_j(k, q, m, \epsilon))^n},$$

with $d\alpha = d\alpha_1 \cdots d\alpha_{n-1}$ and $\alpha_n = 1 - \sum_{j=1}^{n-1} \alpha_j$, and where Δ is the simplex

$$\Delta = \left\{ \alpha : \sum_{j=1}^n \alpha_j = 1, \quad \alpha_j \geq 0 \right\}.$$

The reader is invited to prove (a generalized version of) Feynman's trick in the exercises at the end of this chapter.

It is clear that if the Feynman integral associated to a diagram D has positive dimension it is divergent, and hence needs to be renormalized by subtracting counter-terms. The strategy is to expand the integrand in Taylor series and subtract the part with positive dimension. But even if the dimension is negative, it can also be divergent due to the presence of a subdiagram with positive dimension; i.e. the divergence happens when some of the momenta go to infinity and others remain finite. If a diagram is divergent but does not have any divergent subgraph, we say that it is primitively divergent. Such a diagram is renormalized by subtracting from the integrand the divergent part of the Taylor expansion of the integrand around zero momenta, and then we can use the above Weinberg theorem to conclude that the renormalized integral is absolutely convergent. If we denote the integrand by $\mathcal{I}(D)$ and its Taylor polynomial up to the superficial divergence of the diagram by $T_D\mathcal{I}(D)$ then the renormalized amplitude of the primitively divergent diagram is $I(D) = \int (1 - T_D)\mathcal{I}(D)$. If the diagram is not primitively divergent, we need an inductive procedure that requires a combinatorial analysis.

Definition 8.2 *A* subdiagram *of a Feynman diagram Γ is a diagram γ such that the set of lines of γ is a subset of the set of lines of Γ and the vertices of γ are the end points of its lines. If γ is a subgraph of Γ then the reduced graph Γ/γ is the graph obtained from Γ by collapsing the lines of γ to points.*

Notice that if $\gamma_1, \ldots, \gamma_c$ are the connected components of γ then $\mathcal{V}(\Gamma/\gamma) = (\mathcal{V}(\Gamma) \setminus \mathcal{V}(\gamma)) \cup (\overline{V}_1 \cup \ldots \overline{V}_c)$, where $\overline{V}_j = \{\mathcal{V}(\gamma_j)\}$ and the lines that connect a given vertex of $\mathcal{V}(\Gamma) \setminus \mathcal{V}(\gamma)$ to $\overline{V}_j$ are the lines in $\mathcal{L}(\Gamma) \setminus \mathcal{L}(\gamma)$ that connect this vertex to some vertex in $\{\mathcal{V}(\gamma_j)\}$. An important concept is that of a *one-particle irreducible graph*, 1PI. These are connected graphs that remain connected after the suppression of one internal line. Any graph is a tree whose vertices are 1PI graphs. A *forest* is a family $\mathcal{F}$ of subgraphs of G such that any two graphs in $\mathcal{F}$ are either disjoint or one is strictly contained in the other. The relevant forests for renormalization are the forests $\mathcal{F}_{div}$ of divergent 1PI subgraphs. The final formula for the renormalized amplitude of any diagram D is

$$\text{Renorm } I(D) = \int \left(\sum_{\mathcal{F}_{div}} \prod_{\gamma \in \mathcal{F}_{div}} (-T_\gamma \mathcal{I}(D)) \right).$$

This is known as the *forest formula* of Zimmermann.

Remark 8.3 The combinatorics of the Feynman diagrams has reached a much higher status in the recent development of renormalization by Connes and Kreimer [CK1, CK2].

The proof of the BPHS theorem is quite involved. In [Pol] (see also [Sal]), Polchinski gave a much simpler proof for the ϕ^4 theory using the renormalization group approach developed earlier by Kadanoff and Wilson [Wil] (see also [Sh]) to describe critical exponents in phase transitions of statistical mechanics models. Polchinski's method was extended to cover QED and Yang–Mills in [KK, KM1] . The renormalization group method is based on the physical intuition that a physical theory is scaling (energy, momenta, mass)-dependent. In each scale we should have an effective theory given by an effective Lagrangian, and the renormalization group equation tells us how to transform the effective action from one scale to the other. We may parameterize the scales by a cut-off parameter Λ. The renormalization transformation relates a Lagrangian $\mathcal{L}$ at scale Λ with an effective Lagrangian $\mathcal{L}_{\text{eff}}$ at scale $\Lambda_1 < \Lambda$ that has the same correlation functions. This gives a dynamics in the space of Lagrangians whose evolution, in renormalizable theories, converges to some fixed point. In the original approach by Wilson, what is called a renormalization group was in fact a discrete semi-group generated by a map in the space of theories. In Polchinski's approach, the evolution in the space of theories is given by a first-order differential equation, the flow equation.

Let us explain the main ideas of the flow equations method in the case of the Φ^4 theory. The first step is to regularize the bare propagator with a cutoff Λ_0 and also introduce a flow parameter $0 \leq \Lambda \leq \Lambda_0$, so that

$$C^{\Lambda,\Lambda_0}(p) == \frac{1}{p^2+m^2}\left(e^{-(k^2+m^2)/\Lambda_0^2} - e^{-(k^2+m^2)/\Lambda^2}\right).$$

We denote by μ_{Λ,Λ_0} the Gaussian measure of covariance $\hbar C^{\Lambda,\Lambda_0}$. For $\Lambda < \Lambda_0$, this Gaussian measure converges, as $\Lambda \to \Lambda_0$, to a Dirac measure at $\Phi = 0$. To the interaction part of the bare action we add the counter-terms:

$$\begin{aligned} S_{int}^{\Lambda_0} &= \int dx \left(\frac{g}{4!}\Phi^4(x)\right) \\ &+ \int dx \left(\frac{1}{2}a(\Lambda_0)\Phi(x)^2 + \frac{1}{2}z(\lambda_0)(\partial_\mu\Phi)^2(x) + \frac{1}{4!}b(\Lambda_0)\Phi(x)^4\right). \end{aligned}$$

The generating functionals of the Schwinger functions are given by the expression

$$Z^{\Lambda,\Lambda_0}(J) = \int d\mu_{\Lambda,\Lambda_0}(\Phi) e^{-\frac{1}{\hbar}S_{int}^{\Lambda_0}(\Phi)+\frac{1}{\hbar}\langle\Phi,J\rangle},$$

and the generating functional of the truncated Schwinger functions by

$$e^{\frac{1}{\hbar}W^{\Lambda,\Lambda_0}(J)} = \frac{Z^{\Lambda,\Lambda_0}(J)}{Z^{\Lambda,\Lambda_0}(0)}.$$

Finally, we consider the generating functionals of the amputated truncated Schwinger functions

$$e^{\frac{1}{\hbar}(L^{\Lambda,\Lambda_0}(\phi)+I^{\Lambda,\Lambda_0})} = \int d\mu_{\Lambda,\Lambda_0}(\Phi) e^{-\frac{1}{\hbar}S_{int}^{\Lambda_0}(\Phi+\phi)}.$$

Note that

$$L^{\Lambda,\Lambda_0}(0) = 0 \Rightarrow e^{\frac{1}{\hbar}I^{\Lambda,\Lambda_0}} = Z^{\Lambda,\Lambda_0}(0).$$

Hence

$$L^{\Lambda,\Lambda_0}(\phi) = \frac{1}{2}\langle\phi,(C^{\Lambda,\Lambda_0})^{-1}(\phi)\rangle - W^{\Lambda,\Lambda_0}((C^{\Lambda,\Lambda_0})^{-1}(\phi)).$$

Its n-point function in the momentum space has an expansion as a formal power series in $\hbar$, namely

$$L_n^{\Lambda,\Lambda_0}(p_1,\dots p_n) = \sum_{l=0}^{\infty}\hbar^l.\mathcal{L}_{l,n}^{\Lambda,\Lambda_0}(p_1,\dots p_n),$$

where $\mathcal{L}_{l,n}^{\Lambda,\Lambda_0}(p_1,\dots p_n)$ is given by the sum of values of connected Feynman graphs with n external legs and l loops. Taking the derivative with respect to Λ on both sides of the above equation defining the amputated truncated Schwinger functional, and using the loop expansion, we get the system of equations that we describe below.

Let $\omega = (\omega_{1,1}\dots\omega_{n-1,4})$, where the $\omega_{j,\mu}$ are non-negative integers, let $|\omega| = \sum \omega_{i,\mu}$, and let

$$\partial^\omega = \prod_{i=1}^{n-1}\prod_{\mu=1}^{4}\left(\frac{\partial}{\partial p_{i,\mu}}\right)^{\omega_{i,\mu}} .$$

The system of flow equations is the following:

$$\begin{aligned}
&\partial_\Lambda \partial^\omega \mathcal{L}_{l,n}^{\Lambda,\Lambda_0}(p_1,\dots,p_n)\\
&\quad= \frac{1}{2}\frac{1}{(2\pi)^4}\int dk\;\; \partial_\Lambda C^{\Lambda,\Lambda_0}(k)\cdot \partial^\omega \mathcal{L}_{l-1,n+2}^{\Lambda,\Lambda_0}(k,p_1,\dots,p_n,-k)\\
&\quad\quad -\frac{1}{2}\sum_{n_1=0}^{n}\sum_{n_1+n_2=n}\sum_{l_1+l_2=l}\sum_{\omega_1+\omega_2+\omega_3=\omega}\sum_{i_1<\cdots<i_{n_1}} \partial^{\omega_1}\mathcal{L}_{l_1,n_1+1}^{\Lambda,\Lambda_0}(p_{i_1},\dots,p_{i_{n_1}},p)\\
&\quad\quad \times \partial^{\omega_3}\partial_\Lambda C^{\Lambda,\Lambda_0}(p)\times \partial^{\omega_2}\mathcal{L}_{l_2,n_2+1}^{\Lambda,\Lambda_0}(-p,p_{j_1},\dots,p_{j_{n_2}}) .
\end{aligned} \tag{8.1}$$

Here, we have $p = -p_{i_i} - \cdots - p_{i_{n_1}}$, and $j_1 < \cdots < j_{n_2}$ satisfy

$$\{i_1,\dots i_{n_1}, j_1,\dots j_{n_2}\} = \{1,\dots n\} .$$

In the Φ^4 theory, because of the symmetry $\Phi \mapsto -\Phi$, all the n-point functions for n odd vanish. Thus in the above formula the integers n, n_1, n_2 are all even.

Notice that the equations defining the various generating functionals above are just formal expressions because the support of the Gaussian measure is the space of distributions and, since the product of distributions is not well defined, the integrands do not make sense. Therefore, we have first to discretize the space and restrict to a bounded volume, perform all the manipulations, and later take the continuous limit and the thermodynamic limit. For finite values of Λ_0, we get that the continuous limit of each coefficient of the loop expansion converges to a function that is smooth in the variables and also in the parameters, and the above system of differential equations holds in the limit. As we remarked before, as $\Lambda \to \Lambda_0$ the Gaussian measure $d\mu_{\Lambda,\Lambda_0}$ converges to a delta function at $\Phi = 0$. Therefore $L^{\Lambda_0,\Lambda_0} = S_{int}^{\Lambda_0}$ and also

$$\partial^\omega \mathcal{L}_{l,n} = 0 \quad \text{for } n + |\omega| \geq 5 .$$

That the Φ^4 theory is renormalizable is the content of the following result.

Theorem 8.4 *The coefficients*

$$v(\Lambda_0) = \sum_{l=1}^{\infty} \hbar^l v_l(\Lambda_0) \ldots c(\Lambda_0) = \sum_{l=1}^{\infty} \hbar^l c_l(\Lambda_0)$$

can be adjusted so that the limit

$$\lim_{\Lambda_0 \to \infty} \lim_{\Lambda \to 0} L_{l,n}(p_1, \ldots, p_n) = L_{l,n}(p_1, \ldots, p_n)$$

exists for all n and for all l.

To prove this theorem, we need to bound the solutions of the system of flow equations by an induction procedure, and for that we need some renormalization condition, which we impose as

$$\mathcal{L}_4^{0,\Lambda_0} = g, \quad \mathcal{L}_2^{0,\Lambda_0} = 0, \quad \partial_{p^2} \mathcal{L}_2^{0,\Lambda_0} = 0\,.$$

A direct calculation in the tree level ($l = 0$) gives

$$\mathcal{L}_{0,1}^{\Lambda,\Lambda_0} = 0\,, \qquad \mathcal{L}_{0,2}^{\Lambda,\Lambda_0}(p, -p) = 0\,,$$

$$\mathcal{L}_{0,3}^{\Lambda,\Lambda_0}(p_1, p_2, p_3) = 0\,, \quad \mathcal{L}_{0,4}^{\Lambda,\Lambda_0}(p_1, p_2, p_3, p_4) = g\,.$$

Notice that with the above conditions, for each l the right-hand side of the system of flow equations involves only terms where the second index is smaller than l. Hence, to estimate the left-hand side at the level n, l by induction, we have to have an estimate of the right-hand side at the level $n + 2, l - 1$ and below that. To perform the induction step, we have to integrate the estimate in the derivative with respect to Λ. If $n + |\omega| > 4$, the bounds decrease with increasing Λ. In this case, we integrating from Λ_0 down to Λ using the initial condition $\partial^\omega \mathcal{L}_{l,n}^{\Lambda_0,\Lambda_0}(p_1, \ldots, p_n) = 0$. In the few other cases, we integrate from $\Lambda = 0$ up to Λ first at zero moments using the renormalization condition and extend to all momenta using the Taylor formula (see [Mul]). The theorem follows from the two lemmas below.

Lemma 8.5 *There exist polynomials P_1, P_2 with positive coefficients that depend only on l, n, ω such that*

$$\left|\partial^\omega L_{l,n}^{\Lambda,\Lambda_0}(p_1, \ldots, p_n)\right| \leq (\Lambda + m)^{4-n-|\omega|}\, P_1\left(\log \frac{\Lambda + m}{m}\right) P_2\left(\frac{|p|}{\Lambda + m}\right).$$

This lemma is proved by an inductive procedure starting at $l = 0$ and $n = 4$, where we can take

$$P_1 = 1 \quad \text{and} \quad P_2 = \frac{g}{(\Lambda + m)^{4-n-|\omega|}}\,.$$

From the flow equation, we get from this an estimate of $\partial_\Lambda \partial^\omega \mathcal{L}_{1,2}$. From that we get by integration that the estimate for $\partial^\omega \mathcal{L}_{1,2}^{\lambda,\Lambda_0}$ holds. We proceed by induction that increases $n + 2l$ and for $n + 2l$ constant decreases in n.

Lemma 8.6 *There exist polynomials P_3, P_4 with positive coefficients that depend only on l, n, ω such that*

$$\left|\partial_{\Lambda_0} \partial^\omega L_{l,n}^{\Lambda,\Lambda_0}(p_1, \dots, p_n)\right| \leq \frac{(\Lambda + m)^{5-n-|\omega|}}{(\Lambda_0 + m)^2} P_3\left(\log \frac{\Lambda_0 + m}{m}\right) P_4\left(\frac{|p|}{\Lambda + m}\right).$$

This lemma is proved by the same inductive scheme on the flow equation derived with respect to Λ_0 and using the previous lemma. The theorem is an easy consequence of this lemma.

To prove the renormalizability of gauge-invariant theories following the above ideas, we have to face the problem that gauge invariance is destroyed by the regularization. In fact, the Schwinger functions are not individually gauge-invariant, but the gauge invariance of the theory is expressed by some identities between the different Schwinger functions called the Taylor–Slavnov identities (generalizing the Ward–Takahashi identities in QED). Because the Taylor–Slavnov identities are not satisfied after the cut-off, a careful analysis must be performed to prove that they are restored when the cut-off goes to infinity. This is done in Chapter 4 of [Mul]. See also the short survey [K].

8.2 Constructive field theory

The great success of the perturbative theory discussed in the previous section and the impressive precision of its predictions naturally indicate that it is really a perturbative expansion of a quantum non-linear theory. The program of constructive field theory is to build non-linear examples and analyze the spectrum of the particles. One should look for a Hilbert space, a positive energy representation of the Poincaré group in this Hilbert space, and an operator-valued distribution satisfying natural axioms and having a non-trivial scattering matrix. To be physically relevant, the theory should also exhibit what is called a "mass gap." This means that the spectrum of the Hamiltonian, which has zero as an eigenvalue with the vacuum as an eigenvector, must have a gap between 0 and a positive number m. The importance of this mass gap is that it implies the exponential decay of the correlations and, in the case of QCD, it explains why the nuclear forces are strong but short-ranged. The development of this program in the 1970s involved very hard mathematical estimates in functional analysis, and the final outcome was the construction of families of non-trivial theories for spacetimes of dimensions 2 and 3.

From the intimate connection established by Osterwalder and Schrader between the Minkowski and Euclidean formulations of quantum field theory, one can formulate the existence problem in the realm of Euclidean quantum field theory, where the goal is to construct interesting probability measures on the σ-algebra generated by the cylindrical sets in the space of distributions. Here, a cylinder is a set of the form $\{\Phi \in \mathcal{S}'; (\Phi(f_1), \ldots, \Phi(f_n)) \in B \subset \mathbb{R}^n\}$, where $f_j \in \mathcal{S}$ are test functions and B is a Borel set. The idea is to make sense of the formal definition $d\mu = \frac{1}{Z} e^{-S_{int}(\Phi)} d\mu_C$, where $d\mu_C$ is the measure in the distribution space defined by the quadratic part of the classical action, S_{int} is the higher-order part of the action, for example, $S_{int}(\Phi) = \lambda\Phi^4$, and $Z = \int_{\mathcal{S}'} e^{-S_{int}(\Phi)} d\mu_C$ is the normalizing factor. As we have mentioned before, this expression is formal because the functional S_{int} does not make sense in the space of distributions. A natural strategy would be to use a cut-off Λ such that for the corresponding measure $d\mu_{C(\Lambda)}$ the definition does make sense and we get a measure $d\mu_\Lambda$ in some space of functions. The next step is to renormalize the action by adding a counter-term with the same terms that are in the original Lagrangian but multiplied by coefficients that depend on the cut-off (and will diverge as we remove the cut-off). Then, we have to prove some *a priori* bounds on the renormalized measure that are uniform in the cut-off and use these bounds to prove the convergence of the approximate measures to a measure in the distribution space. In the 1970s, many papers by Jaffe, Glimm, Simon and others were dedicated to this program, and the final outcome was the existence of the measure, and of the mass gap, when the spacetime has dimension 3 with $S_{int}(\Phi) = \lambda\Phi^4$ for small λ and in dimension 2 with $S_{int}(\Phi) = \lambda\mathcal{P}(\Phi)$, where $\mathcal{P}$ is any polynomial bounded from below (the so-called $\mathcal{P}(\Phi)_2$ theory).

The question of compatibility of special relativity and quantum theory in four-dimensional spacetime remains one of the greatest challenges in mathematics and in physics. This question, even for the $\lambda\Phi^4$ theory, is not resolved in dimension 4, although most people believe that the theory does not exist in this dimension. The same problem is expected for quantum electrodynamics. This is probably related to the fact that these theories are not asymptotically free. A theory is said to be asymptotically free if the quantum behavior at short distances (high energy) approximates the classical behavior; i.e. interaction decays with energy and particles start to behave like free particles. On the other hand, one the greatest discoveries of the 1970s is the asymptotic freedom of Yang–Mills theories with non-abelian gauge groups. Politzer, Gross, and Wilczek received the Nobel Prize for this discovery that justifies the use of perturbative expansions in the strong interactions. It also indicates that a quantum theory for Yang–Mills in spacetime of dimension 4 may exist. In fact, the

Clay Mathematics Institute established as one of the Millennium Problems a quantum theory of the four-dimensional Yang–Mills system and the existence of a mass gap, with a one million–dollar prize for its solution. Notice that the possible existence of a mass gap for the Yang–Mills theory is a *quantum effect*. It is not manifested classically, since all the gauge bosons, without the symmetry breaking by the Higgs boson, are massless. See Chapter 9.

Exercises

8.1 Prove the easiest version of Feynman's trick: if a, b are non-zero numbers, then

$$\frac{1}{ab} = \int_0^1 \frac{dx}{[xa + (1-x)b]^2} .$$

8.2 Generalize the previous exercise as follows. Let $a \neq 0 \neq b$ be complex numbers, and let $\alpha, \beta > 0$. Then

$$\frac{1}{a^\alpha b^\beta} = \frac{\Gamma(\alpha+\beta)}{\Gamma(\alpha)\Gamma(\beta)} \int_0^1 \frac{x^{\alpha-1}(1-x)^{\beta-1}}{[xa+(1-x)b]^{\alpha+\beta}}\, dx .$$

To prove this formula, perform the substitution $z = xa/(xa + (1-x)b)$ in the integral on the right-hand side to see that the integral is equal to $a^{-\alpha}b^{-\beta}B(\alpha, \beta)$, where $B(\alpha, \beta)$ is the *beta function*

$$B(\alpha, \beta) = \int_0^1 z^{\alpha-1}(1-z)^{\beta-1}\, dz .$$

Then apply a well-known identity relating the beta and gamma functions.

8.3 Using Exercise 8.2 and induction, prove the following general formula. If $a_1, a_2, \ldots, a_n$ are non-zero complex numbers and $\alpha_1, \alpha_2, \ldots, \alpha_n$ are positive numbers, then

$$\frac{1}{a_1^{\alpha_1} a_2^{\alpha_2} \cdots a_n^{\alpha_n}} = \frac{\Gamma(\alpha_1 + \cdots + \alpha_n)}{\Gamma(\alpha_1)\cdots\Gamma(\alpha_n)}$$

$$\times \int_0^1 \cdots \int_0^1 \frac{\prod x_i^{\alpha_i - 1}}{\left[\sum x_i a_i\right]^{\sum \alpha_i}}\, \delta\left(\sum x_i - 1\right) dx_1 \cdots dx_n .$$

This formula generalizes the Feynman trick used in the present chapter. Why?

9
The Standard Model

In this chapter, we describe the most general Lagrangian that is both Lorentz invariant and renormalizable. This is the Lagrangian of the Standard Model that describes the interaction of all known particles with all known forces, except for gravity. The group of internal symmetries of the Standard Model is $U(1) \times SU(2) \times SU(3)$. All particles predicted by the Standard Model have been detected experimentally except for one, the Higgs boson, which plays an essential role, as we will see below. The elementary matter particles are the leptons (electron, electron neutrino, muon, muon neutrino, tau, and tau neutrino and the corresponding antiparticles), the quarks in six different flavors (up, down, charmed, strange, top, bottom) and each in three different colors (red, blue, green), and their antiparticles. The interaction carriers are the photon for the electromagnetic field, three bosons associated with the weak interaction, corresponding to the internal symmetry $U(1) \times SU(2)$, and eight gluons associated with the strong interaction, corresponding to the group $SU(3)$ (each also coming in three different colors), plus the hypothetical Higgs boson, which is the only one not yet detected experimentally.

We emphasize that the Standard Model as presented here is a *semi-classical* model. After the appropriate Lagrangian is written down, it is still necessary to *quantize* it. No one knows so far how to do this in a mathematically rigorous, constructive way. The next best thing is to use the methods of perturbative QFT (Chapter 7) and renormalization (Chapter 8). This, of course, we will not do. However, we do want to remark that very good results matching experiment can be obtained from just the simplest first-order Feynman graphs! These have no loops, and therefore require no renormalization. It is indeed a remarkable feature of the Standard Model that it produces good predictions even at the semi-classical, pre-quantized level.

Table 9.1 *Interaction carriers.*

Interaction	Boson	Spin
Gravitational	"gravitons" (conjectured)	2
Electromagnetic	photons γ^0	1
Weak force	weak bosons Z^0, W^+, W^-	1
Strong force	gluons G_i (postulated)	1

9.1 Particles and fields

As we already discussed, in particle physics we classify particles into two major groups, according to whether they satisfy or fail to satisfy *Pauli's exclusion principle*. Thus, they can be *fermions*, with half-integral spin, or *bosons*, with integral spin. On the other hand, we also know that in QFT everything physical is defined in terms of *fields*. There are two basic types of fields:

(i) *Interaction carriers*. These are also called *force fields*, and are *gauge* fields. Their particle manifestations are bosons. They subdivide further into
- Photon γ^0: carriers of the electromagnetic interaction.
- Weak bosons W^+, W^-, Z^0: carriers of the weak interaction.
- Gluons G_i ($i = 1, \ldots, 8$): carriers of the strong interaction.

See Table 9.1.

(ii) *Matter fields*. These are fields whose associated particle manifestations are fermions, and can be further subdivided into
- Leptons: these do not "feel" the strong force, their interactions being mediated by the carriers of the electromagnetic and weak interactions only. Examples of leptons are the electron, the muon, the tau particle (and corresponding antiparticles: the positron, etc.), and all neutrinos. See Table 9.2.
- Hadrons: these are subject to the strong interaction (which is very short-range). They subdivide even further into *mesons* (such as the so-called *pions*) and *baryons* (and their corresponding antiparticles). Examples of baryons include the proton and the neutron. See Table 9.3.

From a mathematical standpoint, the two basic types of fields described above are very distinct:

(i) Interaction carriers are *connections* on certain auxiliary vector bundles.
(ii) Matter fields are *sections* of specific vector bundles over spacetime (M, g).

Table 9.2 *The three generations of leptons (spin 1/2).*

Lepton	Mass in MeV/c^2	Lifetime in s
Electron e^-	0.511	∞
Electron neutrino ν_e	$<3\times10^{-6}$	
Muon μ^-	105.658	2.197×10^{-6}
Muon neutrino ν_μ		
Tau τ^-	1777	291×10^{-15}
Tau neutrino ν_τ		

Table 9.3 *The baryon decuplet.*

Baryon	qqq	Charge	Strangeness
Δ^{++}	uuu	2	0
Δ^{+}	uud	1	0
Δ^{0}	udd	0	0
Δ^{-}	ddd	-1	0
Σ^{*+}	uus	1	-1
Σ^{*0}	uds	0	-1
Σ^{*-}	dds	-1	-1
Ξ^{*0}	uss	0	-2
Ξ^{*-}	dss	-1	-2
Ω^{-}	sss	-1	-3

The description above is certainly not complete. In particular, hadrons are not elementary: they are made up of elementary particles called *quarks*. More about that below.

9.2 Particles and their quantum numbers

Certain particle *decays* are never observed in nature or the laboratory. For instance, the proton p^+ and the positron e^+ have the same charge, and the proton mass (energy) equals approximately that of the positron plus the energy of a single photon γ^0; nevertheless,

$$p^+ \nrightarrow e^+ + \gamma^0 .$$

This strongly suggests that some *quantum number* exists that is not preserved under such a putative decay, preventing it from happening. This quantum number is called the *baryon number*, denoted B. It is postulated that $B = 1$ for baryons and $B = 0$ for leptons, and that the total baryon number in a given particle-to-particle interaction should be conserved. These is indeed satisfied in all events occurring in nature, and in all experiments performed in the laboratory. Equivalently, one could assign a *lepton number* $L = 1$ to leptons, and $L = 0$ to all other particles.

In fact, one can actually be more specific and define one quantum number for each type of lepton. For instance, another example of a particle decay that is never observed is muon μ^- decay. Muons are negatively charged leptons just as are electrons, but heavier. Despite charge and energy conservation, we have

$$\mu^- \nrightarrow e^- + \gamma^0 .$$

Again, this is explained by the introduction of yet another quantum number, the so-called *muon number* L_μ, equal to 1 for muons, -1 for anti-muons, and 0 for other particles, and a corresponding conservation law. One can similarly define several other quantum numbers, and corresponding conservation principles.

These ad hoc conservation principles are dictated by experimental observation, and the challenge is to incorporate them into a coherent mathematical model. From a mathematical standpoint, it should be clear after our discussion of Noether's theorem in Chapter 5 that such conservation of various quantum numbers should correspond to *symmetries* of the model. These quantum numbers are defined in terms of various representations of a suitable symmetry group. This is consistent with the mathematical interpretation of particles advanced by Wigner, as we saw in Chapter 4.

9.3 The quark model

The quark model was proposed by Gell-Mann in 1964. It postulates that all hadrons are composite states of more elementary particles called *quarks*, along with their antiparticles (called *anti-quarks*). Quarks and anti-quarks are fermions. According to this model, baryons are composed of three quarks, whereas mesons are made up of one quark and one anti-quark (see Table 9.4). In the original proposed model, quarks came in three different types, or *flavors*, namely the *up quark* u, the *down quark* d, and the *strange quark* s. Empirical facts and symmetry considerations lead to the introduction of three other quarks, namely the *charmed quark* c, the *top quark* t, and the *bottom quark* b. See Table 9.5.

Despite its success in the explanation of hadron "genealogy," including the prediction of new hadrons, the initial quark model suffered from two embarrassing problems. The first trouble was that quarks were not, and have never been, observed in isolation. This was explained by the concept of *quark*

Table 9.4 *The meson nonet.*

Meson	$q\overline{q}$	Charge	Strangeness
π^0	$u\overline{u}$	0	0
π^+	$u\overline{d}$	1	0
π^-	$d\overline{u}$	-1	0
η	$d\overline{d}$	0	0
K^+	$u\overline{s}$	1	1
K^0	$d\overline{s}$	0	1
K^-	$s\overline{u}$	-1	-1
$\overline{K}^0$	$s\overline{d}$	0	-1
η'	$s\overline{u}$	0	0

Table 9.5 *The three generations of quarks (spin* $1/2$*).*

Flavor	Charge	Mass range (in GeV/c^2)
up u	$\frac{2}{3}$	1.5 to 4.0 $\times$ 10^{-3}
down d	$-\frac{1}{3}$	4 to 8 $\times$ 10^{-3}
charmed c	$\frac{2}{3}$	1.15 to 1.35
strange s	$-\frac{1}{3}$	80 to 130 $\times$ 10^{-3}
top t	$\frac{2}{3}$	169 to 174
bottom b	$-\frac{1}{3}$	4.1 to 4.4

confinement, the mechanism of which is still not fully understood. The second trouble was the violation of Pauli's exclusion principle: there are particles such as the Δ^{++} hadron that are made up of *three* identical quarks (uuu, in this case; see Table 9.3). The way out of this second trouble was the introduction of a new quark property called *color*.

Recall that in QED, the interaction between two charged particles, such as electrons, is mediated through the emission and absorbtion of photons (the electromagnetic interaction carriers). In quantum chromo-dynamics (QCD), the basic particles are quarks, and they have "colors." *Color* is a quantum property of nuclear interactions akin to charge in electromagnetism. Quarks come in three colors: R, B, G (for red, blue, and green, respectively). Anti-quarks come in three "anti-colors," or complementary colors: $\overline{R}, \overline{B}, \overline{G}$ (also called cyan, yellow, and magenta, respectively). Quarks and anti-quarks interact with each other by *exchanging colors*. The interactions carriers are called *gluons*. The gluons themselves are said by physicists to be "bi-colored objects."

9.4 Non-abelian gauge theories

The Standard Model is a gauge theory with symmetry group $U(1) \times SU(2) \times SU(3)$. Let us therefore digress a bit and recall some of the basic structure of such theories. We do everything locally, and leave to the reader the task of translating everything into the coordinate-free language of bundles and connections, as presented in Chapter 4. See also Section 9.6 below.

9.4.1 The Yang–Mills Lagrangian

At a semi-classical level, one can formulate a Yang–Mills theory with general non-abelian symmetry group G. The formulation mimics that of electromagnetism, which is an *abelian* gauge theory with group $U(1)$. For concreteness, the reader can think of the case where $G = SU(N)$. We assume that we have a gauge field A_μ, i.e. a connection, which will be coupled with a matter field Ψ, represented by a section of a suitable vector bundle over spacetime. We are given a representation $R : G \to \mathrm{Aut}(V)$, where V is a finite-dimensional complex Hilbert space (e.g. we could have $G = SU(N)$ and R the regular representation of G into $V = \mathbb{C}^N$). Let $\{T_R^a\}$ be the generators of the Lie algebra of G in the representation R, and let us write $A_\mu = A_\mu^a T_R^a$ (when $G = SU(N)$, there are $N^2 - 1$ generators). Recall that the *structure constants* of the Lie algebra of G are implicitly defined by the relations

$$[T_R^a, T_R^b] = i\, f^{abc} T_R^c\,. \tag{9.1}$$

We define a covariant derivative on matter fields by

$$D_\mu \Psi = \left(\partial_\mu - igA_\mu^a T_R^a\right) \Psi\,. \tag{9.2}$$

Here g is a constant, called the coupling constant.

As in the case of electromagnetism, we let the *field strength tensor* be given by

$$F_{\mu\nu} = \partial_\mu A_\nu - \partial_\nu A_\mu - ig[A_\mu, A_\nu]\,. \tag{9.3}$$

From (9.1) and (9.2), a one-line computation left as an exercise shows that

$$F_{\mu\nu}^a = \partial_\mu A_\nu^a - \partial_\nu A_\mu^a + gf^{abc} A_\mu^b A_\nu^c\,. \tag{9.4}$$

Let us now define an appropriate Yang–Mills Lagrangian density. The *kinetic part* of this Lagrangian is given by the curvature of our connection, namely

$$\mathcal{L}_{YM}^{\mathrm{kin}} = -\frac{1}{4}\,\mathrm{Tr}(F_{\mu\nu} F^{\mu\nu})\,. \tag{9.5}$$

This is also sometimes called the pure Yang–Mills Lagrangian. In keeping with the paradigm provided by electromagnetism, we now couple the gauge field with the matter field in the most economic way, the so-called minimal coupling, by means of a *Dirac current*, namely

$$\mathcal{L}_{YM}^{\text{int}} = \overline{\Psi}(i\not{D} - m)\Psi\,, \tag{9.6}$$

where M is a constant and $\not{D}$ is the Dirac operator, defined as follows. We think of each component Ψ^α of Ψ as a spinor ($\Psi^\alpha(x) \in \mathbb{R}^4$ or $\mathbb{C}^2$), for each $\alpha = 1, 2, \ldots, \dim R/4$ (here $\dim R$ denotes the real dimension of the representation of G). We take $\not{D}$ to be the direct sum of the usual Dirac operators on each component,

$$\not{D}\Psi = \sum_\alpha^\oplus \gamma^\mu D_\mu \Psi^\alpha\,,$$

where γ^μ are the usual Dirac matrices. The total Yang–Mills Lagrangian is therefore the sum of (9.5) and (9.6), namely

$$\mathcal{L}_{YM} = -\frac{1}{4}\,F_{\mu\nu}^a F^{\mu\nu\,a} + \overline{\Psi}(i\not{D} - m)\Psi\,, \tag{9.7}$$

where we have taken the trouble of spelling out the trace defining the curvature in terms of the field strength components.

Now, the main thing to observe is that the Yang–Mills Lagrangian as we have just defined is gauge-invariant. Indeed, let $x \mapsto U(x) \in G$ be a local gauge transformation. We may write $U(x) = \exp(ig\theta^a(x)T_R^a)$, where the $\theta^a(x)$ are local functions over spacetime. We have

$$\begin{aligned} \Psi &\mapsto U(x)\Psi \\ A_\mu &\mapsto U(x)A_\mu U^\dagger(x) - \frac{i}{g}\,\big(\partial_\mu U(x)\big)U^\dagger(x)\,. \end{aligned}$$

The reader can now verify as an exercise using these transformations that the covariant derivative transforms as $D_\mu\Psi \mapsto U(x)D_\mu\Psi$. From these facts, it follows at once that the Yang–Mills Lagrangian (9.7) is gauge-invariant, as stated.

9.4.2 Spontaneous symmetry breaking

The formulation of non-abelian gauge theories given above seems fine but, as far as its physical content goes, it suffers from a major drawback: all of its fields are massless! Recall, by analogy with the Klein–Gordon field, that the mass terms in the Lagrangian are those that are quadratic in the fields – but there are no such terms in (9.7). This would not be a problem if the only field we cared about were the electromagnetic field, whose carrier is the photon, which as we know is massless. But it is certainly not acceptable even for other gauge fields such as the weak interaction fields – the weak bosons are known to be massive particles.

It turns out that the reason for this masslessness feature of the Lagrangian (9.7) is its *excess of symmetry*. In the mid-sixties, Higgs created a mechanism for breaking the symmetry, allowing the gauge and matter fields to acquire mass. The idea is to introduce a new (bosonic) field, called the Higgs field, that is subject to a quartic potential that shifts the vacuum to a new place, around which part of the original symmetry is lost. This is known as the *Higgs mechanism* or *spontaneous symmetry breaking*. The role of the Higgs field is to break the symmetry by shifting the classical vacuum (the minimum of the action) away from the origin. If we expand the Lagrangian in powers of the deviation from the shifted vacuum and diagonalize the quadratic part by defining new fields as linear combinations of the old ones, we will see that the coefficients of the diagonal terms are the *masses* of the fields (again, the reader should keep in mind the analogy with the Klein–Gordon field).

Let us explain this idea in more detail, working in perhaps the simplest non-abelian situation, namely the case where the symmetry group is $SU(2)$ and there are no matter fields. In other words, we start with the pure Yang–Mills Lagrangian

$$\mathcal{L}_{YM}^0 = -\frac{1}{4}\,\mathrm{Tr}(F_{\mu\nu}F^{\mu\nu})\,.$$

We will describe the Higgs mechanism that, in the present context, allows the gauge bosons to acquire mass. Let us add to our theory a scalar doublet field

$$\phi = \begin{pmatrix}\phi^+\\ \phi^0\end{pmatrix}\,,$$

subject to a potential of the form

$$V(\phi^\dagger\phi) = \frac{1}{2}\lambda^2\left[(\phi^\dagger\phi)^2 - v^2\right]^2\,,$$

where λ and v are real constants (v will determine the new vacuum state). Our new Lagrangian reads

$$\mathcal{L}_{YM}^H = (D_\mu\phi)^\dagger(D_\mu\phi) - \frac{1}{2}\lambda^2\left[(\phi^\dagger\phi)^2 - v^2\right]^2 - \frac{1}{4}\,\mathrm{Tr}(F_{\mu\nu}F^{\mu\nu})\,. \quad (9.8)$$

Now, the new vacuum states of the theory should correspond to the minima of the potential V. These are degenerate minima, occurring at those fields ϕ such that $\phi^\dagger\phi = v^2$. These vacua are in the gauge-group orbit of the constant field

$$\phi_0 = \begin{pmatrix}0\\ v\end{pmatrix}\,.$$

The idea now is to fix the gauge so that this is our new vacuum state, and to expand the potential around this new vacuum. The original $SU(2)$ symmetry will therefore be lost in the process, but we will achieve our goal of giving mass to the gauge bosons. Note that the field ϕ has four real components (or two complex ones) and because the group $SU(2)$ is a three-dimensional Lie group, we can use a gauge transformation to gauge away three of the four components of ϕ, getting a representative of the form

$$\phi(x) = \begin{pmatrix} 0 \\ v + \frac{1}{\sqrt{2}}h(x) \end{pmatrix},$$

where h is a *real* scalar field. Let us then write the expression of the Lagrangian for this ϕ, in terms of h. Using the fact that the generators of the Lie algebra of $SU(2)$ are given by the Pauli matrices, we see after some computations that

$$\begin{aligned} D_\mu\phi &= \left(\partial_\mu + \frac{ig}{2}A_\mu^a\sigma^a\right)\begin{pmatrix} 0 \\ v + \frac{1}{\sqrt{2}}h \end{pmatrix} \\ &= \begin{pmatrix} -\frac{ig}{2}(A_\mu^1 - iA_\mu^2)(v + \frac{1}{\sqrt{2}}h) \\ \frac{1}{\sqrt{2}}\partial_\mu h - \frac{ig}{2}A_\mu^3\,(v + \frac{1}{\sqrt{2}}h) \end{pmatrix}. \end{aligned}$$

Let us now introduce the linear combinations

$$A_\mu^\pm = \frac{1}{\sqrt{2}}(A_\mu^1 \mp A_\mu^2)\,.$$

Using this notation and the above computation, we can calculate explicitly the first term on the right-hand side of (9.8), obtaining

$$\begin{aligned} &(D_\mu\phi)^\dagger(D_\mu\phi) \\ &= \frac{1}{2}(\partial_\mu\phi)(\partial^\mu\phi) + \frac{g^2}{2}A_\mu^- A_\mu^+\left(v + \frac{h}{\sqrt{2}}\right)^2 + \frac{g^2}{4}A_\mu^3 A^{\mu\,3}\left(v + \frac{h}{\sqrt{2}}\right)^2. \end{aligned} \tag{9.9}$$

Note that, because

$$A_\mu^- A_\mu^+ = \frac{1}{2}(A_\mu^1 A^{\mu\,1} + A_\mu^2 A^{\mu\,2}),$$

the second term on the right-hand side of (9.9) is telling us that the gauge bosons associated with the fields A_ν^1 and A_μ^2 both have the same mass m_A, given by

$$\frac{1}{2}m_A^2 = \frac{g^2}{4}v^2\,;$$

in other words, $m_A = gv/\sqrt{2}$. Therefore, the linear combinations A^-_μ and A^-_μ also have the same mass m_A. The mass of the third gauge boson A^3_μ can be read off from the third term on the right-hand side of (9.9), and we see that it must also be equal to m_A. Thus, in this Yang–Mills theory with symmetry breaking, the gauge bosons have all acquired mass (and the masses are equal). The Lagrangian (9.8) also tells us, of course, that the Higgs boson itself is a massive particle (its mass being $m_H = \lambda$, the parameter in the potential V).

The Higgs field can also be coupled with the matter field Ψ in the full Yang–Mills Lagrangian, giving it mass. We will describe this mechanism directly in the context of the Standard Model below.

9.5 Lagrangian formulation of the standard model

The Standard Model Lagrangian is a gauge-field Lagrangian, with gauge group $U(1) \times SU(2) \times SU(3)$. The basic paradigm leading to the construction of this Lagrangian is provided by Yang–Mills theory, as described in the previous section: the interaction fields are given by connections, and the matter fields by sections of suitable vector or spinor bundles over spacetime. In the sequel we will describe the Lagrangian as usually presented in the physics literature. In the next section we will describe also the intrinsic geometric meaning of this Lagrangian.

9.5.1 The electroweak model of Glashow–Weinberg–Salam

According to the electroweak theory of Glashow–Weinberg–Salam, the interactions between the leptons are mediated by the bosons of the gauge group $U(1) \times SU(2)$. This theory presents a unification of the electromagnetic and weak interactions. Let us progressively describe the ingredients in the construction of the electroweak Lagrangian. This will be a $U(1) \times SU(2)$ gauge-invariant Lagrangian.

9.5.1.1 The kinetic terms Let us first describe the *kinetic* Lagrangian, and later the part corresponding to the interactions with leptons.

(i) We have a $U(1)$ gauge field, or *abelian connection* (B_μ). As we saw in Chapter 5, the corresponding field strength tensor $B_{\mu\nu}$ is given by $B_{\mu\nu} = \partial_\mu B_\nu - \partial_\nu B_\mu$. The kinetic part of the Lagrangian associated with this field is

$$\mathcal{L}^B_{\text{kin}} = -\frac{1}{4} B_{\mu\nu} B^{\mu\nu} \,. \tag{9.10}$$

(ii) To incorporate the weak interaction into the model, one needs the gauge fields (W_μ). These constitute a non-abelian connection with group $SU(2)$. The corresponding covariant derivative is given by

$$D_\mu = \partial_\mu + \frac{ig_2}{2} W_\mu ,$$

where g_2 is the so-called *weak coupling constant*. The field strength tensor must be defined in a covariantly natural way, and the way to do this is to write

$$W_{\mu\nu} = D_\mu W_\nu - D_\nu W_\mu . \tag{9.11}$$

This in turn can be rewritten as

$$W_{\mu\nu} = (\partial_\mu W_\nu - \partial_\nu W_\mu) + \frac{ig_2}{2}(W_\mu W_\nu - W_\nu W_\mu) . \tag{9.12}$$

Now, as dictated by the Yang–Mills paradigm, we define the kinetic part of the Lagrangian density corresponding to the weak field as the curvature of our $SU(2)$ connection, namely

$$\mathcal{L}^w_{\text{kin}} = -\frac{1}{8}\,\text{Tr}(W_{\mu\nu}W^{\mu\nu}) . \tag{9.13}$$

Let us record here the effect of a gauge transformation $\psi \mapsto U\psi$ on a given lepton (fermion) field ψ, where $U \in SU(2)$ is a unitary matrix. We have the transformation rule

$$W_\mu \mapsto W'_\mu = UW_\mu U^\dagger + \frac{2i}{g_2}\,\partial_\mu U \cdot U^\dagger$$

expressing the fact that (W_μ) is a $SU(2)$-connection, and therefore two things happen:

(i) $D_\mu\psi \mapsto UD_\mu\psi$.
(ii) $W_{\mu\nu} \mapsto W'_{\mu\nu} = UW_{\mu\nu}U^\dagger$.

Using these facts, it is an easy exercise to check that the Lagrangian (9.13) has the appropriate $SU(2)$ gauge covariance.

We may now put together the $U(1)$ and $SU(2)$ gauge fields, getting the total kinetic Lagrangian density of the electroweak model. This is simply the sum $\mathcal{L}^{ew}_{\text{kin}} = \mathcal{L}^B_{\text{kin}} + \mathcal{L}^w_{\text{kin}}$; in other words

$$\mathcal{L}^{ew}_{\text{kin}} = -\frac{1}{4}\,B_{\mu\nu}B^{\mu\nu} - \frac{1}{8}\,\text{Tr}(W_{\mu\nu}W^{\mu\nu}) . \tag{9.14}$$

Both kinetic terms making up the right-hand side of (9.14) have clear geometric meanings: they are the curvatures of the corresponding gauge fields (B_μ) and (W_μ).

Next, we wish to recast the above formula (9.14) in a slightly different notation, making the expression in terms of coordinates more explicit. For this purpose, we recall that any element of $SU(2)$ can be written in the form

$$U = \exp(-i\alpha^a \sigma_a) ,$$

where α^a, $a = 1, 2, 3$, are real scalars and the σ_a are the Pauli spin matrices, namely

$$\sigma_1 = \begin{pmatrix} 0 & 1 \\ 1 & 0 \end{pmatrix} , \quad \sigma_2 = \begin{pmatrix} 0 & -i \\ i & 0 \end{pmatrix} , \quad \sigma_3 = \begin{pmatrix} 1 & 0 \\ 0 & -1 \end{pmatrix} ,$$

which, as we know, generate the Lie algebra of $SU(2)$. Thus, each gauge field W_μ can be written as $W_\mu = W_\mu^a \sigma_a$; in other words,

$$W_\mu = \begin{pmatrix} W_\mu^3 & W_\mu^1 - i W_\mu^2 \\ W_\mu^1 + i W_\mu^2 & -W_\mu^3 \end{pmatrix} .$$

It is convenient to introduce the complex fields

$$W_\mu^+ = \frac{1}{\sqrt{2}} (W_\mu^1 - i W_\mu^2) \tag{9.15}$$

$$W_\mu^- = \frac{1}{\sqrt{2}} (W_\mu^1 + i W_\mu^2) \tag{9.16}$$

such that $W_\mu^- = (W_\mu^+)^*$. Similarly, we define

$$W_{\mu\nu}^\pm = \frac{1}{\sqrt{2}} (W_{\mu\nu}^1 \mp i W_{\mu\nu}^2) .$$

One can easily check (exercise) that

$$W_{\mu\nu}^3 = \partial_\mu W_\nu^3 - \partial_\nu W_\mu^3 - i g_2 (W_\mu^- W_\mu^+ - W_\nu^- W_\mu^+) .$$

Using this notation and the above relations, we can rewrite the kinetic Lagrangian of the electroweak model as follows:

$$\mathcal{L}_{\text{kin}}^{ew} = -\frac{1}{4} B_{\mu\nu} B^{\mu\nu} - \frac{1}{4} W_{\mu\nu}^3 W^{3\,\mu\nu} - \frac{1}{2} W_{\mu\nu}^- W^{+\,\mu\nu} . \tag{9.17}$$

The details of this calculation are left as an exercise.

The gauge bosons of the theory are defined in terms of the given gauge fields as follows. The W^+ and W^- vector bosons are the particles associated to the fields $W^\pm$ given in (9.15). We define the field Z_μ by the orthogonal combination

$$Z_\mu = W_\mu^3 \cos\theta_w - B_\mu \sin\theta_w ,$$

where θ_w is a constant, called the *Weinberg angle*. We also define the field A_μ by the orthogonal combination

$$A_\mu = W_\mu^3 \sin\theta_w + B_\mu \cos\theta_w \ .$$

These fields correspond merely to a rotation of W_μ^3, B_μ by the Weinberg angle. The vector boson Z^0 is the particle associated to the Z_μ field, whereas the field A_μ is the electromagnetic field, whose associated particle is our old friend the photon γ. One might be tempted to fancy that the Weinberg angle is a quite arbitrary parameter, but this is not so. It has to be carefully chosen so that, after the $SU(2)$ gauge symmetry of the electroweak model is spontaneously broken (as we explain below), the vector bosons acquire mass, whereas the photon remains massless. In fact, it may be shown that the Weinberg angle and the coupling constants are related as follows:

$$\cos\theta_w = \frac{g_2}{\sqrt{g_1^2 + g_2^2}} \ , \qquad \sin\theta_w = \frac{g_1}{\sqrt{g_1^2 + g_2^2}} \ .$$

9.5.1.2 Leptonic interactions Now we have to describe the interactions of the gauge fields with the leptons. Recall that leptons are fermions, and as such they are represented mathematically by spinor fields. Given a spinor ψ in the Dirac representation, we can consider the pair of Weyl spinors consisting of the right and left components of ψ, namely

$$\psi_L = \frac{1}{2}(1 - \gamma^5)\psi \ , \quad \psi_R = \frac{1}{2}(1 + \gamma^5)\psi \ ,$$

where $\gamma^5 = i\gamma^0\gamma^1\gamma^2\gamma^3$ (the γ^μ being the usual Dirac matrices). Note that the matrices defining ψ_L and ψ_R are projection matrices, because

$$\left[\frac{1}{2}(1 \pm \gamma^5)\right]^2 = \left[\frac{1}{2}(1 \pm \gamma^5)\right] .$$

Thus, we may identify ψ with the pair ψ_L, ψ_R. The decomposition

$$\psi = \begin{pmatrix} \psi_L \\ \psi_R \end{pmatrix}$$

is called the *chiral representation* of the spinor ψ.

Remark 9.1 For use below, we recall the relationship between the three 2×2 Pauli matrices σ^i and the 4×4 Dirac matrices γ^μ, namely

$$\gamma^0 = \begin{pmatrix} 0 & \sigma^0 \\ 1 & 0 \end{pmatrix} \quad \text{and} \quad \gamma^i = \begin{pmatrix} 0 & \sigma^i \\ -\sigma^i & 0 \end{pmatrix} \quad (i = 1, 2, 3) \ ,$$

where we have defined $\sigma^0 = I$ (the identity matrix). In more compact notation, this can be rewritten as

$$\gamma^\mu = \begin{pmatrix} 0 & \sigma^\mu \\ \overline{\sigma}^\mu & 0 \end{pmatrix},$$

where $\overline{\sigma}^0 = \sigma^0$ and $\overline{\sigma}^i = -\sigma^i$ $(i = 1, 2, 3)$.

It is an experimentally observed fact that *parity* is not preserved under weak interactions. This fact has lead physicists to treat the left and right components of the lepton (and quark) fields as quite different objects. Let us consider for instance the first generation of leptons, namely e, ν_e. The basic assumption is that the left-handed components e_L and ν_{eL} form a *doublet*

$$\ell_L^e = \begin{pmatrix} \nu_{eL} \\ e_L \end{pmatrix}$$

that is sensitive to $SU(2)$ gauge transformations, whereas the right-handed components e_R and ν_{eR} are regarded as *singlets*, sensitive only to $U(1)$ gauge transformations. In particular, the electron and electron-neutrino interact with the weak gauge fields *only through their left-handed components*. Working by analogy with the Dirac current for the electromagnetic Lagrangian, the minimal $U(1) \times SU(2)$ gauge–covariant way to define the coupling of these lepton fields with the gauge fields turns out to be (in the chiral representation)

$$\mathcal{L}_{\text{int}}^e = (\ell_L^e)^\dagger \overline{\sigma}^\mu i D_\mu \ell_L^e + (e_R)^\dagger \sigma^\mu i D'_\mu e_R + (\nu_{eR})^\dagger \sigma^\mu i \partial_\mu \nu_{eR} \,. \tag{9.18}$$

Here, there are two covariant derivatives in action, given by

$$D_\mu \psi = \left(\partial_\mu + \frac{ig_2}{2} W_\mu + \frac{ig_1}{2} B_\mu \right) \psi$$

$$D'_\mu \psi = \left(\partial_\mu + \frac{ig_1}{2} B_\mu \right) \psi \,.$$

Note that the first covariant derivative, containing the W_μ fields, only intervenes in the first term on the right-hand side of (9.18). So there is no interaction of the W_μ fields with the right-hand components of the electron and electron-neutrino fields, as expected. Note, however, the following fact: *no right-handed neutrino has ever been found!* Thus, in the Standard Model there is just one singlet in each lepton generation – i.e. three lepton singlets altogether: e_R, μ_R, and τ_R. There are no right-handed neutrino fields. Accordingly, we drop the last term in the interaction Lagrangian (9.18), writing instead

$$\mathcal{L}_{\text{int}}^e = (\ell_L^e)^\dagger \overline{\sigma}^\mu i D_\mu \ell_L^e + (e_R)^\dagger \sigma^\mu i D'_\mu e_R \,. \tag{9.19}$$

One defines the doublets of the remaining two generations

$$\ell_L^\mu = \begin{pmatrix} \nu_{\mu L} \\ \mu_L \end{pmatrix}, \quad \ell_L^\tau = \begin{pmatrix} \nu_{\tau L} \\ \tau_L \end{pmatrix}$$

and the corresponding singlets μ_R, τ_R in a completely similar way, and the partial interaction Lagrangians $\mathcal{L}_{\text{int}}^\mu$ and $\mathcal{L}_{\text{int}}^\tau$ by expressions analogous to (9.19).

9.5.1.3 Symmetry-breaking mechanism The electroweak Lagrangian, as defined so far, has that serious defect endemic to non-abelian gauge theories: all of its particles are massless! The reason is that there are too many symmetries. We already know the recipe for remedying this situation: the Higgs symmetry-breaking mechanism. We will break the $SU(2)$ symmetry of the electroweak model, still keeping its $U(1)$ symmetry. Let us briefly indicate the results, leaving the computational details as an exercise. We introduce the Higgs field as the scalar doublet

$$\phi = \begin{pmatrix} \phi^+ \\ \phi^0 \end{pmatrix}$$

subject to the potential

$$V(\phi^\dagger \phi) = \frac{1}{2}\lambda^2(\phi^\dagger \phi - v^2)^2 .$$

We use the $SU(2)$ gauge invariance to fix the new ground state to

$$\phi_0 = \begin{pmatrix} 0 \\ v \end{pmatrix}$$

and expand the Lagrangian around this new vacuum. Due to this gauge fixing, we only care about fields ϕ that are perturbations of the new vacuum of the form

$$\phi = \begin{pmatrix} 0 \\ v + \frac{1}{\sqrt{2}}h \end{pmatrix},$$

where h is a real field. This time the covariant derivative applied to ϕ gives us

$$D_\mu \phi = \begin{pmatrix} 0 \\ \frac{1}{\sqrt{2}}\partial_\mu h \end{pmatrix} + \frac{ig_1}{2}\begin{pmatrix} 0 \\ B_\mu(v + \frac{1}{\sqrt{2}}h) \end{pmatrix}$$
$$+ \frac{ig_2}{2}\begin{pmatrix} \sqrt{2}W_\mu^+(v + \frac{1}{\sqrt{2}}h) \\ -W_\mu^3(v + \frac{1}{\sqrt{2}}h) \end{pmatrix}.$$

Note that we are employing here the weak boson fields $W_\mu^\pm$. Using the above expression for $D_\mu \phi$ to calculate $(D_\mu \phi)(\dagger D_\mu \phi)$, we deduce after some lengthy computations that the kinetic electroweak Lagrangian with the Higgs field added in is equal to

$$\begin{aligned}\mathcal{L}^{H}_{\text{kin-ew}} &= \frac{1}{2}(\partial_\mu h)(\partial^\mu h) + \frac{g_2^2}{2} W_\mu^- W^{+\mu}\left(v + \frac{1}{\sqrt{2}}h\right)^2 \\ &+ \frac{1}{4}(g_1^2 + g_2^2)\, Z_\mu Z^\mu \left(v + \frac{1}{\sqrt{2}}h\right)^2 \\ &- \lambda^2 v^2 h^2 + \frac{1}{\sqrt{2}}\lambda^2 v h^3 + \frac{1}{8}\lambda^2 h^4 .\end{aligned}$$

From this expression for the Lagrangian, it is not difficult to check that the masses of the gauge bosons $W_\mu^\pm$ and Z_μ are now positive. One can also check that the field combination $A_\mu = W_\mu^3 \sin\theta_w + B_\mu \cos\theta_w$ (where θ_w is the Weinberg angle) remains massless. This field is identified with the electromagnetic field, and its carrier is the photon.

Now, what about the leptons? To give them mass, we need to couple them with the Higgs field. This requires adding new terms to the electroweak Lagrangian, and these terms should be Lorentz and gauge-invariant. The way to do this is to add a "mass Lagrangian" for each lepton generation. For instance, for the first generation we have

$$\mathcal{L}^{e}_{\text{mass}} == -c_e \left[((\ell_L^e)^\dagger \phi) e_R + e_R^\dagger (\phi^\dagger \ell_L^e)\right] , \tag{9.20}$$

where c_e is a coupling constant (it has to be very small: the added mass Lagrangian should not upset the perturbative calculations of QED!). The expressions for the mass Lagrangians of the other two lepton generations are entirely analogous, and we omit them. Note that (9.20) can be rewritten in terms of the components of the left lepton field as follows:

$$\mathcal{L}^{e}_{\text{mass}} = -c_e \left[(v_{eL}^\dagger \phi^+ + e_L^\dagger \phi^0) e_R + e_R^\dagger ((\phi^+)^\dagger v_{eL} + (\phi^0)^\dagger e_L)\right] .$$

After fixing the gauge so the symmetry is broken, we get

$$\mathcal{L}^{e}_{\text{mass}} = -c_e v (e_L^\dagger e_R + e_R^\dagger e_L) - \frac{c_e h}{\sqrt{2}} (e_L^\dagger e_R + e_R^\dagger e_L) .$$

Note that the neutrino field has disappeared from the scene. It continues to be a massless particle, whereas the electron has acquired a mass, as it should. Analogous results hold for the other two lepton generations.

Remark 9.2 There is strong empirical evidence that massive neutrinos exist (albeit with a *very* small mass). The Standard Model has to be slightly modified in order to accommodate such experimental evidence. This can be done, but we will not discuss it here. See [CG, Chapters 19, 20].

9.5.2 Quantum chromo-dynamics

The theory describing strong interactions is known as *quantum chromo-dynamics*, or QCD for short. The interaction carriers are gluons, and the matter fields are quarks. QCD is a non-abelian gauge theory, with gauge group $SU(3)$, the *color* group. Each quark q (a generic notation for any one of the six flavors u, d, c, s, t, b) is a *triplet*

$$q = \begin{pmatrix} q_r \\ q_b \\ q_g \end{pmatrix}$$

and the same is true for the corresponding anti-quarks $\overline{q}$. We require gauge invariance under the group $SU(3)$.

9.5.2.1 The gluon gauge fields The gluon gauge fields (G_μ) determine an $SU(3)$ connection, and the corresponding covariant derivative (using the fundamental representation of $SU(3)$) is given by

$$D_\mu q = (\partial_\mu + i g_3 G_\mu) q \ .$$

The constant g_3 is the so-called *strong coupling constant*. Under a gauge transformation $q \mapsto q' = Uq$ $(U \in SU(3))$, we have the following transformation rule:

$$G_\mu \mapsto G'_\mu = U G_\mu U^\dagger + \frac{i}{g_3} \partial_\mu U \cdot U^\dagger \ .$$

Note that G_μ belongs to the Lie algebra of $SU(3)$, which consists of traceless Hermitian matrices. This Lie algebra is 8-dimensional, and a basis is provided by the Gell-Mann matrices

$$\lambda_1 = \begin{pmatrix} 0 & 1 & 0 \\ 1 & 0 & 0 \\ 0 & 0 & 0 \end{pmatrix}, \quad \lambda_2 = \begin{pmatrix} 0 & -i & 0 \\ i & 0 & 0 \\ 0 & 0 & 0 \end{pmatrix}, \quad \lambda_3 = \begin{pmatrix} 1 & 0 & 0 \\ 0 & -1 & 0 \\ 0 & 0 & 0 \end{pmatrix},$$

$$\lambda_4 = \begin{pmatrix} 0 & 0 & 1 \\ 0 & 0 & 0 \\ 1 & 0 & \mu \end{pmatrix}, \quad \lambda_5 = \begin{pmatrix} 0 & 0 & -i \\ 0 & 0 & 0 \\ i & 0 & 0 \end{pmatrix}, \quad \lambda_6 = \begin{pmatrix} 0 & 0 & 0 \\ 0 & 0 & 1 \\ 0 & 1 & 0 \end{pmatrix},$$

$$\lambda_7 = \begin{pmatrix} 0 & 0 & 0 \\ 0 & 0 & -i \\ 0 & i & 0 \end{pmatrix}, \quad \lambda_8 = \frac{1}{\sqrt{3}} \begin{pmatrix} 1 & 0 & 0 \\ 0 & 1 & 0 \\ 0 & 0 & -2 \end{pmatrix}.$$

Using this basis, one may write

$$G_\mu = \frac{1}{2} G^a_\mu \lambda_a .$$

The field strength tensor associated with this $SU(3)$ connection is given by

$$\begin{aligned} G_{\mu\nu} &= D_\mu G_\nu - D_\nu G_\mu \\ &= \partial_\mu G_\nu - \partial_\nu G_\mu + i g_3 [G_\mu, G_\nu] . \end{aligned}$$

Following the Yang–Mills paradigm, we know that the kinetic Lagrangian of the strong interaction must be defined as the curvature of our $SU(3)$ connection, namely

$$\mathcal{L}^s_{\text{kin}} = -\frac{1}{2} \operatorname{Tr}(G_{\mu\nu} G^{\mu\nu}) .$$

9.5.2.2 Quark interactions So much for the kinetic part of the QCD Lagrangian. Now, we need to take care of the part corresponding to the interactions with the quark fields. Let us first describe the electroweak interaction of quarks. The quark model must be able to explain certain nuclear decays, such as the neutron decay $n \to p + e^- + \overline{\nu_e}$. This decay, at quark level, is simply the decay of a down quark into an up quark plus an electron and an electron neutrino, i.e. $d \to u + e^- + \overline{\nu_e}$. This decay is due to the weak interaction, whose carrier is the W boson. Therefore, working by analogy with the muon decay $\mu^- \to \nu_\mu + e^- + \overline{\nu_e}$, which is *also* mediated by the W boson, one sees that the left-handed components u_L and d_L of the up and down quarks should be put together in a doublet

$$q^1_L = \begin{pmatrix} u_L \\ d_L \end{pmatrix} ,$$

whereas the right-handed components u_R and d_R, just like the right-handed leptons e_R and $\nu_{e,R}$, are unchanged by $SU(2)$ transformations. Similar doublet–singlet arrangements should hold for the other two generations, namely the doublets

$$q^2_L = \begin{pmatrix} c_L \\ s_L \end{pmatrix} , \quad q^3_L = \begin{pmatrix} t_L \\ b_L \end{pmatrix} ,$$

and the corresponding singlets c_R, s_R and t_R, b_R.

Given such symmetry structure of quark fields, it turns out that there is only one way to make the quark-dynamical Lagrangian density invariant under

$U(1) \times SU(2)$ gauge transformations. Again, the guideline is provided by analogy with the Dirac (QED) Lagrangian. We get, for the up–down generation,

$$\begin{aligned}\mathcal{L}^{u,d} &= \overline{q_L}\, i \left[\partial_\mu + \frac{ig_2}{2} W_\mu + \frac{ig_1}{6} B_\mu \right] q_L \\ &+ \overline{u_R}\, i \left[\partial_\mu + \frac{2ig_1}{3} B_\mu \right] u_R \\ &+ \overline{d_R}\, i \left[\partial_\mu - \frac{ig_1}{3} B_\mu \right] d_R \,.\end{aligned} \tag{9.21}$$

The fractions $2/3$ and $-1/3$ multiplying $ig_1 B_\mu$ in the second and third lines above reflect the fact that the up quark carries an electromagnetic charge of $2/3$, whereas the down quark carries an electromagnetic charge of $-1/3$. Completely similar expressions $\mathcal{L}^{c,s}$ and $\mathcal{L}^{t,b}$ hold for the charmed–strange and top–bottom quark generations. Hence, the dynamical part of the total quark Lagrangian is

$$\mathcal{L}^{\text{quark}}_{\text{dyn}} = \mathcal{L}^{u,d} + \mathcal{L}^{c,s} + \mathcal{L}^{t,b} \,.$$

9.5.2.3 Coupling with the Higgs field Let us now quickly describe how the quarks acquire mass. The procedure is essentially the same as in the case of leptons. There is an important difference, however. In our model, the left lepton doublets have the neutrino fields as their first components, and these are massless. By contrast, both components of the quark doublets, which we write as

$$q_L^i = \begin{pmatrix} u_L^i \\ d_L^i \end{pmatrix} \qquad i = 1, 2, 3 \,,$$

should have mass. The interaction Lagrangian corresponding to quarks coupled with the Higgs field can be written as

$$\mathcal{L}^H_{\text{quark}} = \mathcal{L}^u_H + \mathcal{L}^d_H \,,$$

where

(i) The interaction of the *down quarks* d, s, b with the Higgs field is given by

$$\mathcal{L}^d_H = -\sum \left\{ G^d_{ij} [(q_L^i)^\dagger \phi] d_R^j + G^{d*}_{ij} (d_R^j)^\dagger [\phi^\dagger q_L^i] \right\} .$$

The coefficients G^d_{ij} determine a 3×3 matrix, which *a priori* is quite arbitrary.

(ii) The interaction of the *up quarks* u, c, t with the Higgs field is given by

$$\mathcal{L}^u_H = -\sum \left\{ G^u_{ij} [(q_L^i)^\dagger i\sigma^2 \phi^*] u_R^j + G^{u*}_{ij} (u_R^j)^\dagger [\phi^T i\sigma^2 q_L^i] \right\} .$$

Here, the matrix

$$i\sigma^2 = \begin{pmatrix} 0 & 1 \\ -1 & 0 \end{pmatrix}$$

interchanges up and down components of the Higgs field. The choice of sign is made so that the above partial Lagrangian is $SU(2)$-invariant.

Upon $SU(2)$ symmetry-breaking, following the Higgs strategy already described, the partial Lagrangian (i) becomes

$$\mathcal{L}^d_{\text{mass}} = -v \sum \left\{ G^d_{ij} (d^i_L)^\dagger d^j_R + G^{d*}_{ij} (d^j_R)^\dagger d^i_L \right\} .$$

Likewise, (ii) becomes

$$\mathcal{L}^u_{\text{mass}} = -v \sum \left\{ G^u_{ij} (u^i_L)^\dagger u^j_R + G^{u*}_{ij} (u^j_R)^\dagger u^i_L \right\} .$$

From these last two expressions, it is possible to verify that all quarks have acquired mass. We will not do this here. However, we will make the following remarks. The 3×3 matrices $G^d = (G^d_{ij})$ and $G^d = (G^d_{ij})$ are arbitrary, but they can both be reduced to diagonal form by multiplication on the left and on the right by distinct unitary matrices, say

$$G^d = D^\dagger_L M^d D_R ,$$
$$G^u = U^\dagger_L M^u U_R .$$

Here, the matrices M^d, M^u are diagonal matrices (whose diagonal terms correspond to the fermion masses). This requires a change of basis in the quark fields, namely

$$d'^i_L = D^{ij}_L d^j_L , \quad d'^i_R = D^{ij}_R d^j_L$$
$$u'^i_L = U^{ij}_L u^j_L , \quad u'^i_R = U^{ij}_R u^j_L .$$

The fields in this new basis are called *true quark fields*. In this basis the quadratic form corresponding to the quark fields becomes diagonal, and one can read off the quark masses from the diagonal terms.

Now, it turns out that, in terms of these true quark fields, the part of the quark Lagrangian corresponding to the interaction with the weak fields can be written as

$$\mathcal{L}_{q,w} = \frac{1}{\sqrt{2}\sin\theta_w} (u'^\dagger_L, c^\dagger_L, t^\dagger_L) \begin{pmatrix} V_{ud} & V_{us} & V_{ub} \\ V_{cd} & V_{cs} & V_{cb} \\ V_{td} & V_{ts} & V_{tb} \end{pmatrix} \begin{pmatrix} \overline{\sigma}^\mu d'_L \\ \overline{\sigma}^\mu s'_L \\ \overline{\sigma}^\mu b'_L \end{pmatrix} W^+_\mu .$$

Here, we have reverted to using the flavor names as indices. The angle θ_w is the Weinberg angle. The 3×3 matrix V appearing on the right-hand side of the above expression is equal to the unitary matrix $U_L D_L^\dagger$. This matrix is called the *Kobayashi–Maskawa* matrix. As we can see, it mixes the three generations of quarks.

9.5.3 The final Standard Model Lagrangian

To write the final Lagrangian of the Standard Model, we have to collect together all the contributions in the previous sections, including the Higgs field. From Noether's theorem (see Chapter 5), each remaining symmetry of the model generates conserved currents and charges. The charges of the $SU(3)$ symmetries are the quark colors. The photons, which are the quantum particles that intermediate the electromagnetic interaction, do not have an electrical charge because the corresponding symmetry group is abelian and there is no interaction between photons. Because $SU(3)$ is non-abelian, the gluons do carry color charges, even though they have no mass. We must of course include fields corresponding to the three families of leptons and quarks that we mentioned before. This gives rise to a rather complicated expression for the Lagrangian, which is the following:

$$\begin{aligned}
\mathcal{L}_{SM} = & -\frac{1}{4}\left(G^a_{\mu\nu}G^{\mu\nu a} + W^\alpha_{\mu\nu}W^{\mu\nu\alpha} + B_{\mu\nu}B^{\mu\nu}\right) \\
& + D_\mu\phi^\dagger D^\mu\phi + \mu^2\phi^\dagger\phi - \lambda(\phi^\dagger\phi)^2 \\
& + i\overline{q_L}\gamma^\mu\left(\partial_\mu - \frac{i}{2}g_2 W^\alpha_\mu\sigma^\alpha - \frac{i}{6}g_1 B_\mu - \frac{i}{2}g_3 G^a_\mu\lambda_a\right)q_L \\
& + i\overline{d_R}\gamma^\mu\left(\partial_\mu + \frac{i}{3}g_1 B_\mu - \frac{i}{2}g_3 G^a_\mu\lambda_a\right)d_R \\
& + i\overline{u_R}\gamma^\mu\left(\partial_\mu - \frac{2i}{3}g_1 B_\mu - \frac{i}{2}g_3 G^a_\mu\lambda_a\right)u_R \\
& + i\overline{\ell_L}\gamma^\mu\left(\partial_\mu - \frac{i}{2}g_2 W^\alpha_\mu\sigma^\alpha + \frac{i}{2}g_1 B_\mu\right)\ell_L \\
& + i\overline{e_R}\gamma^\mu\left(\partial_\mu + i g_1 B_\mu\right)e_R \\
& + k^d\overline{q_L}\phi d_R + k^u\overline{q_L}\tau^2\phi u_R + k^e\overline{\ell_L}\phi e_R \\
& + \text{h.c}\,. \qquad (9.22)
\end{aligned}$$

The abbreviation "h.c." stands for the sum of *Hermitian conjugates* of the terms displayed, so the above formula is in fact twice as long when written in full.

Remark 9.3 This is just the classical description. In quantizing the theory, two other developments played a fundamental role. The first was the discovery of *asymptotic freedom* by Gross, Wilczek, and Politzer (Nobel Prize in physics 2004), which allows the use of perturbation theory in QCD when the scale of energy is high or distances are small; see [GW, P]. This means that the strong force that acts on the quarks decreases with the distances between the quarks so that at very small distances they behave as free particles, whereas if the distances increase, it increases strongly, accounting for *confinement* of quarks in QCD. The second important step was the proof by t'Hooft and Veltman (Nobel Prize in physics 1999) that non-abelian gauge theories are renormalizable; see [tHV].

Remark 9.4 The Standard Model also contains 19 free parameters (masses, coupling constants, etc.) that cannot be computed by the theory: they have to be obtained experimentally. This stimulated the search for unification models with a small number of parameters where the Standard Model, because be imbedded. The biggest challenge is to include gravitation, whose Lagrangian is highly non-renormalizable, in the quantum theory. String theory, see [Polc], which has just one parameter, is a serious candidate for this model, because it contains both the standard model and gravitation. However there are a huge number of string theory models, due to different compactifications of part of the higher-dimensional spacetime, giving different physical predictions, and there are no experimental data to select one out of them all. More recently, Connes and co-workers [CCM] proposed a new unification model based on non-commutative geometry.

Remark 9.5 A very lucid description of the Standard Model at the level presented in this chapter can be found in the book by Cottingham and Greenwood [CG]. A more sophisticated treatment is given by Chamseddine in [Ch]. Note that, apart from a a slight change in notation, the SM Lagrangian expression presented in (9.22) is exaclty the same as that given in [Ch]. For a very interesting phenomenological description of the model, see Veltman's account in [V].

9.6 The intrinsic formulation of the Lagrangian

Having presented the semi-classical expression of the Lagrangian of the Standard Model as given in the physics literature, we give a brief explanation of the intrinsic (i.e. coordinate-free) mathematical meaning of such Lagrangian. The Lagrangian itself will depend only on the 1-jet of the fields involved.

There will be several ingredients. First we will need certain connections corresponding to the interaction fields. We will also need several vector or spinor (Hilbert) bundles over spacetime, whose sections will be the matter fields. These vector bundles are obtained as associated bundles of the principal bundles where the interaction fields live, via suitable group representations. The various kinetic terms of the Lagrangian arise directly from the covariant derivatives induced by the interaction field connections on these associated vector bundles.

Here are the building blocks.

(i) *Spin bundle and Clifford connection*: All bundles will be defined over a spacetime M. As always, M is assumed to be a spin 4-manifold, with a Lorentzian metric that in general will be non-flat. We start with a Lorentz principal bundle over M and an associated vector bundle

$$\begin{array}{ccc} \mathbb{R}^4 & \hookrightarrow & E \\ & & \downarrow \\ & & M \end{array}$$

We also consider a *Clifford connection* on this vector bundle, i.e. a connection $\nabla^c : \Gamma(TM) \times \Gamma(E) \to \Gamma(E)$ having the following properties:

(a) It is compatible with the metric: for all sections $\psi, \theta \in \Gamma(E)$ we have

$$\partial_\mu \langle \psi, \theta \rangle = \langle \nabla^c_\mu \psi, \theta \rangle + \langle \psi, \nabla^c_\mu \theta \rangle \ ,$$

where $\langle \cdot, \cdot \rangle$ denotes the inner product on fibers.

(b) It is compatible with Clifford multiplication: for all $X \in \Gamma(TM)$, all $\lambda \in \Gamma(T^*M)$, and all $\psi \in \Gamma(E)$, we have

$$\nabla^c_X(\lambda \cdot \psi) = (\nabla_X \lambda) \cdot \psi + \lambda \cdot (\nabla_X \psi) \ ,$$

where $\cdot$ denotes Clifford multiplication.

It is a *theorem* that we shall not prove here that, given a spin bundle (E, M) as above, there exists a unique Clifford connection defined over E. See the book by Lawson and Michelson [LM].

Let us briefly describe how one can write down an expression for the Clifford connection in local coordinates. Let ω denote the Levi-Civita (Lorentz) connection on M. Recall (see Chapter 4) that ω is a matrix $\omega = (\omega^\mu_\nu)$ of 1-forms, which can be written in local coordinates as

$$\omega^\mu_\nu = \omega^\mu_{\sigma\nu}\, dx^\sigma \ .$$

The matrix ω is anti-symmetric: $\omega^\mu_\nu = -\omega^\nu_\mu$. By a slight abuse of notation, let γ^μ denote the Dirac matrices transported via the representation of $SL(2, \mathbb{C})$ defining the vector (spinor) bundle E to matrices acting on the fibers of E. Then the Clifford connection on E can be written as follows:

$$\Omega = \frac{1}{4}\,\omega^{\mu}_{\nu}\,\gamma_{\mu}\gamma^{\nu}\,.$$

Here, as usual, we write $\gamma_{\mu} = g_{\mu\nu}\gamma^{\nu}$, where $g_{\mu\nu} = g^{\mu\nu}$ are the components of the Lorentzian metric tensor on M. Thus, the Clifford covariant derivative of any section $\psi \in \Gamma(E)$ is given in local coordinates by

$$\nabla^{c}_{\mu}\psi = \left(\partial_{\mu} + \frac{1}{4}\omega^{\nu}_{\sigma\mu}\gamma_{\nu}\gamma^{\sigma}\right)\psi\,.$$

Note that we have used Clifford multiplication on the fibers of E, as we should. Therefore, the local expression of the corresponding Dirac operator is simply

$$D\!\!\!/\psi = \left(\gamma^{\mu}\partial_{\mu} + \frac{1}{4}\omega^{\nu}_{\sigma\mu}\gamma^{\mu}\gamma_{\nu}\gamma^{\sigma}\right)\psi\,.$$

(ii) *Basic Hermitian bundles*: The matter fields are constructed from the following auxiliary Hermitian vector bundles:

(a) Structure group $U(1)$, with representation **1**:

$$\begin{array}{ccc} \mathbb{C} & \hookrightarrow & E^{1} \\ & & \downarrow \\ & & M \end{array}$$

(b) Structure group $SU(2)$, with representation **1**:

$$\begin{array}{ccc} \mathbb{C} & \hookrightarrow & \tilde{E}^{2} \\ & & \downarrow \\ & & M \end{array}$$

(c) Structure group $SU(2)$, with representation **2**:

$$\begin{array}{ccc} \mathbb{C}^{2} & \hookrightarrow & E^{2} \\ & & \downarrow \\ & & M \end{array}$$

(d) Structure group $SU(3)$, with representation $\mathbf{3}$:

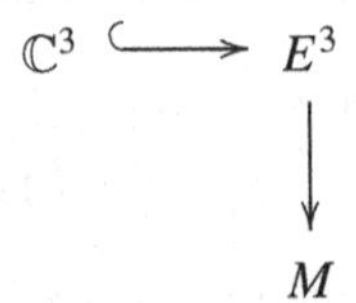

(e) Structure group $SU(3)$, with representation $\overline{\mathbf{3}}$:

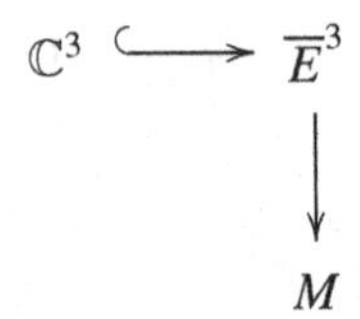

(f) Structure group $SU(3)$, with the trivial representation

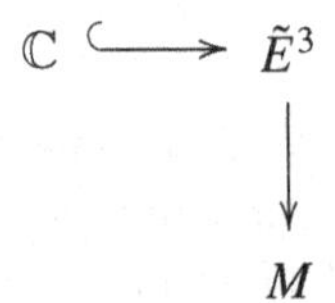

The Hermitian structure on each of these bundles is pretty obvious.

(iii) *The matter bundles*: These are constructed from the above auxiliary vector bundles by suitable tensor products:

(a) Left-quark doublet bundle $\mathcal{Q}_L = E^1 \otimes E^2 \otimes E^3 \otimes E$. Here the representation is $\mathbf{1} \otimes \mathbf{2} \otimes \mathbf{3}$, and the vector bundle is

$$\begin{array}{ccc} \mathbb{C} \otimes \mathbb{C}^2 \otimes \mathbb{C}^3 \otimes \mathbb{R}^4 & \hookrightarrow & \mathcal{Q}_L \\ & & \downarrow \\ & & M \end{array}$$

(b) Up-quark singlet bundle $\mathcal{U}_R = E^1 \otimes \tilde{E}^2 \otimes \overline{E}^3 \otimes E$. Here, the representation is $\mathbf{1} \otimes \mathbf{1} \otimes \overline{\mathbf{3}}$, and the vector bundle is

$$\begin{array}{ccc} \mathbb{C} \otimes \mathbb{C} \otimes \mathbb{C}^3 \otimes \mathbb{R}^4 & \hookrightarrow & \mathcal{U}_R \\ & & \downarrow \\ & & M \end{array}$$

(c) Down-quark singlet bundle $\mathcal{D}_R = E^1 \otimes \tilde{E}^2 \otimes \overline{E}^3 \otimes E$. Here, the representation is $\mathbf{1} \otimes \mathbf{1} \otimes \overline{\mathbf{3}}$, and the vector bundle is

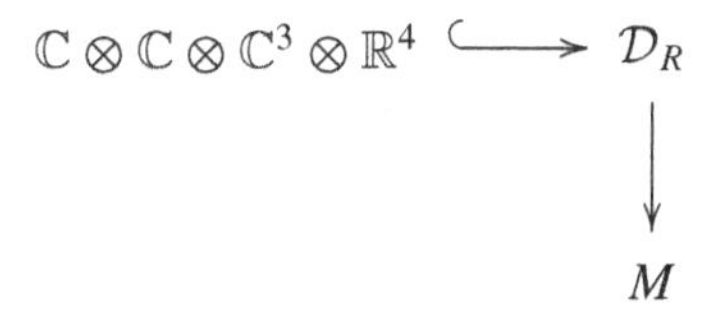

(d) Left-lepton doublet bundle $\mathcal{L}_L = E^1 \otimes E^2 \otimes \tilde{E}^3 \otimes E$. Here, the representation is $\mathbf{1} \otimes \mathbf{2} \otimes \mathbf{1}$, and the vector bundle is

$$\begin{array}{ccc} \mathbb{C} \otimes \mathbb{C}^2 \otimes \mathbb{C} \otimes \mathbb{R}^4 & \hookrightarrow & \mathcal{L}_L \\ & & \downarrow \\ & & M \end{array}$$

(e) Right-electron singlet bundle $\mathcal{E}_R = E^1 \otimes \tilde{E}^2 \otimes \tilde{E}^3 \otimes E$. Here, the representation is $\mathbf{1} \otimes \mathbf{1} \otimes \mathbf{1}$, and the vector bundle is

$$\begin{array}{ccc} \mathbb{C} \otimes \mathbb{C} \otimes \mathbb{C} \otimes \mathbb{R}^4 & \hookrightarrow & \mathcal{E}_R \\ & & \downarrow \\ & & M \end{array}$$

(f) Higgs bundle $\mathcal{H} = E^1 \otimes E^2 \otimes \tilde{E}^3 \otimes E$. Here, the representation is $\mathbf{1} \otimes \mathbf{2} \otimes \mathbf{1}$, and the vector bundle is

$$\begin{array}{ccc} \mathbb{C} \otimes \mathbb{C}^2 \otimes \mathbb{C} \otimes \mathbb{R}^4 & \hookrightarrow & \mathcal{H} \\ & & \downarrow \\ & & M \end{array}$$

(iv) *Charge conjugation*: On each building-block vector bundle described above, we can define an anti-linear bundle involution. Taken together, these yield a *charge conjugation operator* on the space of fields.

(v) *Kobayashi–Maskawa matrix*: As we saw, there are three generations of quarks and leptons. The coupling of the Higgs field with the matter fields mixes these three generations. The way the mixing is accomplished is through the so-called Kobayashi–Maskawa matrix (also called the Cabibbo–Kobayashi–Maskawa matrix).

(vi) *Higgs morphism*: To incorporate the symmetry breaking mechanism that assigns masses to leptons and quarks, we define a bundle morphism

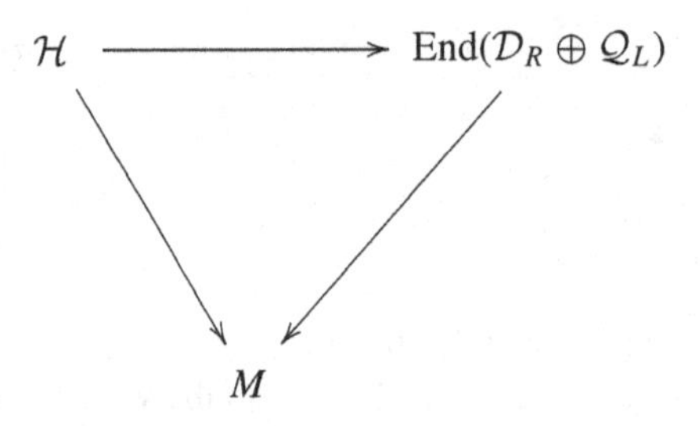

In coordinates, its expression is

$$(x, h) \mapsto (x, T_{x,h}),$$

where

$$T_{x,h} : \begin{pmatrix} d_R \\ q_L \end{pmatrix} \mapsto \begin{pmatrix} 0 \\ h d_R \end{pmatrix}.$$

One easily checks (exercise) that this definition does not depend on the choice of trivialization.

One should also take into account that there are three generations of leptons and quarks. With these ingredients in place, one can define the Lagrangian of the Standard Model intrinsically as a map

$$\mathcal{L}_{SM} : \mathcal{D} \to \Gamma(M \times \mathbb{R}),$$

where the domain of $\mathcal{L}_{SM}$, namely $\mathcal{D}$, is the product space

$$\Gamma(\mathcal{C}) \times \prod_{i=1}^{3} \Gamma(\mathcal{Q}_L^i) \times \prod_{i=1}^{3} \Gamma(\mathcal{U}_R^i) \times \prod_{i=1}^{3} \Gamma(\mathcal{D}_R^i)$$
$$\times \prod_{i=1}^{3} \Gamma(\mathcal{L}_L^i) \times \prod_{i=1}^{3} \Gamma(\mathcal{E}_R^i) \times \Gamma(\mathcal{H}).$$

The definitions have been set up so that the formula (9.22) now makes sense. Note that $\Gamma(M \times \mathbb{R}) \equiv C^\infty(M)$, so the Lagrangian evaluated at a 17-tuple of fields produces a function over M that can therefore be integrated over M to yield a number (the value of the *action* at that 17-tuple of fields). This Lagrangian is, by construction, invariant under the gauge group $G = U(1) \times SU(2) \times SU(3)$ (it is also Lorentz invariant, so the gauge group can be enlarged by taking the product of G with the Lorentz group), and it respects charge conjugation.

Exercises

9.1 Give a detailed proof of the $U(1) \times SU(2)$ gauge invariance of the electroweak Lagrangian.

9.2 Do the same for the $U(1) \times SU(2) \times SU(3)$ gauge invariance of the Standard Model Lagrangian.

9.3 Using the Euler–Lagrangian equations for the action functional associated to the QCD Lagrangian, show that the components of the gluon field strength tensor satisfy the equations

$$\partial_\mu G^{a\,\mu\nu} = j^{a\,\nu} ,$$

where the current components are given by

$$j^{a\,\nu} = g_3 \left[f^a_{bc} G^b_\mu G^{c\,\mu\nu}_\mu + \sum_f \overline{q_f} \gamma^\nu \frac{1}{2} \lambda^a q_f \right] .$$

Here, f^a_{bc} are the structure constants of the group $SU(3)$.

9.4 Prove that the definition of the Higgs morphism given in Section 9.6 is indeed independent of the choices of trivializations of the bundles involved.

Appendix A
Hilbert spaces and operators

In this Appendix, we present a crash course in functional analysis on Hilbert spaces. We limit our presentation to those facts and results that are essential for the proper foundations of quantum theory. These results include the spectral theorem for unbounded, self-adjoint operators, the functional calculus for such operators, Stone's theorem, and the Kato–Rellich theorem, all of which are presented here with complete proofs. The literature on Hilbert space theory is incredibly vast. Among the references we found most useful are [RS1], [RS2], [Th], [L], and [AJP].

A.1 Hilbert spaces

We start with the definition of inner product space, or pre-Hilbert space. We are interested only in *complex* Hilbert spaces.

Definition A.1 *A (complex)* inner product space $(V, \langle\cdot,\cdot\rangle)$ *consists of a (complex) vector space V together with a map $\langle\cdot,\cdot\rangle : V \times V \to \mathbb{C}$, called an* inner product, *satisfying*

(i) $\langle v, v\rangle \geq 0$ *for all* $v \in V$, *and* $\langle v, v\rangle = 0$ *if and only if* $v = 0$;

(ii) $\langle\cdot,\cdot\rangle$ *is bilinear and skew-symmetric:*

- $\langle \alpha u + \beta v, w\rangle = \alpha \langle u, w\rangle + \beta \langle v, w\rangle$ *for all* $u, v, w \in V$ *and all* $\alpha, \beta \in \mathbb{C}$;
- $\langle v, w\rangle = \overline{\langle w, v\rangle}$ *for all* $v, w \in V$.

An inner product on V induces a norm on V, namely $\|v\| = \sqrt{\langle v, v\rangle}$, and this norm gives rise, of course, to a metric: $d(v, w) = \|v - w\|$.

Definition A.2 *A* Hilbert space $\mathcal{H}$ *is a complex inner-product space that is* complete *with respect to the metric induced by its inner product.*

Example A.3 *Let (X, μ) be a measure space, and consider $L^2_\mu(X)$, the complex vector space of all measurable functions $f : X \to \mathbb{C}$ that are square-integrable, i.e. satisfy*

$$\int_X |f|^2 \, d\mu < \infty .$$

We regard two functions in $L^2_\mu(X)$ that agree μ-almost everywhere as equal. Define an inner product in $L^2_\mu(X)$ as follows:

$$\langle f, g \rangle = \int_X f \bar{g} \, d\mu .$$

Then $L^2_\mu(X)$, endowed with this inner product, is a Hilbert space (exercise).

Example A.4 *Let I be a (finite or infinite) index set, and let $\ell^2(I) \subset \mathbb{C}^I$ denote the complex vector space of all sequences $x = (x_i)_{i \in I}$ such that*

$$\sum_{i \in I} |x_i|^2 < \infty .$$

It is implicit that convergence of the series entails in particular that all but at most countably many terms are non-zero. Then $\ell^2(I)$ is an inner-product space under the inner product

$$\langle x, y \rangle = \sum_{i \in I} x_i \overline{y_i} .$$

This inner product is complete (exercise), so $\ell^2(I)$ is a Hilbert space. When I is finite, say with cardinality n, then $\ell^2(I) \equiv \mathbb{C}^n$.

A.2 Linear operators

Let us introduce the basics on linear functionals and operators in Hilbert spaces.

Let $T : \mathcal{H} \to \mathcal{H}'$ be a linear operator between Hilbert spaces. If $\mathcal{H}' = \mathbb{C}$, we call T a *linear functional*. As usual in linear algebra, we denote by $\ker T$ the *kernel* of T, i.e. the subspace of $\mathcal{H}$ consisting of all $\xi \in \mathcal{H}$ such that $T(\xi) = 0$. The operator T is said to be *bounded* if $\|T(\xi)\| \leq C\|\xi\|$ for all $\xi \in \mathcal{H}$, for some constant $C > 0$. The smallest such constant is called the *operator norm* of T, denoted $\|T\|$. A linear operator T is bounded if and only if it is continuous. The kernel $\ker T$ of a bounded operator T is a closed subspace of $\mathcal{H}$. We are interested primarily in the case of operators mapping a Hilbert space $\mathcal{H}$ into itself. The space of all such bounded linear operators $T : \mathcal{H} \to \mathcal{H}$ will be denoted $B(\mathcal{H})$. Endowed with the operator norm, $B(\mathcal{H})$ is a *Banach* space (a complete normed linear space).

Let us first deal with linear functionals. The fundamental fact about bounded linear functionals is known as *Riesz's representation lemma*. In the proof, we will need the notion of the *orthogonal complement* of a given subspace $V \subset \mathcal{H}$. This is simply the subspace $V^\perp = \{\xi \in \mathcal{H} : \langle \xi, v \rangle = 0 \text{ for all } v \in V\}$.

Lemma A.5 (Riesz) *If $\phi : \mathcal{H} \to \mathbb{C}$ is a bounded linear functional, then there exists a unique $\eta \in \mathcal{H}$ such that $\phi(\xi) = \langle \xi, \eta \rangle$ for all $\xi \in \mathcal{H}$.*

Proof. Let $V = \ker \phi$. Because ϕ is bounded, V is closed in $\mathcal{H}$. If $V = \mathcal{H}$, then $\phi = 0$ and there is nothing to prove. If $V \neq \mathcal{H}$, then $V^\perp$ is one-dimensional. Indeed, if $u, v \in V^\perp$ are both non-zero, then

$$w = u - \frac{\phi(u)}{\phi(v)} v$$

is in $V^\perp$ and satisfies $\phi(w) = 0$, so $w \in V \cap V^\perp = \{0\}$. Now let $u \in V^\perp$ be such that $\phi(u) = 1$ and take $\eta = \|u\|^{-2} u \in V^\perp$, so that

$$\phi(\eta) = \frac{\phi(u)}{\|u\|^2} = \|\eta\|^2 .$$

Let $L : \mathcal{H} \to \mathbb{C}$ be the linear functional given by $L(\xi) = \phi(\xi) - \langle \xi, \eta \rangle$. If $\xi \in V$, then $\xi \perp \eta$, and therefore $L(\xi) = 0$. If instead $\xi \in V^\perp$, then $\xi = \lambda \eta$ for some $\lambda \in \mathbb{C}$ and

$$L(\xi) = \lambda \phi(\eta) - \lambda \langle \eta, \eta \rangle = \lambda(\phi(\eta) - \|\eta\|^2) = 0 .$$

Because $\mathcal{H} = V \oplus V^\perp$, this shows that $L \equiv 0$, and we are done. □

The Riesz lemma has many applications. Here is one of the most basic. Let $T : \mathcal{H} \to \mathcal{H}$ be a bounded linear operator, and let $\eta \in \mathcal{H}$. Then the correspondence $\xi \mapsto \langle T\xi, \eta \rangle$ defines a bounded linear functional on $\mathcal{H}$. By the Riesz lemma, there exists a unique element $T^*\eta \in \mathcal{H}$ such that $\langle T\xi, \eta \rangle = \langle \xi, T^*\eta \rangle$ for all $\xi \in \mathcal{H}$. The map $T : \eta \mapsto T^*\eta$ defines a bounded linear operator on $\mathcal{H}$, called the *adjoint* of T.

Definition A.6 *Several types of operators occur naturally in applications.*

(i) An operator $T \in B(\mathcal{H})$ is said to be self-adjoint *if $T = T^*$.*

(ii) An operator $T \in B(\mathcal{H})$ is an isometry *if $\langle T\xi, T\eta \rangle = \langle \xi, \eta \rangle$ for all $\xi, \eta \in \mathcal{H}$.*

(iii) An operator $U \in B(\mathcal{H})$ is said to be unitary *if $UU^* = U^*U = I$ (I being the identity operator).*

It is an easy exercise to prove that $U \in B(\mathcal{H})$ is unitary if and only if U is a *surjective* isometry. In particular, a simple example of an isometry that is *not* unitary is provided by the *right shift operator* $T : \ell^2(\mathbb{N}) \to \ell^2(\mathbb{N})$ given by $T(x_1, x_2, \ldots) = (0, x_1, x_2, \ldots)$, which is clearly not surjective.

A.3 Spectral theorem for compact operators

A linear operator $T : \mathcal{H} \to \mathcal{H}$ on a Hilbert space $\mathcal{H}$ is *compact* if it maps bounded sets onto relatively compact sets. Equivalently, T is compact if the image of the unit ball is relatively compact in $\mathcal{H}$.

Example A.7 *Here is a prototypical example. Let $\mathcal{H}$ be a separable Hilbert space, and let $\{e_n\}_{n\geq 1}$ be an orthonormal basis for $\mathcal{H}$. Given a sequence $\{\lambda_n\}_{n\geq 1}$ of real numbers, say with $|\lambda_n| \geq |\lambda_{n+1}|$ for all n and $\lambda_n \to 0$ as $n \to \infty$, let $T : \mathcal{H} \to \mathcal{H}$ be the linear operator with $T(e_n) = \lambda_n e_n$ for all n (and extended by linearity). Then T is compact and self-adjoint. The proof is left as an exercise.*

The spectral theorem for compact, self-adjoint operators states that the above example is the *only* example up to unitary equivalence. Before we carefully state and prove this result, a definition and an auxiliary lemma are in order.

We define the *spectrum* of $T : \mathcal{H} \to \mathcal{H}$ to be the set

$$\sigma(T) = \{\lambda \in \mathbb{C} : \ker(T - \lambda I) \neq 0\} .$$

The spectrum of a compact operator is always non-empty, as the following result shows.

Lemma A.8 *If $T : \mathcal{H} \to \mathcal{H}$ is compact and self-adjoint, then T has an eigenvalue.*

Proof. We shall prove that either $\|T\|$ or $-\|T\|$ is an eigenvalue of T. We assume that $\|T\| \neq 0$; otherwise there is nothing to prove.

First we claim that $\ker(T^2 - \|T\|^2 I) \neq 0$. To see why, let (u_n) be a sequence of unit vectors such that $\|Tu_n\| \to \|T\|$. Because T is compact, we may assume also that $Tu_n \to w$, for some $w \in \mathcal{H}$. Now we have, as $n \to \infty$,

$$\begin{aligned} \|T^2 u_n - \|T\|^2 u_n\|^2 &= \|T^2 u_n\|^2 - 2\|T\|^2 \|Tu_n\|^2 + \|T\|^4 \\ &\leq 2\|T\|^4 - 2\|T\|^2\|Tu_n\|^2 \to 0 . \end{aligned} \tag{A.1}$$

Note that self-adjointness was used in (A.1). But $T^2 u_n \to Tw$, because T is continuous. Because (A.1) implies

$$\|u_n - \|T\|^{-2} T^2 u_n\| \to 0$$

as $n \to \infty$, it follows that $u_n \to v = \|T\|^{-2} Tw$. Therefore $T^2 v - \|T\|^2 v = 0$, and this proves our claim.

The rest is now very easy, because

$$0 = (T^2 - \|T\|^2 I)v = (T - \|T\| I)(T + \|T\| I)v$$

implies that *either* $(T + \|T\| I)v = 0$, in which case v is an eigenvector with eigenvalue $-\|T\|$, *or* $w = (T + \|T\| I)v \neq 0$, in which case w is an eigenvector with eigenvalue $\|T\|$. □

We are now ready for the first version of the spectral theorem.

Theorem A.9 *Let $T : \mathcal{H} \to \mathcal{H}$ be compact and self-adjoint. Then*

(i) the spectrum $\sigma(T)$ is at most countable, and contained in $\mathbb{R}$;
(ii) the subspaces $\ker(T - \lambda I) \subset \mathcal{H}$ with $\lambda \in \sigma(T)$ are pairwise orthogonal;
(iii) we have $\mathcal{H} = \bigoplus_{\lambda \in \sigma(T)} \ker(T - \lambda I)$;
(iv) for each $\epsilon > 0$, $\sum_{\lambda \in \sigma(T),\, |\lambda| \geq \epsilon} \dim \ker(T - \lambda I) < \infty$.

Proof. First note that, if $\lambda \in \sigma(T)$, then λ is real, because T is self-adjoint. Next, if $\lambda, \mu \in \sigma(T)$ are distinct, then, given $v \in \ker(T - \lambda I)$ and $w \in \ker(T - \mu I)$, we have

$$\begin{aligned} \langle Tv, w\rangle = \langle v, Tw\rangle &\Rightarrow \langle \lambda v, w\rangle = \langle v, \mu w\rangle \\ &\Rightarrow (\lambda - \mu)\langle v, w\rangle = 0 \\ &\Rightarrow \langle v, w\rangle = 0\,. \end{aligned}$$

This proves (ii).

Now, for each $\lambda \in \sigma(T)$, let $\mathcal{H}_\lambda = \ker(T - \lambda I)$. Note that, whenever $\lambda \in \sigma(T)$ is $\neq 0$, $\mathcal{H}_\lambda$ must be finite-dimensional. Indeed, the restriction $T|_{\mathcal{H}_\lambda}$ maps $\mathcal{H}_\lambda$ into itself, and because $Tv = \lambda v$ for each $v \in \mathcal{H}_\lambda$, this restriction is in fact a multiple of the identity, which can never be compact unless $n_\lambda = \dim \mathcal{H}_\lambda$ is finite.

But more is true. If $\epsilon > 0$ is given, let

$$\mathcal{H}^\epsilon = \bigoplus_{\lambda \in \sigma(T),\, |\lambda| \geq \epsilon} \mathcal{H}_\lambda\,.$$

Then $T(\mathcal{H}^\epsilon) \subset \mathcal{H}^\epsilon$. Let $e_\lambda \in \mathcal{H}_\lambda$ be a unit vector, for each $\lambda \in \sigma(T)$. Then $\{Te_\lambda : |\lambda| \geq \epsilon\}$ is a relatively compact set, so we can find a sequence (λ_n) with $|\lambda_n| \geq \epsilon$ such that (Te_{λ_n}) converges. But if $\lambda \neq \mu$ are any two elements of the spectrum, then

$$\|Te_\lambda - Te_\mu\|^2 = \|\lambda e_\lambda - \mu e_\mu\|^2 = \lambda^2 + \mu^2 \geq 2\epsilon^2 > 0\,,$$

which is a contradiction unless $\mathcal{H}^\epsilon$ is finite-dimensional. This proves (iv), and it also proves (i) (why?).

Finally, let $V = \bigoplus_{\lambda \in \sigma(T)} \mathcal{H}_\lambda$. Then V is a closed subspace, and it is T-invariant. Hence, so is its orthogonal complement. The operator $T|_{V^\perp} : V^\perp \to V^\perp$ is compact and self-adjoint. But every such operator must have an eigenvalue, by Lemma A.8. This contradicts the very definition of V, unless $V = \mathcal{H}$. This proves (ii). □

This theorem implies that every compact, self-adjoint operator can be *diagonalized*. More precisely, we have the following result.

Corollary A.10 *A compact, self-adjoint operator $T : \mathcal{H} \to \mathcal{H}$ admits an orthonormal basis of eigenvectors.*

Proof. For each $\lambda \in \mathbb{C}$ such that $\mathcal{H}_\lambda = \ker(T - \lambda I) \neq 0$, let B_λ be an orthonormal basis of $\mathcal{H}_\lambda$. Each $e \in B_\lambda$ is an eigenvector with eigenvalue λ, and $\cup B_\lambda$ is a basis for $\mathcal{H}$ because, as we have seen, $\mathcal{H} = \bigoplus \mathcal{H}_\lambda$. □

Remark A.11 If $\{e_n\}_{n\geq 1}$ is a basis of eigenvectors for T, with $Te_n = \lambda_n e_n$ for all n ($\lambda_n \in \mathbb{R}$), then for each $v = \sum a_n e_n \in \mathcal{H}$ we have $Tv = \sum a_n \lambda_n e_n$. Because T is compact, we know that $|\lambda_n| \to 0$ as $n \to \infty$. This justifies the remark we made immediately after Example A.3.

Remark A.12 The function $\varphi_T : \lambda \mapsto \dim \ker(T - \lambda I)$ characterizes T completely up to unitary equivalence. In other words, if $T_i : \mathcal{H}_i \to \mathcal{H}_i$ ($i = 1, 2$) are both compact and self-adjoint, then T_1 is unitarily equivalent to T_2 if and only if $\varphi_{T_1} = \varphi_{T_2}$. The proof is left as an exercise.

A.4 Spectral theorem for normal operators

A bounded operator $T : \mathcal{H} \to \mathcal{H}$ is said to be *normal* if it commutes with its adjoint: $TT^* = T^*T$. This wide class of operators includes of course all self-adjoint operators, as well as all unitary operators. It also includes all multiplication operators, when $\mathcal{H} = L^2_\mu(X)$ (where (X, μ) is a finite measure space): given any bounded measurable function $g : X \to \mathbb{C}$, the operator $M_g : L^2_\mu(X) \to L^2_\mu(X)$ given by $M_g\psi = g \cdot \psi$ is bounded, and $M_g^* = M_{\bar{g}}$; in particular, $M_g^* M_g = M_{\bar{g}g} = M_{g\bar{g}} = M_g M_g^*$, so M_g is normal.

As it turns out, multiplication operators as above are the only normal operators up to unitary equivalence. This is the content of the following spectral theorem.

Theorem A.13 (spectral theorem) *Let $T : \mathcal{H} \to \mathcal{H}$ be a bounded, normal operator on a separable Hilbert space. Then T is unitarily equivalent to a multiplication operator. In other words, there exist a σ-finite measure space*

(X, μ), a bounded measurable function $g : X \to \mathbb{C}$, and a unitary isometry $U : \mathcal{H} \to L^2_\mu(X)$ such that the diagram

$$\begin{array}{ccc} \mathcal{H} & \xrightarrow{T} & \mathcal{H} \\ {\scriptstyle U}\downarrow & & \downarrow{\scriptstyle U} \\ L^2_\mu(X) & \xrightarrow[M_g]{} & L^2_\mu(X) \end{array}$$

commutes (i.e. $T = U^* M_g U$).

The proof of this theorem will be given in Appendix B, after we talk about C^* algebras and their representations. As we shall see there, the above theorem holds true for any finite set $T_1, \ldots, T_n$ of commuting normal operators; in this case, the same unitary isometry U conjugates each T_i to a multiplication operator $M_{g_i} \in B(L^2_\mu(X))$, where $g_i : X \to \mathbb{C}$ is bounded and measurable.

A.5 Spectral theorem for unbounded operators

A.5.1 Unbounded operators

The vast majority of operators (in Hilbert space) that appear in applications ranging from classical to quantum physics are *unbounded* operators. These operators are usually only densely defined. Let us formulate the appropriate definitions.

We consider operators of the form $T : V \to \mathcal{H}$, where $V \subset \mathcal{H}$ is a (not necessarily closed) linear subspace, which are *linear*. We say that T is *densely defined* if V is dense in $\mathcal{H}$. We say that T is *closed* if the graph $\mathrm{Gr}(T) = \{(v, Tv) : v \in V\}$ is a closed subspace of $\mathcal{H} \oplus \mathcal{H}$.

Example A.14 *Let (X, μ) be a measure space, and let $g : X \to \mathbb{C}$ be a measurable function. Define $V_g = \{\varphi \in L^2_\mu(X) : g\varphi \in L^2_\mu(X)\}$. This is a dense subspace of $L^2_\mu(X)$. The multiplication operator $M_g : V_g \to L^2_\mu(X)$ given by $M_g(\varphi) = g\varphi$ is therefore densely defined, and it is also closed. The details are left as an exercise.*

We shall frequently write $T : \mathrm{dom}\,(T) \to \mathcal{H}$, sometimes also $(T, \mathrm{dom}\,(T))$, when we want to specify a linear operator T; it is implicit here that $\mathrm{dom}\,(T)$ is the *domain* of T. The *image* of T will be denoted by $\mathrm{ran}\,(T)$, as is customary among functional analysts. The *graph norm* on $\mathrm{dom}\,(T)$ is the norm defined by

$$\|v\|_T^2 = \|v\|^2 + \|Tv\|^2$$

for all $v \in \mathrm{dom}\,(T)$. That this is indeed a norm, deriving from an inner product, is left as an exercise. One can show that the operator T is closed if and only if the graph norm is complete for $\mathrm{dom}\,(T)$ (i.e. iff $\mathrm{dom}\,(T)$ is a Hilbert space under this norm).

Let us now define the adjoint of a densely defined operator $T : \mathrm{dom}\,(T) \to \mathcal{H}$. First, we define its domain $\mathrm{dom}\,(T^*)$ to be the set of all $\eta \in \mathcal{H}$ such that the correspondence $\xi \mapsto \langle T\xi, \eta\rangle$ extends to a bounded linear functional $\varphi_\eta : \mathcal{H} \to \mathbb{C}$. This extension is unique because $\mathrm{dom}\,(T)$ is dense in $\mathcal{H}$. Then $\mathrm{dom}\,(T^*) \subset \mathcal{H}$ is a linear subspace, and by Riesz's representation theorem, there exists $v_\eta \in \mathcal{H}$ such that $\varphi_\eta(\xi) = \langle \xi, v_\eta\rangle$. We define $T^* : \mathrm{dom}\,(T^*) \to \mathcal{H}$ by $T^*(\eta) = v_\eta$. It is clear that T^* is linear and satisfies $\langle T\xi, \eta\rangle = \langle \xi, T^*\eta\rangle$ for each $\xi \in \mathrm{dom}\,(T)$ and each $\eta \in \mathrm{dom}\,(T^*)$. The operator T^* is the *adjoint* of T. The reader is invited to prove that T^* is always a *closed* operator.

A few more definitions are in order. Given two linear operators T, S on $\mathcal{H}$, we say that T is an *extension* of S, written $S \subset T$, if $\mathrm{dom}\,(S) \subset \mathrm{dom}\,(T)$ and $S\xi = T\xi$ for all $\xi \in \mathrm{dom}\,(S)$. An operator $T : \mathrm{dom}\,(T) \to \mathcal{H}$ is *symmetric* if for all $\xi\eta \in \mathrm{dom}\,(T)$ we have $\langle T\xi, \eta\rangle = \langle \xi, T\eta\rangle$. It is easy to see (exercise) that T is symmetric iff $T \subset T^*$. We say that T is *self-adjoint* if $T = T^*$; in other words, T is self-adjoint if $\mathrm{dom}\,(T) = \mathrm{dom}\,(T^*)$ and T is symmetric.

A.5.2 The Cayley transform

Let $T : \mathrm{dom}\,(T) \to \mathcal{H}$ be a densely defined, symmetric operator. The Cayley transform of T is the linear operator W_T defined as follows:

(i) $\mathrm{dom}\,(W_T) = \{(T + i)v : v \in \mathrm{dom}\,(T)\}$;

(ii) $W_T((T + i)v) = (T - i)v$, for all $v \in \mathrm{dom}\,(T)$.

To see that W_T is well-defined, note that if $w \in \mathrm{dom}\,(W_T)$ then $w = (T + i)v$ for a unique $v \in \mathrm{dom}\,(T)$. Indeed, if we had $(T + i)v_1 = (T + i)v_2$ then $\xi = v_1 - v_2$ would satisfy $T\xi = -i\xi$, but this is impossible unless $\xi = 0$, because T is symmetric:

$$-i\,\|\xi\|^2 = \langle T\xi, \xi\rangle = \langle \xi, T\xi\rangle = i\,\|\xi\|^2 .$$

Hence W_T is well-defined, and clearly linear. Moreover, an easy calculation yields

$$\|(T \pm i)v\|^2 = \|Tv\|^2 + \|v\|^2$$

for all $v \in \mathrm{dom}\,(T)$. Hence $\|(T + i)v\|^2 = \|(T - i)v\|^2$, and therefore $\|W_T((T + i)v)\| = \|(T + i)v\|$ for all $v \in \mathrm{dom}\,(T)$. This shows that W_T is a *partial isometry*; i.e. it is a linear operator that is an isometry wherever it is defined. Note also that W_T is onto the subspace $\mathrm{ran}\,(W_T) = \{(T - i)v : v \in \mathrm{dom}\,(T)\}$. Thus, we can view the Cayley transform as a correspondence

$\kappa : T \mapsto W_T$ that sends densely defined symmetric operators to partial isometries of $\mathcal{H}$. We shall see that this correspondence is one-to-one and onto. This requires the following three lemmas.

Lemma A.15 *Let $U : \mathrm{dom}\,(U) \to \mathcal{H}$ be a partial isometry such that $I - U$ has a dense image in $\mathcal{H}$. Then*

(i) $I - U : \mathrm{dom}\,(U) \to \mathrm{ran}\,((I - U))$ is bijective;

(ii) the linear operator A_U, defined by taking $\mathrm{dom}\,(A_U) = \{(I - U)\xi : \xi \in \mathrm{dom}\,(U)\}$ and $A_U((I - U)\xi) = i(I + U)\xi$ for all $\xi \in \mathrm{dom}\,(U)$, is symmetric and densely defined.

Proof.

(i) Let $\xi \in \ker(I - U)$; then for each $\eta \in \mathrm{dom}\,(U)$ we have

$$\begin{aligned}\langle \xi, (I - U)\eta \rangle &= \langle \xi, \eta \rangle - \langle \xi, U\eta \rangle \\ &= \langle \xi, \eta \rangle - \langle U\xi, U\eta \rangle \\ &= \langle \xi, \eta \rangle - \langle \xi, \eta \rangle = 0\,.\end{aligned}$$

Hence, $\xi \perp \mathrm{ran}\,(I - U)$, and because $\mathrm{ran}\,(I - U)$ is dense in $\mathcal{H}$, it follows that $\xi = 0$. This shows that $I - U$ is injective (hence a bijection onto its image).

(ii) The operator A_U is clearly densely defined, for its domain is $\mathrm{ran}\,(I - U)$. Now suppose $v = (I - U)\xi \in \mathrm{dom}\,(A_U)$ and $w = (I - U)\eta \in \mathrm{dom}\,(A_U)$. Then on one hand we have

$$\begin{aligned}\langle A_U v, w \rangle &= \langle i(\xi + U\xi), \eta - U\eta \rangle \\ &= i\,[\langle U\xi, \eta \rangle - \langle \xi, U\eta \rangle]\,, \end{aligned} \tag{A.2}$$

and on the other hand

$$\begin{aligned}\langle v, A_U w \rangle &= \langle \xi - U\xi, i(\eta + U\eta) \rangle \\ &= -i\,[-\langle U\xi, \eta \rangle + \langle \xi, U\eta \rangle]\,. \end{aligned} \tag{A.3}$$

From (A.2) and (A.3), it follows that $\langle A_U v, w \rangle = \langle v, A_U w \rangle$, so A_U is symmetric.

□

Lemma A.16 *Let $T : \mathrm{dom}\,(T) \to \mathcal{H}$ be densely defined and symmetric, and let $U : \mathrm{dom}\,(U) \to \mathcal{H}$ be a partial isometry with $\mathrm{ran}\,(I - U)$ dense in $\mathcal{H}$. Then we have $W_{A_U} = U$ and $A_{W_T} = T$.*

Proof. For each $\xi \in \mathrm{dom}\,(A_U)$, we have $W_{A_U}(A_U\xi + i\xi) = A_U\xi - i\xi$. But $\mathrm{dom}\,(A_U) = \mathrm{ran}\,(I - U)$, where $\xi = (I - U)\eta$ for some $\eta \in \mathrm{dom}\,(U)$. This means that

$$A_U\xi - i\xi = i(I+U)\eta + i(I-U)\eta = 2i\eta\ ,$$

and also

$$A_U\xi + i\xi = i(I+U)\eta - i(I-U)\eta = 2iU\eta\ .$$

Therefore, $W_{A_U}(2i\eta) = 2iU\eta$; i.e. $W_{A_U} = U$. This proves the first equality in the statement. The second equality follows from a similar argument. □

Lemma A.17 *Let* $T : \mathrm{dom}\,(T) \to \mathcal{H}$ *be a symmetric operator. Then the following statements are equivalent:*

(i) T *is closed;*
(ii) $\mathrm{ran}\,(T+i)$ *is closed;*
(iii) $\mathrm{ran}\,(T-i)$ *is closed.*

Proof. The equivalence of all three assertions is an easy consequence of the fact that the operators $(\xi, T\xi) \mapsto (T \pm i)\xi$ from $\mathrm{Gr}(T)$ onto $\mathrm{ran}\,(T \pm i)$ are both unitary. The details are left as an exercise. □

Lemma A.18 *Let* $T : \mathrm{dom}\,(T) \to \mathcal{H}$ *be symmetric and densely defined. Let* $N_{\pm} = \mathrm{ran}\,(T \pm i)^{\perp}$. *Then we have*

$$N_+ = \ker(T^* - i) \quad \textit{and} \quad N_- = \ker(T^* + i)\ . \tag{A.4}$$

Moreover, endowing $\mathrm{dom}\,(T^*)$ *with the graph norm, we have the following orthogonal decomposition:*

$$\mathrm{dom}\left(T^*\right) = \overline{\mathrm{dom}\,(T)} \oplus N_+ \oplus N_-\ . \tag{A.5}$$

Proof. Let $\xi \in N_+$, and take any $\eta \in \mathrm{dom}\,(T)$. Then $\langle (T+i)\eta, xi\rangle = 0$, i.e. $\langle T\eta, \xi\rangle + i\,\langle \eta, \xi\rangle = 0$. Hence $\xi \in \mathrm{dom}\,(T^*)$ and $\langle \eta, T^*\xi - i\xi\rangle = 0$, so $(T^* - i)\xi$ is orthogonal to $\mathrm{dom}\,(T)$. Because $\mathrm{dom}\,(T)$ is dense in $\mathcal{H}$, it follows that $(T^* - i)\xi = 0$, i.e. $\xi \in \ker(T^* - i)$. This shows that $N_+ \subset \ker(T^* - i)$. For the reverse inclusion, note that the argument just given is itself reversible. This proves the first equality in (A.4). The proof of the second equality is similar.

To prove (A.5), we first show that N_+ and N_- are orthogonal subspaces, relative to the graph inner product $\langle \cdot, \cdot \rangle$ on $\mathrm{dom}\,(T^*)$. Given $\xi_{\pm} \in N_{\pm}$, we have, using (A.4),

$$\begin{aligned} \langle \xi_+, \xi_- \rangle_{T^*} &= \left\langle T^*\xi_+, T^*\xi_- \right\rangle + \langle \xi_+, \xi_- \rangle \\ &= \langle i\xi_+, -i\xi_- \rangle + \langle \xi_+, \xi_- \rangle \\ &= 0\ . \end{aligned}$$

Next, we show that $N_+ \perp \overline{\mathrm{dom}\,(T)}$. It suffices of course to show that $N_+ \perp \mathrm{dom}\,(T)$. Let $\xi \in N_+ = \ker(T^* - i)$ and $\eta \in \mathrm{dom}\,(T)$. Then, using the fact that T is symmetric, we have

$$\begin{aligned}\langle \xi, \eta \rangle_{T^*} &= \left\langle T^*\xi, T^*\eta \right\rangle + \langle \xi, \eta \rangle = \langle i\xi, T\eta \rangle + \langle \xi, \eta \rangle \\ &= \left\langle iT^*\xi, \eta \right\rangle + \langle \xi, \eta \rangle \\ &= i\left\langle -i\xi + T^*\xi, \eta \right\rangle = 0\,.\end{aligned}$$

The proof that $N_- \perp \overline{\mathrm{dom}\,(T)}$ is entirely similar.

Finally, in order to establish equality in (A.5), it suffices to show that the orthogonal complement of $\mathrm{dom}\,(T)$ in $\mathrm{dom}\,(T^*)$ with respect to the graph inner product falls within $N_+ \oplus N_-$. Let v belong to this orthogonal complement. Then $\langle v, \eta \rangle_{T^*} = 0$ for all $\eta \in \mathrm{dom}\,(T)$; that is

$$\langle v, \eta \rangle + \left\langle T^*v, T^*\eta \right\rangle = 0\,,$$

or (using symmetry)

$$\left\langle v + (T^*)^2 v, \eta \right\rangle = 0\,,$$

for all $\eta \in \mathrm{dom}\,(T)$. Because $\mathrm{dom}\,(T)$ is dense in $\mathcal{H}$, it follows that $(T^*)^2 v = -v$. This can also be written as

$$(T^* + i)(T^* - i)v = (T^* - i)(T^* + i)v = 0\,,$$

and this implies that $(T^* - i)v \in \ker(T^* + i) = N_-$ and that $(T^* + i)v \in \ker(T^* - i) = N_+$. But because

$$v = \frac{1}{2i}\left[(T^* + i)v - (T^* - i)v\right]\,,$$

it follows that $v \in N_+ \oplus N_-$. This completes the proof. □

The following theorem is a key result concerning the Cayley transform and will be used in the proof of the spectral theorem for unbounded self-adjoint operators.

Theorem A.19 *The Cayley transform κ is a bijection between the set of densely defined, symmetric operators $T : \mathrm{dom}\,(T) \to \mathcal{H}$ and the set of partial isometries $U : \mathrm{dom}\,(U) \to \mathcal{H}$ such that $\mathrm{ran}\,(I - U)$ is dense. Moreover, T is self-adjoint iff $W_T = \kappa(T)$ is unitary.*

Proof. Given $T : \mathrm{dom}\,(T) \to \mathcal{H}$ densely defined and symmetric, let $U = W_T = \kappa(T)$. Then $\mathrm{ran}\,(I - U) = \mathrm{dom}\,(T)$. Indeed, $\mathrm{dom}\,(U) = \mathrm{ran}\,(T + i)$, so if $\xi \in \mathrm{dom}\,(U)$ then $\xi = (T + i)\eta$ for some $\eta \in \mathrm{dom}\,(T)$, and therefore

$$(I - U)\xi = (T + i)\eta - W_T(T + i)\eta = (T + i)\eta - (T - i)\eta = 2i\eta \in \mathrm{dom}\,(T)\ .$$

This shows that $\mathrm{ran}\,(I - U) \subset \mathrm{dom}\,(T)$, and the argument is reversible, so $\mathrm{dom}\,(T) \subset \mathrm{ran}\,(I - U)$ as well. Hence, W_T is a densely defined partial isometry with $\mathrm{ran}\,(I - W_T)$ a dense subspace. Applying Lemmas A.15 and A.16, we see at once that $\kappa : T \mapsto W_T$ is surjective and injective (for there we constructed the inverse κ^{-1} quite explicitly). This proves the first assertion of our theorem.

To prove the second assertion, let T be self-adjoint. Then, by (A.5) in Lemma A.18, we have $N_+ = N_- = \{0\}$. Thus, we have that $\mathrm{dom}\,(W_T) = \mathrm{ran}\,(T + i)$ is dense (because its orthogonal complement in $\mathcal{H}$ is N_+) and $\mathrm{ran}\,(W_T) = \mathrm{ran}\,(T - i)$ is also dense (because its orthogonal complement in $\mathcal{H}$ is N_-). But because T is self-adjoint, T is closed, and therefore $\mathrm{ran}\,(T \pm i) = \mathcal{H}$ (we are using Lemma A.17 here), which shows that W_T is unitary. Conversely, if W_T is unitary then $\mathrm{ran}\,(T \pm i) = \mathcal{H}$. Hence, T is closed (again by Lemma A.17) and we have $N_+ = N_- = \{0\}$. Now Lemma A.18 tells us that $\mathrm{dom}\,(T^*) = \overline{\mathrm{dom}\,(T)} = \mathrm{dom}\,(T)$. This proves that T is self-adjoint, and we are done. □

Let us now examine Cayley transforms of multiplication operators. We consider a measure space (X, μ) and for each measurable function $\varphi : X \to \mathbb{C}$ we consider the multiplication operator $M_\varphi : D_\varphi \to L^2_\mu(X)$ given by $M_\varphi(f) = \varphi f$, where $D_\varphi = \{f \in L^2_\mu(X) : \varphi f \in L^2_\mu(X)\}$.

Lemma A.20 *Let $\varphi : X \to \mathbb{R}$ be a real-valued measurable function. Then the Cayley transform of M_φ is M_ψ, where $\psi = (\varphi - i)(\varphi + i)^{-1}$. Moreover, M_φ is self-adjoint.*

Proof. For each $\xi \in D_\varphi$ we have

$$W_{M_\varphi} : (M_\varphi + i)\xi \ \mapsto\ (M_\varphi - i)\xi\ .$$

Because $|\varphi(x) + i)^{-1}| \leq 1$ for all $x \in X$, it follows that $(W_{M_\varphi}) = L^2_\mu(X)$ (check!) and in addition we have, for all $\eta \in (W_{M_\varphi})$,

$$\begin{aligned} W_{M_\varphi}\eta &= W_{M_\varphi}\left[(\varphi + i)(\varphi + i)^{-1}\eta\right] \\ &= (\varphi - i)(\varphi + i)^{-1}\eta = \psi\eta \\ &= M_\psi\eta\ . \end{aligned}$$

Hence, $W_{M_\varphi} = M_\psi$ as asserted. Now, because

$$|\psi(x)| = |\varphi - i| \cdot |\varphi + i|^{-1} = 1$$

for all $x \in X$, we see that ψ maps X into the unit circle. This shows that $\sigma(M_\psi) \subset \mathbb{T}^1$, and therefore M_ψ is unitary. Because $M_\psi = \kappa(M_\varphi)$, it follows from Theorem A.19 that M_φ is self-adjoint. □

A.5.3 Unitary equivalence

Just as in the case of bounded operators, the suitable notion of equivalence between densely defined, unbounded operators is *unitary* equivalence. Given Hilbert spaces $\mathcal{H}$, $\mathcal{G}$ and two operators $T : \operatorname{dom}(T) \to \mathcal{H}$ and $S : \operatorname{dom}(S) \to \mathcal{G}$, with $\operatorname{dom}(T) \subset \mathcal{H}$ dense in $\mathcal{H}$ and $\operatorname{dom}(S) \subset \mathcal{G}$ dense in $\mathcal{G}$, we say that T and S are *unitarily equivalent* if there exists a unitary isometry $U : \mathcal{H} \to \mathcal{G}$ such that $UT\xi = SU\xi$ for all $\xi \in \operatorname{dom}(T)$. The isometry U is called by functional analysts an *intertwining operator* between T and S.[1]

Lemma A.21 *Let $T_1 : \operatorname{dom}(T_1) \to \mathcal{H}_1$ and $T_2 : \operatorname{dom}(T_2) \to \mathcal{H}_2$ be densely defined, symmetric operators. Then T_1 and T_2 are unitarily equivalent if and only if W_{T_1} and W_{T_2} are unitarily equivalent.*

Proof. Let $U : \mathcal{H}_1 \to \mathcal{H}_2$ be a unitary operator intertwining W_{T_1} and W_{T_2}. Let us write $W_i = W_{T_i}$ for simplicity of notation. Using Lemma A.15 we see that

$$U(\operatorname{dom}(T_1)) = U(\operatorname{ran}(I - W_1)) = \operatorname{ran}(I - W_2) = \operatorname{dom}(T_2) \ .$$

Moreover, using the fact that $\operatorname{ran}(I - W_1)$ is dense in $\mathcal{H}_1$ (again by Lemma A.15), we see that $UT_1 = T_2U$, because

$$\begin{aligned} UT_1(I - W_1)\xi &= U(i(I + W_1)\xi) = i(U + W_2U)\xi \\ &= i(I + W_2)U\xi = T_2(I - W_2)U\xi \\ &= T_2U(I - W_1)\xi \ . \end{aligned}$$

This shows that U intertwines T_1 and T_2. The converse is easier, and is left as an exercise. □

A.5.4 The spectral theorem

We are finally in a position to state and prove the spectral theorem for unbounded self-adjoint operators.

Theorem A.22 *Let $T : \operatorname{dom}(T) \to \mathcal{H}$ be a densely defined operator. Then T is self-adjoint if and only if T is unitarily equivalent to a multiplication operator M_φ, where φ is a measurable and real-valued function on some measure space.*

Proof. Let $T : \operatorname{dom}(T) \to \mathcal{H}$ be self-adjoint. Then its Cayley transform W_T is unitary, hence normal, and so by the spectral theorem for *bounded normal*

[1] Dynamicists (such as the authors of this book) prefer the term *conjugacy*.

operators (Theorem A.13), we know that there exist a measure space (X, μ) and a measurable function $\psi : X \to \mathbb{T}^1$ such that W_T is unitarily equivalent to $M_\psi : L^2_\mu(X) \to L^2_\mu(X)$. In other words, there exists a unitary operator $U : \mathcal{H} \to L^2_\mu(X)$ such that $UW_T = M_\psi U$. We claim that ψ cannot be equal to 1 on a set of positive μ-measure. If $E \subset X$ were such a set, then we would have $(I - M_\psi)\mathbf{1}_E = 0$ (check!). However, $I - W_T$ is injective, so $I - M_\psi$ is injective as well. This contradiction shows that $\psi \neq 1$ μ-almost everywhere. Hence we can define $\varphi : X \to \mathbb{R}$ by

$$\varphi(x) = i(1 + \psi(x))(1 - \psi(x))^{-1} .$$

This is a well-defined measurable function, and applying Lemma A.20 we easily see that $W_{M_\varphi} = M_\psi$. Thus, we now know that U intertwines W_T and W_{M_φ}. Therefore, by Lemma A.21, it follows that T and M_φ are unitarily equivalent. The converse is much easier and is left as an exercise. □

A.6 Functional calculus

In this section, we develop a version of the so-called *functional calculus* for unbounded, self-adjoint operators as a consequence of the spectral theorem. Given a metric space X, we denote by $BM(X)$ the space of all bounded measurable functions $X \to \mathbb{C}$. Endowed with the sup-norm $\|\cdot\|_\infty$, $BM(X)$ is a Banach space, and in fact a Banach algebra (see Appendix B). The goal of functional calculus, for a self-adjoint operator T in a Hilbert space $\mathcal{H}$, is to make sense of $f(T)$ for all $f \in BM(\mathbb{R})$. This is tantamount to finding a special type of representation of $BM(\mathbb{R})$ into $B(\mathcal{H})$ called a *spectral homomorphism.*

Definition A.23 *A* spectral homomorphism *is a map $\pi : BM(X) \to B(\mathcal{H})$ with the following properties:*

(i) π is a continuous representation of the Banach algebra $BM(X)$;

(ii) for each $\xi \in \mathcal{H}$, the function $\nu_\xi : \mathcal{B}_X \to \mathbb{C}$ given by $\nu_\xi(E) = \langle \pi(\mathbf{1}_E)\xi, \xi\rangle$ defines a complex measure on X (here $\mathcal{B}_X$ denotes the Borel σ-algebra of X).

Given a representation $\pi_0 : C(X) \to B(\mathcal{H})$, let us agree to call a vector $\xi \in \mathcal{H}$ *cyclic* if $\{\pi_0(f)\xi : f \in C(X)\}$ is *dense* in $\mathcal{H}$. If a representation has a cyclic vector, we call it *cyclic*. Cyclic representations are obviously *irreducible*. As it turns out, every representation of $C(X)$ into $B(\mathcal{H})$ can be written as a direct sum of cyclic representations. This fact, which we simply assume here, will be proved later (Appendix B) in the much more general context of C^* algebras.

The important abstract fact about spectral homomorphisms is the following.

Theorem A.24 *Let X be a compact metric space, and let $\pi_0 : C(X) \to B(\mathcal{H})$ be a representation. Then there exists a unique spectral homomorphism $\pi : BM(X) \to B(\mathcal{H})$ extending π_0 (i.e. such that $\pi|_{C(X)} \equiv \pi_0$).*

Proof. Due to the observation just preceding the statement, it suffices to present the proof when π_0 has a cyclic vector $\xi \in \mathcal{H}$. The correspondence $f \mapsto \langle \pi_0(f)\xi, \xi \rangle$ defines a complex linear functional on $C(X)$. By the Riesz–Markov theorem, there exists a finite, complex Borel measure μ on X such that

$$\langle \pi_0(f)\xi, \xi \rangle = \int_X f \, d\mu$$

for all $f \in C(X)$. Let us consider the Hilbert space $L^2_\mu(X)$, of which $C(X)$ is a dense subspace. Given $f, g \in C(X) \subset L^2_\mu(X)$, we have

$$\begin{aligned} \langle f, g \rangle &= \int_X \bar{g} f \, d\mu = \langle \pi_0(\bar{g} f)\xi, \xi \rangle \\ &= \langle \pi_0(\bar{g})\pi_0(f)\xi, \xi \rangle \\ &= \langle \pi_0(f)\xi, \pi_0(g)\xi \rangle \ . \end{aligned}$$

This shows that the map $f \mapsto \pi_0(f)$ extends continuously to a linear isometry $U : L^2_\mu(X) \to \mathcal{H}$. This isometry is *onto* $\mathcal{H}$, because ξ is cyclic. Hence, U is unitary. This unitary map gives rise to an isometry $\widehat{U} : B(\mathcal{H}) \to B(L^2_\mu(X))$ given by $\widehat{U}(T) = U^{-1}TU$. Now, $\widehat{U}$ conjugates the representation π_0 to a representation $\tilde{\pi}_0 : C(X) \to B(L^2_\mu(X))$. In fact, for each $f \in C(X)$ and each $\varphi \in L^2_\mu(X)$, we have

$$\tilde{\pi}_0(f)\varphi = U^{-1}\pi_0(f)U(\varphi) = U^{-1}(\pi_0(f\varphi)\xi) = f\varphi = M_f\varphi \ .$$

Hence, $\tilde{\pi}_0(f) = M_f$, a multiplication operator, for all $f \in C(X)$. But this has an obvious extension to a representation $\tilde{\pi} : BM(X) \to B(L^2_\mu(X))$: one simply defines $\tilde{\pi}(f) = M_f$ for all $f \in BM(X)$. Finally, let $\pi : BM(X) \to B(\mathcal{H})$ be given by $\pi(f) = U\tilde{\pi}(f)U^{-1}$. This clearly extends π, and one verifies at once that it enjoys properties (i) and (ii) of Definition A.23. Thus, existence of π is established. Uniqueness is in fact easier to prove; it is left as an exercise. □

Remark A.25 This theorem remains true if we replace $C(X)$ with $C_0(X)$, the space of continuous functions vanishing at ∞ on a locally compact space X.

Combining this remark, the above theorem, and the spectral theorem for unbounded self-adjoint operators, we arrive at the following version of functional calculus.

Theorem A.26 *Let $T : \mathrm{dom}(T) \to \mathcal{H}$ be a self-adjoint operator. Then there exists a unique spectral homomorphism $\pi : BM(\mathbb{R}) \to B(\mathcal{H})$ such that $\pi((t+i)^{-1}) = (T+i)^{-1}$.*

Proof. By the spectral theorem A.22, T is unitarily equivalent to $M_\varphi : f \mapsto \varphi f$, where $\varphi : X \to \mathbb{R}$ is measurable ((X, μ) some measure space). Hence we may assume, in fact, that $T = M_\varphi$. Let $\pi : BM(\mathbb{R}) \to B(L^2_\mu(X))$ be given by $\pi(\psi) = M_{\psi\circ\varphi}$. This is well defined, because $\psi \circ \varphi$ is bounded and measurable, for each $\psi \in BM(\mathbb{R})$. We clearly have

$$\begin{aligned}\pi(\psi_1\psi_2) &= M_{(\psi_1\psi_2)\circ\varphi} = M_{(\psi_1\circ\varphi)\cdot(\psi_2\circ\varphi)} \\ &= M_{\psi_1\circ\varphi} \cdot M_{\psi_2\circ\varphi} = \pi(\psi_1)\pi(\psi_2)\,.\end{aligned}$$

One also shows quite easily that $\pi(\psi_1 + \psi_2) = \pi(\psi_1) + \pi(\psi_2)$. Thus, π is a spectral homomorphism. Moreover, letting $\psi(t) = (t+i)^{-1}$, we have $\psi \circ \varphi = (\varphi + i)^{-1}$, and therefore

$$\pi((t+i)^{-1}) = M_{(\varphi+i)^{-1}} = (M_\varphi + i)^{-1} = (T+i)^{-1}\,.$$

This establishes the *existence* of π with the desired properties.

Now for *uniqueness*. Suppose π_1, π_2 are spectral homomorphisms into $B(\mathcal{H}_1)$ and $B(\mathcal{H}_2)$ respectively and that

$$\pi_1((t+i)^{-1}) = (T+i)^{-1} = \pi_2((t+i)^{-1})\,.$$

Let $\mathcal{A} \subset BM(\mathbb{R})$ be the algebra of all $f \in BM(\mathbb{R})$ such that $\pi_1(f) = \pi_2(f)$. The sub-algebra of $\mathcal{A}$ generated by the constants and $(t+i)^{-1} \in \mathcal{A}$ separates points, because $(t+i)^{-1}$ already does, and it is also closed under conjugation. By the Stone–Weierstrass theorem, this sub-algebra is dense in $C_0(\mathbb{R})$, and therefore $\mathcal{A} \supset C_0(\mathbb{R})$. By Theorem A.24 and the remark following it, $\pi_0 = \pi_1|_{C_0(\mathbb{R})} = \pi_2|_{C_0(\mathbb{R})}$ has a unique extension to $BM(\mathbb{R})$. But then it follows that $\pi_1 = \pi_2$. □

A.7 Essential self-adjointness

Let $T : \mathrm{dom}(T) \to \mathcal{H}$ be a densely defined operator. We say that T is *essentially self-adjoint* if T is symmetric and has a unique self-adjoint extension. Equivalently, T is essentially self-adjoint if it is symmetric and $\bar{T}$ is self-adjoint. The following result gives us a useful criterion for essential self-adjointness.

Theorem A.27 *Let $T : \mathrm{dom}(T) \to \mathcal{H}$ be a symmetric, densely defined operator. Then the following are equivalent:*

(i) *T is essentially self-adjoint;*
(ii) $\mathrm{ran}\,(T \pm i) \subset \mathcal{H}$ *are dense subspaces;*
(iii) $\ker(T^* \pm i) = \{0\}$.

Proof. Recall from Lemma A.18 that

$$\mathrm{dom}\,\left(T^*\right) = \overline{\mathrm{dom}\,(T)} \oplus N_+ \oplus N_- \tag{A.6}$$

(the closure being with respect to the graph norm of $\mathrm{dom}\,(T^*)$), where

$$N_\pm = \mathrm{ran}\,(T \pm i)^\perp = \ker(T^* \pm i)\,. \tag{A.7}$$

The equivalence between (i) and (ii) follows easily from (A.6), whereas the equivalence between (ii) and (iii) is immediate from (A.7). □

The following example is of fundamental importance in quantum mechanics.

Example A.28 (the Laplace operator) *Perhaps the most important example of a self-adjoint operator is the Laplacian, $L = -\Delta$. The minus sign is chosen because in this way L is a positive operator. We define this operator on Euclidean d-dimensional space $\mathbb{R}^d$ as follows. First we look at $L = -\Delta$ on the Schwarz space $\mathcal{S}(\mathbb{R}^d) \subset L^2(\mathbb{R}^d)$, a dense subspace of $L^2(\mathbb{R}^d)$, defining it directly by the formula*

$$-\Delta\varphi = -\sum_{j=1}^{d} \frac{\partial^2 \varphi}{\partial x_j^2}\,, \quad \textit{for all } \varphi \in C_0^\infty(\mathbb{R}^d)\,. \tag{A.8}$$

Let us verify directly that $-\Delta$ is essentially self-adjoint on $\mathcal{S}(\mathbb{R}^d)$. To do this, we use the Fourier transform $\mathcal{F} : L^2(\mathbb{R}^d) \to L^2(\mathbb{R}^d)$ given by

$$(\mathcal{F}\varphi)(\xi) = \frac{1}{(2\pi)^{n/2}} \int_{\mathbb{R}^d} \varphi(x) e^{-i\langle \xi, x\rangle}\,dx\,.$$

We assume that the reader knows the basic properties of the Fourier transform, among them (i) the fact that $\mathcal{F}$ is a unitary isometry; (ii) the fact that $\mathcal{F}(\partial_{x_j}\varphi) = -i\xi_j\mathcal{F}(\varphi)$; (iii) the fact that $\mathcal{F}$ maps $\mathcal{S}(\mathbb{R}^d)$ into itself.

The fact that $-\Delta$ is symmetric is a consequence of one of Green's identities, and is left as an exercise. Hence, in order to show that $-\Delta$ is essentially self-adjoint, it suffices to show, by Theorem A.27 above, that $\mathrm{ran}\,(-\Delta \pm i) \subset L^2(\mathbb{R}^d)$ *are dense subspaces. Let us show that* $\mathrm{ran}\,(-\Delta + i) \supset \mathcal{S}(\mathbb{R}^d)$. *Given $f \in \mathcal{S}(\mathbb{R}^d)$, we need to solve the PDE*

$$(-\Delta + i)\varphi = f\,. \tag{A.9}$$

The Fourier transform is tailor-made for such problems! Applying it to both sides of (A.9)*, we get*

$$(-|\xi|^2 + i)\hat{\varphi}(\xi) = \hat{f}(\xi) \quad (\xi \in \mathbb{R}^n)\,,$$

where the circumflex denotes the Fourier transform. Hence we have

$$\hat{\varphi}(\xi) = \frac{\hat{f}(\xi)}{-|\xi|^2 + i} \in \mathcal{S}(\mathbb{R}^d)\,,$$

and by Fourier inversion we deduce that

$$\varphi(x) = \frac{1}{(2\pi)^{n/2}} \int_{\mathbb{R}^d} \frac{\hat{f}(\xi) e^{i\langle x,\xi\rangle}}{-|\xi|^2 + i}\, d\xi \ \in \mathcal{S}(\mathbb{R}^d)$$

is the desired solution. This shows that ran $(-\Delta + i)$ *is dense, and the proof that* ran $(-\Delta - i)$ *is dense is the same. Thus, the Laplacian is essentially self-adjoint on* $\mathcal{S}(\mathbb{R}^d)$*, as claimed.*

In fact, the unique self-adjoint extension of $-\Delta$ *can be defined directly via the Fourier transform in the following way. Let* $P : \mathbb{R}^d \to \mathbb{R}$ *be the polynomial* $P(\xi) = -\sum_{j=1}^{d} \xi_j^2$ *(the symbol of* $-\Delta$*), and consider the multiplication operator* $M_P : \varphi \mapsto P\varphi$*, with domain* $\mathrm{dom}\,(M_P) = \{\varphi \in L^2(\mathbb{R}^d) :\ P\varphi \in L^2(\mathbb{R}^d)\}$. *Note that* $\mathrm{dom}\,(M_P)$ *contains* $\mathcal{S}(\mathbb{R}^d)$ *as a dense subspace. We let* $\mathrm{dom}\,(-\Delta) = \mathcal{F}^{-1}(\mathrm{dom}\,(M_P))$ *and define the extension*

$$-\Delta :\ \mathrm{dom}\,(-\Delta) \to L^2(\mathbb{R}^d)$$

by $-\Delta = \mathcal{F}^{-1} \circ M_P \circ \mathcal{F}$*. Because* M_P *is self-adjoint in its domain* $\mathrm{dom}\,(M_P)$ *and* $\mathcal{F}$ *is unitary, it follows that* $-\Delta$ *is self-adjoint on* $\mathrm{dom}\,(-\Delta) \supset \mathcal{S}(\mathbb{R}^d)$.

A.8 A note on the spectrum

The spectrum of an operator $T : \mathrm{dom}\,(T) \to \mathcal{H}$ is the set $\sigma(T)$ of all $\lambda \in \mathbb{C}$ such that $T - \lambda : \mathrm{dom}\,(T) \to \mathcal{H}$ does not have a bounded inverse. The complement of $\sigma(T)$ in $\mathbb{C}$ is called the *resolvent set* of T, and for each λ in the resolvent set, $(T - \lambda)^{-1}$ is called the *resolvent* of T. Certainly every eigenvalue of T is in the spectrum, but not every element of $\sigma(T)$ needs to be an eigenvalue; in fact, T may have no eigenvalues at all!

We distinguish two subsets of the spectrum whose elements exhibit very different behavior. If $\lambda \in \sigma(T)$, then we call $\dim\ker(T - \lambda)$ the *multiplicity* of λ. We define the *point spectrum* $\sigma_p(T)$ to be

$$\sigma_p(T) = \{\lambda \in \sigma(T) :\ \lambda \text{ is isolated and has finite multiplicity}\}\,.$$

We also define the *continuous spectrum* (also called the *essential spectrum*) of T to be the set $\sigma_c(T)$ whose elements are *approximate eigenvalues* of T in the following sense. An element $\lambda \in \sigma(T)$ is an approximate eigenvalue if there exists a sequence $\xi_n \in \mathrm{dom}\,(T)$ with $\|\xi_n\| = 1$ such that (i) $\|(T - \lambda)\xi_n\| \to 0$ as $n \to \infty$, and (ξ_n) converges weakly to 0, i.e. $\langle \xi_n, \eta \rangle \to 0$ as $n \to \infty$, for each $\eta \in \mathcal{H}$. It turns out that for self-adjoint operators the point and continuous spectra are a dichotomy.

Theorem A.29 (Weyl) *If $T : \mathrm{dom}\,(T) \to \mathcal{H}$ is a self-adjoint operator, then $\sigma(T) = \sigma_p(T) \cup \sigma_c(T)$*

For a proof of this result, see [RS1]. We note that the spectrum, point spectrum, and continuous spectrum of an operator $T : \mathrm{dom}\,(T) \to \mathcal{H}$ are all invariant under conjugacies by unitary isometries; in other words, if $U \in B(\mathcal{H})$ is unitary and $S = U^{-1}TU$, then $\sigma_p(S) = \sigma_p(T)$, $\sigma_c(S) = \sigma_c(T)$, and $\sigma(S) = \sigma(T)$.

A.9 Stone's theorem

A major application of the functional calculus developed in Section A.6 is a proof of Stone's theorem. Stone's theorem states that every *evolution group* has an *infinitesimal generator*. As we saw in Chapter 2, this theorem is crucial in the Heisenberg formulation of quantum mechanics.

Definition A.30 *An* evolution group *on a Hilbert space $\mathcal{H}$ is a one-parameter family $(U_t)_{t\in\mathbb{R}}$ of unitary operators $U_t : \mathcal{H} \to \mathcal{H}$ such that*

(i) $U_{t+s} = U_t U_s$ for all $t, s \in \mathbb{R}$;

(ii) $t \mapsto U_t$ is strongly continuous.

As the following result shows, every self-adjoint operator in Hilbert space gives rise to an evolution group by exponentiation.

Theorem A.31 *Let $A : \mathrm{dom}\,(A) \to \mathcal{H}$ be a self-adjoint operator. Then $U_t = \exp itA$, given by functional calculus applied to $f_t(x) = e^{itx}$, is an evolution group. Moreover,*

(i) $U_t(\mathrm{dom}\,(A)) \subset \mathrm{dom}\,(A)$ for all $t \in \mathbb{R}$;

(ii) for each $\xi \in \mathrm{dom}\,(A)$ we have

$$\lim_{t\to 0} \frac{1}{t}(U_t\xi - \xi) = iA\xi \; ; \tag{A.10}$$

(iii) conversely, if $\xi \in \mathcal{H}$ is such that the limit in the left-hand side of (A.10) exists, then $\xi \in \mathrm{dom}\,(A)$.

Proof. The fact that (U_t) is well-defined and a one-parameter group is an easy consequence of the properties of $x \mapsto e^{itx}$. Property (ii) follows from functional calculus applied to the functions $x \mapsto (e^{itx} - 1)/t$ (for $t \neq 0$) and $x \mapsto ix$ (for $t = 0$). In order to prove (iii), we define an operator B in $\mathcal{H}$ as follows. First we take

$$\operatorname{dom}(B) = \left\{ \xi \in \mathcal{H} : \lim_{t \to 0} \frac{1}{t}(U_t \xi - \xi) \text{ exists} \right\} .$$

Then we let, for each $\xi \in \operatorname{dom}(B)$,

$$B\xi = \lim_{t \to 0} \frac{1}{t}(U_t \xi - \xi) . \tag{A.11}$$

One easily checks that $-iB$ is a symmetric operator, and that $-iB \supset A$ (exercise). But because $A = A^*$, we know that A is closed, and therefore we have $iA = B$. Now (i) follows as well, for if $\xi \in \operatorname{dom}(A)$ and $s \in \mathbb{R}$, then $U_s \xi$ will belong to $\operatorname{dom}(A)$ provided we can show that

$$\lim_{t \to 0} \frac{1}{t}(U_t(U_s \xi) - U_s \xi)$$

exists; but because U_s is continuous, we have

$$\begin{aligned} \lim_{t \to 0} \frac{1}{t}(U_t U_s \xi - U_s \xi) &= U_s \left[\lim_{t \to 0} \frac{1}{t}(U_t \xi - \xi) \right] \\ &= U_s(iA\xi) . \end{aligned}$$

This finishes the proof. □

Remark A.32 The operator $B = iA$ is called the *infinitesimal generator* of $(U_t)_{t \in \mathbb{R}}$. In general, given an evolution group $(U_t)_{t \in \mathbb{R}}$, we define its infinitesimal generator B to be the operator defined as the limit (A.11).

Stone's theorem is the *converse* of Theorem A.31.

Theorem A.33 (Stone) *Let $(U_t)_{t \in \mathbb{R}}$ be an evolution group in Hilbert space $\mathcal{H}$. Then there exists a unique self-adjoint operator $A : \operatorname{dom}(A) \to \mathcal{H}$ such that $B = iA$ is the infinitesimal generator of $(U_t)_{t \in \mathbb{R}}$.*

Proof. We shall define A on a certain dense subspace of $\mathcal{H}$, prove that A is essentially self-adjoint, and finally verify that $U_t = e^{it\bar{A}}$.

For each $\varphi \in C_0^\infty(\mathbb{R})$ and each $\xi \in \mathcal{H}$, let

$$\xi_\varphi = \int_{\mathbb{R}} \varphi(t) U_t \xi \, dt \in \mathcal{H} .$$

We define $D = \{\xi_\varphi : \varphi \in C_0^\infty(\mathbb{R}),\ \xi \in \mathcal{H}\}$. Then $D \subset \mathcal{H}$ is a linear subspace. We claim that D is dense in $\mathcal{H}$. To see why, let $\xi \in \mathcal{H}$ be arbitrary, and for each $n \geq 1$ let $\varphi_n \in C_0^\infty(\mathbb{R})$ be such that $\operatorname{supp}\varphi_n \subset [-n^{-1}, n^{-1}]$, $\varphi_n \geq 0$, and $\int_{\mathbb{R}} \varphi_n = 1$. Then we have

$$\begin{aligned}\|\xi_{\varphi_n} - \xi\| &= \left\| \int_{\mathbb{R}} \varphi_n(t)(U_t\xi - \xi)\,dt \right\| \\ &\leq \sup_{|t|\leq n^{-1}} \|U_t\xi - \xi\| \to 0\end{aligned}$$

as $n \to \infty$, so $\xi_{\varphi_n} \to \xi$ and the claim is proved.

Next, we define $A : D \to \mathcal{H}$ as follows. If $\eta = \xi_\varphi \in D$, let

$$\begin{aligned}A\eta &= -i \lim_{s\to 0} \frac{1}{s}(U_s\eta - \eta) \\ &= -i \lim_{s\to 0} \frac{1}{s} \int_{\mathbb{R}} \varphi(t)[U_{t+s}\xi - U_t\xi]\,dt \\ &= -i \lim_{s\to 0} \int_{\mathbb{R}} \frac{\varphi(\tau - s) - \varphi(\tau)}{s} U_\tau\xi\,d\tau \\ &= -i \int_{\mathbb{R}} \varphi'(\tau)\,U_\tau\xi\,d\tau\,,\end{aligned}$$

where in the last step we have used Lebesgue's dominated convergence theorem. It is clear that A defined in this way is linear. We claim that A is essentially self-adjoint. The fact that A is symmetric is an easy exercise. Thus, by Theorem A.27, it suffices to show that $\ker(A^* \pm i) = \{0\}$. Let $\xi \in \ker(A^* - i)$. Because for all $\eta \in D$ we have $U_t\eta \in D$ for all t, we see that

$$\frac{d}{dt}\langle \xi, U_t\eta\rangle = \langle \xi, iAU_t\eta\rangle = -i\left\langle A^*\xi, U_t\eta\right\rangle = \langle \xi, U_t\eta\rangle\ . \tag{A.12}$$

Solving the resulting elementary ODE, we get $\langle \xi, U_t\eta\rangle = \langle \xi, \eta\rangle\, e^t$ for all $t \in \mathbb{R}$. But because

$$|\langle \xi, U_t\eta\rangle| \leq \|\xi\|\,\|U_t\eta\| = \|\xi\|\,\|U_t\eta\| < \infty\,,$$

we see that (A.12) is possible iff $\langle \xi, \eta\rangle = 0$. Since D is dense in $\mathcal{H}$, it follows that $\xi = 0$. The proof that $\ker(A^* + i) = \{0\}$ is entirely analogous. This establishes the claim that A is essentially self-adjoint. Therefore the *closure* $\bar{A}$ is self-adjoint.

The final step of the proof is to show that $U_t = e^{it\bar{A}}$ for all $t \in \mathbb{R}$. Let $\xi \in D \subset \bar{D} = \operatorname{dom}\left(\bar{A}\right)$. Then on the one hand, by Theorem A.31 (i) and (ii), we have

$$e^{it\bar{A}}\xi \in D \quad \text{and} \quad \frac{d}{dt}(e^{it\bar{A}}\xi) = iAe^{it\bar{A}}\xi\ . \tag{A.13}$$

On the other hand,

$$\frac{d}{dt}(U_t\xi) = iAU_t\xi \ . \tag{A.14}$$

Writing $\psi(t) = U_t\xi - e^{it\bar{A}}\xi$ and using (A.13) and (A.14), we see that

$$\psi'(t) = iAU_t\xi - iAe^{it\bar{A}}\xi = iA\psi(t) \ .$$

Therefore, because $\bar{A}$ is self-adjoint, we have

$$\begin{aligned}\frac{d}{dt}\langle\psi(t),\psi(t)\rangle &= \langle\psi'(t),\psi(t)\rangle + \langle\psi(t),\psi'(t)\rangle \\ &= i\langle\bar{A}\psi(t),\psi(t)\rangle - i\langle\psi(t),\bar{A}\psi(t)\rangle \\ &= 0\ .\end{aligned}$$

This shows that $\|\psi(t)\|^2 =$ constant. But $\psi(0) = 0$, so $\psi(t) = 0$ for all t. Hence, $U_t\xi = e^{it\bar{A}}\xi$ for all t, and the proof is complete. □

Remark A.34 Stone's theorem characterizes one-parameter *groups*, and as we saw in Chapter 2, it is essential in the study of the dynamical evolution of *closed* quantum systems with a finite number of particles. It is, however, insufficient for the study of the dynamics of *open* quantum systems (such as a quantum gas in the grand canonical ensemble). For the study of such open systems, one needs results about one-parameter *semigroups* of operators, the most fundamental of which is the so-called *Hille–Yoshida theorem*. See [AJP] for more on this subject.

A.10 The Kato–Rellich theorem

As we saw in Chapter 2, the time evolution of a non-relativistic quantum system having a fixed number of particles is determined by the Schrödinger operator $H = -\Delta + V$ in $L^2(\mathbb{R}^d)$, where V is the interacting potential. The Schrödinger operator yields a unitary group in Hilbert space, via Stone's theorem, *provided we know that H is self-adjoint*. Thus, we need good criteria for self-adjointness.

The simplest such criterion is provided by the *Kato–Rellich* theorem. In a nutsehll, this theorem states that a small linear and symmetric perturbation of a self-adjoint operator is still self-adjoint. As we saw in Section A.7, the Laplacian $-\Delta$ is self-adjoint (in the Sobolev space $H^2(\mathbb{R}^d)$, say). We can regard the potential V as a kind of perturbation and deduce the required self-adjointness of H if the right conditions are met.

Before we can give the statement (and proof) of the Kato–Rellich theorem, we need a definition. Let $A : \text{dom}\,(A) \to \mathcal{H}$ be a self-adjoint operator, and let

$B : \mathrm{dom}(A) \to \mathcal{H}$ be a linear operator. We call B an A-*bounded* operator if there exist constants $\alpha, \beta > 0$ such that

$$\|B\xi\|^2 \leq \alpha\|A\xi\|^2 + \beta\|\xi\|^2 , \tag{A.15}$$

for all $\xi \in \mathrm{dom}(A)$. The infimum over all $\alpha > 0$ with the property that (A.15) holds true for some $\beta > 0$ (and all $\xi \in \mathrm{dom}(A)$) is called the A-*norm* of B, and it is denoted $N_A(B)$.

Theorem A.35 (Kato–Rellich) *Let* $A : \mathrm{dom}(A) \to \mathcal{H}$ *be a self-adjoint operator, and let* $B : \mathrm{dom}(A) \to \mathcal{H}$ *be an* A*-bounded symmetric operator with* $N_A(B) < 1$*. Then* $T = A + B : \mathrm{dom}(A) \to \mathcal{H}$ *is self-adjoint.*

Proof. Because B is symmetric, it is clear that $A + B$ is symmetric. Hence, it suffices to show (see Theorem A.27) that there exists $\lambda > 0$ such that the operators $A + B \pm \lambda i : \mathrm{dom}(A) \to \mathcal{H}$ are both surjective. Because $N_A(B) < 1$, we know that there exist $0 < \alpha < 1$ and $\beta > 0$ such that, for all $\xi \in \mathrm{dom}(A)$, we have

$$\begin{aligned}\|B\xi\|^2 &\leq \alpha\|A\xi\|^2 + \beta\|\xi\|^2 = \alpha\left(\|A\xi\|^2 + \beta\alpha^{-1}\|\xi\|^2\right) \\ &= \alpha\left\|(A \pm (\beta\alpha^{-1})^{\frac{1}{2}} i)\right\|^2 .\end{aligned} \tag{A.16}$$

Let us then take $\lambda = (\beta\alpha^{-1})^{\frac{1}{2}}$. Because A is self-adjoint, the operators $A \pm \lambda i$ are both invertible. Hence, writing $\xi = (A \pm \lambda i)^{-1}\eta$ for $\eta \in \mathcal{H}$ and plugging it in (A.16), we see that

$$\|B(A \pm \lambda i)^{-1}\eta\|^2 \leq \alpha\|\eta\|^2 .$$

Because this holds for all η, we deduce that $B(A \pm \lambda i)^{-1}$ are operators with norm $\leq \alpha < 1$. The usual Neumann series trick (geometric series) now tells us that $I + B(A \pm \lambda i)^{-1}$ are *invertible* operators in $\mathcal{H}$, with bounded inverses. Therefore, because

$$A + B \pm \lambda i = \left(I + B(A \pm \lambda i)^{-1}\right)(A \pm \lambda i) ,$$

we see that the operators $A + B \pm \lambda i : \mathrm{dom}(A) \to \mathcal{H}$ are both *bijective*. This proves that $A + B$ is self-adjoint. □

Exercises

A.1 Show that an inner-product space is a Hilbert space iff every bounded linear functional on $\mathcal{H}$ is representable (in the sense of Riesz).

A.2 Let $T \in B(\mathcal{H})$. Prove that $\|T\|^2 = \|T^*T\|$.

A.3 Let $S, T \in B(\mathcal{H})$ be self-adjoint operators. Show that ST is self-adjoint iff $ST = TS$.

A.4 Show that the linear operator of Example A.3 is indeed compact and self-adjoint as claimed.

A.5 Let $T : \mathcal{H} \to \mathcal{H}$ be a compact operator. Prove that T^* is compact.

A.6 Show that if $T : \mathcal{H} \to \mathcal{H}$ is Hermitian then $T + i$ is self-adjoint.

A.7 Let $T : \mathrm{dom}\,(T) \to \mathcal{H}$ be a linear operator in Hilbert space. Show that the graph norm of $\mathrm{dom}\,(T)$ derives from an inner product. Show also that this norm is complete if and only if T is closed.

A.8 Let $T : \mathrm{dom}\,(T) \to \mathcal{H}$ be a linear operator, and let $J : \mathcal{H} \oplus \mathcal{H} \to \mathcal{H} \oplus \mathcal{H}$ be the operator $J(\xi, \eta) = (-\eta, \xi)$.
(a) Show that $\mathrm{Gr}(T^*) = J(\mathrm{Gr}(T))^{\perp}$.
(b) Deduce that T^* is closed.

A.9 Supply the details of the proof of Lemma A.17.

A.10 Show that, as claimed in the first paragraph of Section A.7, a densely defined operator $T : \mathrm{dom}\,(T) \to \mathcal{H}$ is essentially self-adjoint if and only if T is symmetric and $\bar{T}$ is self-adjoint.

A.11 Using Green's second identity,

$$\int_{B_R} (u\Delta v - v\Delta u)\, dx = \int_{\partial B_R} \left(u\frac{\partial v}{\partial n} - v\frac{\partial u}{\partial n} \right) d\sigma(x)\,,$$

applied to $u, v \in \mathcal{S}(\mathbb{R}^d)$ and to the ball $B_R \subset \mathbb{R}^d$ of radius R centered at 0, and letting $R \to \infty$, prove that $\langle -\Delta u, v \rangle = \langle u, -\Delta v \rangle$; in other words, that $-\Delta$ is a symmetric operator on the Schwarz space $\mathcal{S}(\mathbb{R}^d)$.

A.12 Show that the spectrum of $-\Delta$ is equal to $[0, \infty)$.

For the following two exercises, we need a definition. Let $T : \mathrm{dom}\,(T) \to L^2(\mathbb{R}^d)$ be a linear operator and let $\lambda \in \mathbb{C}$. A sequence $\{\psi_n\}$ in $L^2(\mathbb{R}^d)$ is *spreading* for (T, λ) if (i) $\|\psi_n\| = 1$ for all n; (ii) $\mathrm{supp}(\psi_n) \subset \mathbb{R}^d$ is compact, and moves off to infinity as $n \to \infty$; and (iii) $\|(T - \lambda)\psi_n\| \to 0$ as $n \to \infty$.

A.13 Show that if $T : \mathrm{dom}\,(T) \to L^2(\mathbb{R}^d)$ and $\lambda \in \sigma_c(T)$ (the continuous spectrum of T), then (T, λ) has a spreading sequence.

A.14 Let $H = -\Delta + V$ be a Schrödinger operator on $L^2(\mathbb{R}^3)$, where $V : \mathbb{R}^3 \to \mathbb{R}$ is a *confining* potential, i.e. V is continuous and non-negative and satisfies $V(x) \to \infty$ as $|x| \to \infty$. One knows that H is self-adjoint on its maximal domain (the Sobolev space $H^2(\mathbb{R}^3)$). The purpose of this exercise is to show that H has *discrete* spectrum.
(a) Let $\lambda \in \sigma_c(H)$ and let $\{\psi_n\}$ be a spreading sequence for (H, λ). Show that $\langle \psi_n, (H - \lambda)\psi_n \rangle \to 0$ as $n \to \infty$.
(b) Show that

$$\langle \psi_n, (H - \lambda)\psi_n \rangle = \int_{\mathbb{R}^3} |\nabla \psi_n|^2\, dx + \int_{\mathbb{R}^3} V|\psi_n|^2\, dx - \lambda\,.$$

(c) Show that (b) contradicts (a), whence $\sigma_c(H) = \emptyset$.

(d) Deduce from Weyl's theorem that the spectrum of the Schrödinger operator H is discrete.

(e) If $\sigma(H) = \{\lambda_1, \lambda_2, \ldots\}$ with $|\lambda_n| \leq |\lambda_{n+1}|$, show that $|\lambda_n| \to \infty$ as $n \to \infty$.

A.15 Recall the following fact, which is at the root of Heisenberg's uncertainty principle (see Chapter 2). If A, B are two self-adjoint operators in Hilbert space and $\psi \in \mathrm{dom}\,(A) \cap \mathrm{dom}\,(B)$, then

$$\langle \psi, i[A, B]\psi \rangle = -2\mathrm{Im}\,\langle A\psi, B\psi \rangle \,.$$

This exercise outlines the proof of a refined version of the uncertainty principle for the Laplace operator, namely that on $L^2(\mathbb{R}^d)$ we have

$$-\Delta \geq \frac{(d-2)^2}{4|x|^2} \,. \tag{EA.1}$$

Let us denote by P_j the jth momentum operator $\psi \mapsto -i\hbar\partial_j\psi$, and as usual by M_φ the multiplication operator $\psi \mapsto \varphi\psi$. We leave to the reader the task of identifying suitable domains of self-adjointness for the operators appearing in this problem.

(a) Prove that, for each $\psi \in \mathcal{S}(\mathbb{R}^d)$,

$$\sum_{j=1}^{d} \|P_j\psi\|^2 = \hbar^2 \langle \psi, -\Delta\psi \rangle \,.$$

(b) Prove that

$$\sum_{j=1}^{d} i \left[M_{|x|^{-1}} P_j M_{|x|^{-1}} \,,\, M_{x_j} \right] = d\hbar M_{|x|^{-2}} \,.$$

(c) Deduce from (b) and the Heisenberg principle that

$$d\hbar \|M_{|x|^{-1}}\psi\|^2 = -2\,\mathrm{Im} \sum_{j=1}^{d} \left\langle M_{|x|^{-1}} P_j M_{|x|^{-1}}\psi, M_{x_j}\psi \right\rangle \,.$$

(d) Applying the identity $P_j M_\varphi = M_{P_j\varphi} + M_\varphi P_j$ (with $\varphi = |x|^{-1}$) to (c), deduce that

$$(d-2)\hbar \|M_{|x|^{-1}}\psi\|^2 = -2\,\mathrm{Im} \sum_{j=1}^{d} \left\langle P_j\psi, M_{x_j|x|^{-2}}\psi \right\rangle \,.$$

(e) Using (a) and applying the Cauchy–Schwarz inequality (twice), show that

$$\mathrm{Im} \sum_{j=1}^{d} \left| \left\langle P_j\psi, M_{x_j|x|^{-2}}\psi \right\rangle \right| \leq \hbar \,\langle \psi, -\Delta\psi \rangle^{1/2} \,\|M_{|x|^{-1}}\psi\| \,.$$

(f) Deduce from (d) and (e) that

$$\langle \psi, -\Delta\psi \rangle \geq \frac{(d-2)^2}{4} \|M_{|x|^{-1}}\psi\|^2 = \frac{(d-2)^2}{4} \left\langle \psi, M_{|x|^{-2}}\psi \right\rangle \,,$$

and verify that this is the exact meaning of (EA.1).

A.16 *Stability of matter.* Apply the result of the previous exercise to prove that the hydrogen atom is stable, as follows. The Schrödinger operator for the hydrogen atom is

$$H = -\frac{\hbar^2}{2m}\Delta - \frac{e^2}{|x|}\,.$$

Here, e denotes the charge of the electron, m its mass, $\hbar$ is Planck's constant, and the term $V(x) = -e^2/|x|$ is Coulomb's potential.

(a) Using the previous exercise, show that

$$H \geq \frac{\hbar^2}{8m|x|^2} - \frac{e^2}{|x|}\,.$$

(b) Analyze the expression on the right-hand side and deduce that in fact

$$H \geq -\frac{2me^4}{\hbar^2}\,.$$

(c) Verify that this means that the energy of the hydrogen atom is bounded from below.

In particular, the electron cannot fall onto the nucleus, and the atom is stable. This was one of the early triumphs of quantum mechanics.

Appendix B

C^* algebras and spectral theory

As we saw in Chapter 2, the language of C^* algebras provides a natural framework for laying down the mathematical foundations of quantum mechanics. This appendix is dedicated to the elementary theory of such algebras. In particular, we shall present a complete proof of the spectral theorem for bounded self-adjoint (or normal) operators in Hilbert space. This is but one of many important applications of this beautiful and powerful theory. We shall also attempt to explain in a nutshell how *nets* of C^* algebras can be used to build the foundations of quantum field theory.

B.1 Banach algebras

A *normed algebra* is an algebra $\mathcal{A}$ over $K = \mathbb{R}$ or $\mathbb{C}$ that has a norm $|\cdot|$ satisfying the usual properties:

(i) $|x| \geq 0$ for all $x \in \mathcal{A}$, with equality iff $x = 0$;
(ii) $|\alpha x| = |\alpha| \cdot |x|$ for all $\alpha \in K$ and all $x \in \mathcal{A}$;
(iii) $|x + y| \leq |x| + |y|$ for all $x, y \in \mathcal{A}$;
(iv) $|xy| \leq |x| \cdot |y|$ for all $x, y \in \mathcal{A}$.

The norm $|\cdot|$ generates a metric topology in $\mathcal{A}$. If this metric is complete, we say that $\mathcal{A}$ is a *Banach algebra*.

Example B.1 *Examples of Banach algebras abound.*

(i) The algebra of all continuous functions over a compact space X with values in K, $C(X, K)$, under the sup-norm, is a Banach algebra.

(ii) Let E be a Banach (or Hilbert) space over K. Then the algebra of all bounded K-linear operators over E, denoted $L(E)$ or $B(E)$, is a Banach algebra under the usual operator norm.

(iii) *The algebra $L^\infty(X, \mu)$ of all complex L^∞ functions over a measure space (X, μ) is a Banach algebra, the norm being the L^∞ norm. Note that if X is a compact Hausdorff space and μ is a finite Borel measure, then $C(X) = C(X, \mathbb{C})$ is a (dense) sub-algebra of $L^\infty(X, \mu)$.*

Note that if $\mathcal{A}$ is a Banach algebra, it may or may not have a unit $e \in \mathcal{A}$. If it does, we can always rescale the norm so that $|e| = 1$ (note that, in any case, property (iv) already implies that $|e| \geq 1$). The *spectrum* of an element $x \in \mathcal{A}$ is the set of all $z \in K$ such that $x - ze$ is not invertible as an element of $\mathcal{A}$. If $\mathcal{A}$ does not have a unit, we may adjoin one if necessary by a simple construction called *unitization*. The idea is to take an element e not in $\mathcal{A}$, declare it to be a unit, and let $\mathcal{A}^+ = \mathcal{A} \oplus \mathbb{C}e$ (direct sum as Banach spaces), extending the multiplication to $\mathcal{A}^+$ so that the distributive law still holds. It is not difficult to extend the norm of $\mathcal{A}$ to a norm in $\mathcal{A}^+$. For a more intrinsic construction in the context of C^* algebras, see Exercise B.1. A Banach algebra with unit will be called *unital*.

Lemma B.2 *The unit e of a unital Banach algebra $\mathcal{A}$ has a neighborhood all of whose elements are invertible.*

Proof. Given any $v \in \mathcal{A}$ with $|v| < 1$, let $y = e - v$. Then, one sees that y is invertible, with $y^{-1} = e + v + v^2 + \dots$. The series converges because $|v| < 1$ and $\mathcal{A}$ is complete. □

From this point onward, we assume that all algebras are over $K = \mathbb{C}$.

Proposition B.3 *The spectrum of $x \in \mathcal{A}$ is always closed, bounded, and non-empty. If $z \in \mathbb{C}$ is in the spectrum, then $|z| \leq |x|$.*

Proof. Let $z \in \mathbb{C}$ be such that $|z| > |x|$. Then $y = x - ze$ is invertible, with

$$y^{-1} = -z^{-1}\left(e + z^{-1}x + z^{-2}x^2 + \cdots\right) .$$

Hence, if z is in the spectrum of x, then $|z| \leq |x|$. This shows that the spectrum of x is bounded. Now, Lemma B.2 implies that the set of all invertible elements in $\mathcal{A}$ is *open*. Therefore, the spectrum of x must be *closed* (if $z_0 \in \mathbb{C}$ is such that $x - z_0 e$ is invertible, then $x - ze$ is invertible for all z sufficiently close to z_0). It remains to prove that the spectrum of x is non-empty. We argue by contradiction. Suppose $x - ze$ is invertible for all $z \in \mathbb{C}$, and let $f : \mathbb{C} \to \mathcal{A}$ be given by $f(z) = (x - ze)^{-1}$. This is certainly not constant. Hence we can find a continuous linear functional $\lambda : \mathcal{A} \to \mathbb{C}$ such that $F = \lambda \circ f : \mathbb{C} \to \mathbb{C}$ is also not constant. But, it is an easy matter to check that F is an *analytic* function. Moreover, we have $|F(z)| \to 0$ as $|z| \to \infty$ (use (iv) here). This shows that

f is bounded, and hence constant by Liouville's theorem. This contradiction shows that the spectrum of x cannot be constant. $\square$

Notation B.4 We shall write $\sigma(x)$ for the spectrum of x. The complement $\mathbb{C} \setminus \sigma(x)$ is sometimes denoted $\rho(x)$ and is usually called the *resolvent set* of x.

Here is an important consequence of the above.

Theorem B.5 (Gelfand–Mazur) *Let K be a normed algebra over $\mathbb{R}$ which is also a field. If K is complete, then either $K = \mathbb{R}$ or $K = \mathbb{C}$.*

Proof. Let $e \in K$ be the unit element. First, suppose that there exists $j \in K$ such that $j^2 = -e$. Then, we see that K is also an algebra over $\mathbb{C}$, for we can define, for each $z = x + iy \in \mathbb{C}$ and each $v \in K$,

$$z \cdot v = xv + y(jv).$$

Using the norm $|z| = |x| + |y|$ in $\mathbb{C}$, we see that K becomes in fact a Banach algebra over $\mathbb{C}$. Now, if $v \in K$, then $\sigma(v) \neq \emptyset$ by Proposition B.3, so let $z \in \sigma(v)$. Then $v - ze$ is non-invertible in K, but because K is a field we must have $v - ze = 0$, i.e. $v = ze$. This shows at once that $\sigma(v) = \{z\}$, and that the map $\varphi : \mathbb{C} \to K$ given by $\varphi(z) = ze$ is a field isomorphism, so $K = \mathbb{C}$ in this case.

If, however, there is no element in K whose square is equal to $-e$, we adjoin a new element $j \notin K$ and look at $K + jK$, declaring that $j^2 = -e$ and making $K + jK$ into an algebra (over $\mathbb{C}$) in the obvious way. We also define a norm in $K + jK$ by $|v + jw| = |v| + |w|$. The reader can check as an exercise that $K + jK$ is a Banach algebra with this norm. By what we proved above, we have $K + jK = \mathbb{C}$, so $K = \mathbb{R}$ in this case. $\square$

Let us now consider *ideals* in a Banach algebra $\mathcal{A}$. A linear subspace $I \subset \mathcal{A}$ is said to be a *left ideal* if $x \subset I$ for all $x \in \mathcal{A}$. If $Ix \subset I$ for all $x \in \mathcal{A}$, then I is called a *right* ideal. If $I \subset \mathcal{A}$ is both a left and a right ideal, we say that I is a two-sided ideal. Of course, these distinctions are immaterial when $\mathcal{A}$ is a commutative algebra. An ideal $I \neq \mathcal{A}$ is said to be *maximal* if it is not properly contained in any other ideal. The reader will have no trouble in checking the usual properties of ideals in the present context. In particular, the topological closure $\overline{I}$ of an ideal $I \subset \mathcal{A}$ is also an ideal.

Proposition B.6 *Let $\mathcal{A}$ be a Banach algebra, and let $I \subset \mathcal{A}$ be a closed ideal that is also proper and two-sided. Then $\mathcal{A}/I$ is a Banach algebra.*

Proof. That the quotient of an algebra by a two-sided ideal is an algebra is standard (exercise). As a norm in $\mathcal{A}/I$, define

$$\|x + I\| = \inf_{v \in I} |x + v|$$

(here $|\cdot|$ is the norm in $\mathcal{A}$). We leave it as an exercise for the reader to show that this is indeed a norm in $\mathcal{A}/I$ and that this norm is complete (this of course requires the completeness of $\mathcal{A}$ and the closedness of I). $\square$

We note that a maximal ideal $M \subset \mathcal{A}$ is necessarily closed (exercise). The following is an important fact concerning complex Banach algebras and maximal ideals. All ideals to be considered will be two-sided.

Proposition B.7 *If $\mathcal{A}$ is a complex Banach algebra and $M \subset \mathcal{A}$ is a maximal ideal, then $\mathcal{A}/M \cong \mathbb{C}$.*

Proof. First, we claim that $\mathcal{A}/M$, which we know is a Banach algebra by Proposition B.6, is in fact a *field*. Let $x \in \mathcal{A}$ be such that $x \notin M$, so that $x + M \neq M$. We claim that there exists $y \in \mathcal{A}$ such that $(x + M)(y + M) = e + M$ (which is the unit in $\mathcal{A}/M$, as the reader can check). Because M is maximal, the ideal generated by $M \cup \{x\}$ must be equal to $\mathcal{A}$. Hence, there exist $y \in \mathcal{A}$ and $m \in M$ such that $xy + m = e$. This shows that $xy + M = e + M$, and because $(x + M)(y + M) = xy + M$, this proves the claim. We have thus shown that for each x in $\mathcal{A}$ but not in M, the element $x + M$ is invertible in $\mathcal{A}/M$. Hence $\mathcal{A}/M$ is indeed a field. By the Gelfand–Mazur theorem, either $\mathcal{A}/M \cong \mathbb{C}$ or $\mathcal{A}/M \cong \mathbb{R}$. But $\mathcal{A}/M$ contains $\{ze + M : z \in \mathbb{C}\} \cong \mathbb{C}$, which rules out the latter. Therefore $\mathcal{A}/M \cong \mathbb{C}$. $\square$

A *character* (or multiplicative linear functional) of a complex Banach algebra $\mathcal{A}$ is an algebra homomorphism $\varphi : \mathcal{A} \to \mathbb{C}$; thus, φ is $\mathbb{C}$-linear and $\varphi(xy) = \varphi(x)\varphi(y)$ for all x, y in $\mathcal{A}$. We also require characters to be continuous.[1] The set of all non-zero characters of $\mathcal{A}$ is usually denoted $\widehat{\mathcal{A}}$. Note that for all $\varphi \in \widehat{\mathcal{A}}$ we have $\varphi(e) = 1$, provided $\mathcal{A}$ has a unit e. The set $\widehat{\mathcal{A}}$ is called the *Gelfand spectrum* of $\mathcal{A}$. We put a topology on the Gelfand spectrum of $\mathcal{A}$ as follows. Note that $\widehat{\mathcal{A}}$ is contained in $\mathcal{A}^*$, the dual of $\mathcal{A}$ as a Banach space. Let $\mathcal{A}^*$ be given the weak* topology, and give $\widehat{\mathcal{A}}$ the induced topology.

Lemma B.8 *If $\varphi : \mathcal{A} \to \mathbb{C}$ is a character of a complex Banach algebra $\mathcal{A}$ with unit, then $\|\varphi\| = 1$.*

Proof. Let $x \in \mathcal{A}$ be such that $|x| \leq 1$, and suppose $|\varphi(x)| > 1$. Then, although $|x^n| \leq |x|^n \leq 1$ for all $n \geq 1$, we have $|\varphi(x^n)| = |\varphi(x)^n| = |\varphi(x)|^n \to \infty$ as $n \to \infty$. This contradicts the continuity of φ. Hence $|\varphi(x)| \leq 1$ for all $x \in \mathcal{A}$ with $|x| \leq 1$, so $\|\varphi\| \leq 1$. Because $\varphi(e) = 1$, we see that in fact $\|\varphi\| = 1$. $\square$

The above lemma shows that $\widehat{\mathcal{A}}$ is in fact contained in the unit sphere of $\mathcal{A}^*$. Now we have the following result.

[1] For C^* algebras continuity is automatic!

Theorem B.9 *The Gelfand spectrum $\widehat{\mathcal{A}}$ is a compact Hausdorff space under the weak* topology.*

Proof. The weak* topology is Hausdorff, and by Alaoglu's theorem (see [RS1]) the unit ball $\mathcal{A}_1^*$ is compact in this topology. Hence, it suffices to show that $\widehat{\mathcal{A}}$ is closed in $\mathcal{A}_1^*$. Suppose we have a net $\{\varphi_\lambda\}_{\lambda\in\Lambda}$ of elements of $\widehat{\mathcal{A}}$, and that $\varphi \in \mathcal{A}_1^*$ is a limit point of this net, i.e. for some totally ordered set $\mathcal{O} \subset \Lambda$ we have $\lim_{\mathcal{O}\ni\lambda} \varphi_\lambda(x) = \varphi(x)$ for all $x \in \mathcal{A}$. Then, from $\varphi_\lambda(xy) = \varphi_\lambda(x)\varphi_\lambda(y)$, we deduce that $\varphi(xy) = \varphi(x)\varphi(y)$ for all $x, y \in \mathcal{A}$, whence φ is a character, i.e. $\varphi \in \widehat{\mathcal{A}}$. This shows that $\widehat{\mathcal{A}}$ is closed as claimed, and therefore compact. □

There is a very close relationship between maximal ideals and characters, in any Banach algebra $\mathcal{A}$. This relationship is even closer when $\mathcal{A}$ has a unit, as the following result shows.

Proposition B.10 *Let $\mathcal{A}$ be a complex Banach algebra with unit. There exists a one-to-one correspondence $\Phi : \widehat{\mathcal{A}} \to \mathcal{M}$, where $\mathcal{M}$ is the set of all maximal ideals of $\mathcal{A}$.*

Proof. Given a character $\varphi \in \widehat{\mathcal{A}}$, define $\Phi(\varphi) = \ker\varphi$. This kernel is easily seen to be a maximal ideal in $\mathcal{A}$ (exercise), so we have a well-defined map $\Phi : \widehat{\mathcal{A}} \to \mathcal{M}$. We claim that Φ is one-to-one. Suppose $\varphi_1, \varphi_2 \in \widehat{\mathcal{A}}$ are such that $\ker\varphi_1 = \ker\varphi_2$. Given $x \in \mathcal{A}$, we have $x - \varphi_1(x)e \in \ker\varphi_1$, because $\varphi_1(e) = 1$. Hence, $x - \varphi_1(x)e \in \ker\varphi_2$ as well. Thus, we have $\varphi_2(x - \varphi_1(x)e) = 0$, where $\varphi_2(x) = \varphi_1(x)\varphi_2(e) = \varphi_1(x)$ (because $\varphi_2(e) = 1$). This shows that $\varphi_1 = \varphi_2$ and so Φ is injective as claimed. To see that Φ is onto, let $M \subset \mathcal{A}$ be a maximal ideal, and consider the canonical projection $\pi_M : \mathcal{A} \to \mathcal{A}/M$, an algebra homomorphism. By the Gelfand–Mazur theorem, there exists an isomorphism $\alpha_M : \mathcal{A}/M \to \mathbb{C}$. But then $\varphi_M = \alpha_M \circ \pi_M : \mathcal{A} \to \mathbb{C}$ is a homomorphism, and clearly $\varphi_M(e) = 1$, so φ_M is a character. Because α_M is an isomorphism, we have $\ker\varphi_M = \ker\pi_M = M$. Hence, we have $\Phi(\varphi_M) = M$, which shows that Φ is onto. □

This proposition allows us to carry the topology of $\widehat{\mathcal{A}}$ over to $\mathcal{M}$. Thus, the space of maximal ideals in $\mathcal{A}$ is a compact Hausdorff space in a natural way.

B.2 C^* algebras

In applications, especially to quantum mechanics, we are not so much interested in arbitrary Banach algebras as we are in those that carry some additional structure. Thus, we are usually interested in algebras of self-adjoint operators in some Hilbert space, or more generally in algebras of operators in Hilbert space which are closed under taking adjoints. This motivates the following definitions.

Definition B.11 *A Banach $*$-algebra is a complex Banach algebra $\mathcal{A}$ together with an involution $x \mapsto x^*$ on $\mathcal{A}$ satisfying*

(i) $(x+y)^* = x^* + y^*$;

(ii) $(\alpha x)^* = \overline{\alpha} x^*$;

(iii) $(xy)^* = y^* x^*$;

(iv) $\|x^*\| = \|x\|$, *for all $x, y \in \mathcal{A}$ and all $\alpha \in \mathbb{C}$ (here and throughout, $\|\cdot\|$ is the norm in $\mathcal{A}$).*

Definition B.12 *A C^* algebra $\mathcal{A}$ is a Banach $*$-algebra such that $\|x^*x\| = \|x\|^2$ for all $x \in \mathcal{A}$.*

Here is some additional terminology. The elements $x \in \mathcal{A}$ (a Banach $*$-algebra or C^* algebra) with $x^* = x$ are called *self-adjoint*. If $x \in \mathcal{A}$ is such that $x^*x = xx^*$, then x is called *normal*. When $\mathcal{A}$ has a unit e, we define $\sigma(x) = \{\alpha \in \mathbb{C} : \alpha e - x \text{ is not invertible}\}$ to be the spectrum of x, as for general Banach algebras. From now on, we shall deal exclusively with C^* algebras.

Example B.13 *Let us now present some important examples of C^* algebras.*

(i) Let X be a compact or locally compact Hausdorff space, and let $\mathcal{A} = C_0(X)$ be the space of continuous functions $f : X \to \mathbb{C}$ that vanish at infinity, in the sense that for each $\varepsilon > 0$ the set $\{x \in X : |f(x)| \geq \varepsilon\}$ is compact. Of course, $C_0(X) = C(X) = C(X, \mathbb{C})$ when X is compact. Let $\mathcal{A}$ be endowed with the sup norm $\|f\| = \sup_{x \in X} |f(x)|$, and let $f^* : X \to \mathbb{C}$ be given by $f^*(x) = \overline{f(x)}$, for all $x \in X$. With addition and multiplication defined pointwise, we see at once that $\mathcal{A}$ is a C^*-algebra, in fact a commutative unital C^*-algebra, with unit given by the constant function $f \equiv 1$.

(ii) Let $\mathcal{A}$ be any sub-algebra of $B(\mathcal{H})$, the space of bounded operators in a Hilbert space $\mathcal{H}$, which is closed under the operations of taking adjoints and also closed in the norm topology of $B(\mathcal{H})$. Then $\mathcal{A}$ is a C^* algebra, and it is usually non-commutative. This class of examples includes, of course, finite-dimensional algebras such as $M_n(\mathbb{C})$, the algebra of $n \times n$ complex matrices.

(iii) Once again, consider $L^\infty(X, \mu)$, the algebra of L^∞ functions over a finite measure space (X, μ). With the involution defined by conjugation, and the operations of pointwise addition and multiplication as in (1), $L^\infty(X, \mu)$ is a commutative C^* algebra with unit. Unlike (1), however, $L^\infty(X, \mu)$ displays the extra feature of possessing a pre-dual as a Banach space, namely $L^1(X, \mu)$. A C^* algebra with this property is called a von Neumann algebra.

We shall see in due time that the above examples (1) and (2) constitute *all* C^* algebras up to the appropriate notion of isomorphism.

Lemma B.14 *Let a be a self-adjoint element of a C^* algebra $\mathcal{A}$. Then we have $\|a^n\| = \|a\|^n$ for all $n \geq 1$.*

Proof. First, we prove the statement for powers of 2. Because $a^* = a$, we have $\|a^2\| = \|a^*a\| = \|a\|^2$. By induction on k, it follows that $\|a^{2^k}\| = \|a\|^{2^k}$, because the powers of self-adjoint elements are self-adjoint. Let us now prove the statement for arbitrary n. Again, we proceed by induction, on the (unique) k such that $2^k \leq n < 2^{k+1}$. Suppose the statement holds true for all m such that $m \leq 2^k$, i.e. $\|a^m\| = \|a\|^m$ for all $m \leq 2^k$. Given n such that $2^k \leq n < 2^{k+1}$, let $k_n = 2^{k+1} - n \leq 2^k$. Then, we have

$$\|a\|^{n+k_n} = \|a^{n+k_n}\| \leq \|a^n\| \cdot \|a^{k_n}\| = \|a^n\| \cdot \|a\|^{k_n} .$$

Here, we have used $\|a^{k_n}\| = \|a\|^{k_n}$, which is true by the induction hypothesis. Hence, $\|a\|^n \leq \|a^n\|$. But $\|a^n\| \leq \|a\|^n$ always, so $\|a^n\| = \|a\|^n$. This completes the induction and finishes the proof. □

Definition B.15 *Given a C^* algebra $\mathcal{A}$ (or more generally a Banach algebra) with unit, the* spectral radius *of $a \in \mathcal{A}$ is $r(a) = \sup\{|\lambda| : \lambda \in \sigma(a)\}$*

Lemma B.16 *The spectral radius of $a \in \mathcal{A}$ is given by $r(a) = \lim_{n\to\infty} \|a^n\|^{1/n}$.*

Proof. The reader is invited to supply a proof as an exercise. The existence of the above limit is a simple consequence of the fact that the sequence $\|a^n\|$ is sub-multiplicative (i.e. $\|a^{m+n}\| \leq \|a^m\| \cdot \|a^n\|$, for all $m, n \geq 1$). □

Note that we always have $r(a) \leq \|a\|$. The following lemma tells us that for self-adjoint elements equality holds.

Lemma B.17 *If $a \in \mathcal{A}$ is a self-adjoint element of a C^* algebra $\mathcal{A}$, then $r(a) = \|a\|$.*

Proof. This is an obvious consequence of Lemmas B.14 and B.16. □

More generally, Lemma B.17 holds true for *normal* elements. This is left as an exercise.

We have yet to introduce a suitable notion of morphism between C^* algebras.

Definition B.18 *A $*$-morphism between two C^* algebras $\mathcal{A}$, $\mathcal{B}$ is an algebra homomorphism $\phi : \mathcal{A} \to \mathcal{B}$ that commutes with the $*$-involutions, i.e. $\phi(a^*) = \phi(a)^*$, for all $a \in \mathcal{A}$. If both algebras are unital and ϕ maps the unit of $\mathcal{A}$ to the unit of $\mathcal{B}$, then ϕ is said to be a unital $*$-morphism.*

A $*$-morphism of C^* algebras will sometimes also be called a C^* homomorphism.

Lemma B.19 *Let $\phi : \mathcal{A} \to \mathcal{B}$ be a unital $*$-morphism between two unital C^* algebras. Then $r(\phi(a)) \leq r(a)$ for all $a \in \mathcal{A}$.*

Proof. It suffices to show that $\sigma_{\mathcal{B}}(\phi(a)) \subset \sigma_{\mathcal{A}}(a)$. Let $\lambda \in \sigma(\phi(a))$. Then $\phi(a) - \lambda e_{\mathcal{B}}$ is non-invertible in $\mathcal{B}$, where $e_{\mathcal{B}}$ is the unit of $\mathcal{B}$. Note that ϕ carries invertible elements in $\mathcal{A}$ to invertible elements in $\mathcal{B}$. Because

$$\phi(a) - \lambda e_{\mathcal{B}} = \phi(a - \lambda e_{\mathcal{A}}) ,$$

it follows that $a - \lambda e_{\mathcal{A}}$ is non-invertible in $\mathcal{A}$, and therefore $\lambda \in \sigma_{\mathcal{A}}(a)$. □

We leave it as an exercise for the reader to remove the hypothesis in Lemma B.19 that both algebras and the $*$-morphism are unital.

In the definition of $*$-morphism given above, we did *not* require that $*$-morphisms be continuous. It is a consequence of the result below that such an assumption is unnecessary.

Proposition B.20 *Let $\phi : \mathcal{A} \to \mathcal{B}$ be a $*$-morphism between C^* algebras. Then ϕ is norm-contracting, i.e. $\|\phi(a)\|_{\mathcal{B}} \leq \|a\|_{\mathcal{A}}$.*

Proof. Let $x \in \mathcal{A}$ be self-adjoint. Then $\phi(x)$ is self-adjoint in $\mathcal{B}$, whence by Lemma B.17 we have

$$\|\phi(x)\|_{\mathcal{B}} = r_{\mathcal{B}}(\phi(x)) \leq r_{\mathcal{A}}(x) = \|x\|_{\mathcal{A}} ,$$

where we have used Lemma B.19 as well. Hence $\|\phi(x)\|_{\mathcal{B}} \leq \|x\|_{\mathcal{A}}$ whenever x is self-adjoint. For arbitrary $a \in \mathcal{A}$, the element a^*a is self-adjoint, so

$$\|\phi(a)\|_{\mathcal{B}}^2 = \|\phi(a^*a)\|_{\mathcal{B}} \leq \|a^*a\|_{\mathcal{A}} = \|a\|_{\mathcal{A}}^2 ,$$

i.e. $\|\phi(a)\|_{\mathcal{B}} \leq \|a\|_{\mathcal{A}}$, and we are done. □

We are drawing closer to the *commutative Gelfand–Naimark theorem*. Our next lemma justifies the name *Gelfand spectrum* that we gave to $\widehat{\mathcal{A}}$.

Lemma B.21 *Let $\mathcal{A}$ be a commutative C^* algebra with unit. Given $a \in \mathcal{A}$, we have $\lambda \in \sigma(a)$ iff there exists $\varphi \in \widehat{\mathcal{A}}$ such that $\varphi(a) = \lambda$.*

Proof. If $\lambda \in \sigma(a)$, then $a - \lambda e$ is not invertible, so the ideal $I = (a - \lambda e)\mathcal{A}$ is proper. Let $J \supset I$ be a maximal ideal. By Proposition B.10, there exists $\varphi \in \widehat{\mathcal{A}}$ such that $\ker \varphi = J$. In particular φ vanishes over I, and so $\varphi(a - \lambda e) = 0$, i.e. $\varphi(a) = \lambda$. Conversely, if $\varphi \in \widehat{\mathcal{A}}$ is such that $\varphi(a) = \lambda$, then $a - \lambda e$ cannot be invertible: if $b \in \mathcal{A}$ were an inverse, then $1 = \varphi((a - \lambda e)b) = \varphi(a - \lambda e) \cdot \varphi(b) = 0$, which is absurd. Hence $\lambda \in \sigma(a)$. □

From now on, we shall write $\sigma(\mathcal{A})$ instead of $\widehat{\mathcal{A}}$ for the Gelfand spectrum of a C^* algebra $\mathcal{A}$.

There is a natural way to associate to each element $a \in \mathcal{A}$ of a C^* algebra with unit an element $\hat{a} \in C(\sigma(\mathcal{A}))$, namely the functional given by $\hat{a}(\varphi) = \varphi(a)$, for each $\varphi \in \sigma(\mathcal{A})$. The map $a \mapsto \hat{a}$ is called the *Gelfand transform.*

Theorem B.22 (Gelfand–Naimark) *The Gelfand transform $\mathcal{A} \to C(\sigma(\mathcal{A}))$ is a $*$-isomorphism, and it is also an isometry, provided $\mathcal{A}$ is a commutative C^*-algebra with unit.*

Proof. It is clear that $a \mapsto \hat{a}$ is a homomorphism of algebras, and that it sends the unit of $\mathcal{A}$ to the unit of $C(\sigma(\mathcal{A}))$ (the constant function equal to 1). We claim that it preserves the $*$-involutions of both algebras. To prove this claim, we use the fact that $\varphi(a)$ is *real* for each self-adjoint $a \in \mathcal{A}$ and each $\varphi \in \sigma(\mathcal{A})$. Given an arbitrary element $a \in \mathcal{A}$, let us write $a = b + ic$ with b, c self-adjoint, namely,

$$b = \frac{1}{2}(a + a^*) \quad \text{and} \quad c = \frac{1}{2}(a - a^*) .$$

Then, for each $\varphi \in \sigma(\mathcal{A})$, we see that

$$\begin{aligned} \varphi(a^*) &= \varphi(b - ic) = \varphi(b) - i\varphi(c) \\ &= \overline{\varphi(b) + i\varphi(c)} = \overline{\varphi(b + ic)} = \overline{\varphi(a)} . \end{aligned}$$

This shows that $\widehat{a^*} = \hat{a}^*$, as claimed. Now, the Gelfand transform is also clearly injective. We claim that it is an isometry. Indeed, it follows from B.21 that for each $a \in \mathcal{A}$ we have $\|\hat{a}\|_\infty = r(a)$ (check!). Therefore we have

$$\|\hat{a}\|_\infty^2 = \|\widehat{a^*a}\|_\infty = r(a^*a) = \|a^*a\| = \|a\|^2 ,$$

where we have used Lemma B.17 together with the fact that a^*a is self-adjoint. Thus the Gelfand transform is an isometry as claimed. In particular, it is continuous. Our final claim is that it is *onto* $C(\sigma(\mathcal{A}))$. Indeed, the image in $C(\sigma(\mathcal{A}))$ of $\mathcal{A}$ under the Gelfand transform is a closed algebra that contains the constant functions, separates points and is closed under complex conjugation. Therefore, it is equal to the whole $C(\sigma(\mathcal{A}))$, by the Stone–Weierstrass theorem. □

Let us extract some important consequences of the Gelfand–Naimark theorem.

Corollary B.23 *Let $\psi : \mathcal{A} \to \mathcal{B}$ be an injective homomorphism between two C^* algebras. Then ψ is an isometry.*

Proof. We already know that ψ is norm-contracting. We may assume that both algebras are unital, passing to $\psi^+ : \mathcal{A}^+ \to \mathcal{B}^+$ if necessary (details are left as an exercise). In order to prove that $\|\psi(a)\|_{\mathcal{B}} = \|a\|_{\mathcal{A}}$ for a given $a \in \mathcal{A}$, we may restrict ψ to $\psi : C^*(a, e_{\mathcal{A}}) \to C^*(\phi(a), e_{\mathcal{B}})$, where the restricted domain $C^*(a, e_{\mathcal{A}})$ is the C^* subalgebra of $\mathcal{A}$ generated by a and $e_{\mathcal{A}}$, and similarly for the restricted range. Both $C^*(a, e_{\mathcal{A}})$ and $C^*(\phi(a), e_{\mathcal{B}})$ are *abelian* C^* algebras. Applying the Gelfand–Naimark Theorem B.22, we deduce that $C^*(a, e_{\mathcal{A}}) \cong C(X)$ and $C^*(\phi(a), e_{\mathcal{B}}) \cong C(Y)$, where X and Y are compact Hausdorff spaces (the Gelfand spectra of both abelian algebras). We are therefore reduced to proving that if $\psi : C(X) \to C(Y)$ is an injective C^* homomorphism, then it is an isometry. But every such homomorphism is of the form $\psi(\varphi) = \varphi \circ f$, where $f : Y \to X$ is a surjective continuous map (this is an easy exercise). Therefore, $\|\psi(\varphi)\|_\infty = \|\varphi \circ f\|_\infty = \|\varphi\|_\infty$, for all $\varphi \in C(X)$. This finishes the proof. □

Another consequence of Theorem B.22, is the following result.

Theorem B.24 *Let X and Y be compact metric spaces. Then X and Y are homeomorphic if and only if $C(X)$ and $C(Y)$ are isomorphic C^* algebras.*

Proof. One implication ($\Rightarrow$) is obvious. To prove the converse implication, we first claim that $X \cong \sigma(C(X))$. Indeed, let $\Phi : X \to \sigma(C(X))$ be the map $x \in \hat{x}$, where $\hat{x} : C(X) \to \mathbb{C}$ is the multiplicative linear functional $\hat{x}(\varphi) = \varphi(x)$. Then Φ is injective and continuous. We shall presently see that it is also *onto*. Note that any $\lambda \in \sigma(C(X))$ is also a *positive* linear functional.[2] By the Riesz–Markov theorem, there exists a Borel probability measure μ on X such that $\lambda(\varphi) = \int_X \varphi \, d\mu$ for all $\varphi \in C(X)$. Because λ is multiplicative, we have

$$\int_X \varphi\psi \, d\mu = \lambda(\varphi\psi) = \lambda(\varphi)\lambda(\psi) \tag{B.1A}$$

$$= \left(\int_X \varphi \, d\mu\right) \cdot \left(\int_X \psi \, d\mu\right) \tag{B.1B}$$

for all $\varphi, \psi \in C(X)$. Let $U, V \subset X$ be any two disjoint open sets with $\mu(U) > 0$ and $\mu(V) > 0$. Choose $\varphi \in C(X)$ so that $\varphi > 0$ on U and $\varphi = 0$ everywhere else; choose $\psi \in C(X)$ likewise with respect to V. Then $\varphi \cdot \psi \equiv 0$, so $\lambda(\varphi\psi) = 0$. But $\int_X \varphi \, d\mu > 0$ and $\int_X \psi \, d\mu > 0$ by construction, and from (B.1B) we deduce that $\lambda(\varphi\psi) > 0$, which is a contradiction. From this, it is an easy matter

[2] If $\varphi \in C(X)$ is ≥ 0, write $\varphi = |f|^2 = f\overline{f}$ for some $f \in C(X)$; then $\lambda(\varphi) = \lambda(f\overline{f}) = |\lambda(f)|^2 \geq 0$, because λ is $*$-preserving.

(exercise) to see that μ is an atomic measure supported at a single point $x_0 \in X$; in other words, $\mu = \delta_{x_0}$, the Dirac measure concentrated at x_0. Hence for each $\varphi \in C(X)$ we have

$$\lambda(\varphi) = \int_X \varphi \, d\delta_{x_0} = \varphi(x_0) = \hat{x_0}(\varphi) .$$

This shows that the map Φ is onto as claimed, hence a homeomorphism.

Thus, now we know that $X \cong \sigma(C(X))$ and $Y \cong \sigma(C(Y))$. Hence, if $C(X)$ and $C(Y)$ ar isomorphic as C^* algebras (so that they are isometric, by the above corollary) then by Lemma B.25 below, $\sigma(C(X))$ and $\sigma(C(Y))$ are homeomorphic. Therefore $X \cong Y$, and this finishes the proof of our theorem. □

The lemma referred to in the above proof is the following.

Lemma B.25 *If $\Psi : \mathcal{A} \to \mathcal{B}$ is a C^* isomorphism then $\hat{\Psi} : \sigma(\mathcal{B}) \to \sigma(\mathcal{A})$ given by $\hat{\Psi}(\varphi) = \varphi \circ \Psi$ is a homeomorphism.*

Proof. An exercise for the reader. □

B.3 The spectral theorem

The main application of the commutative Gelfand–Naimark theorem, for our purposes, is to a fairly simple proof of the spectral theorem for bounded, self-adjoint operators. We shall first formulate and prove a C^* version of the spectral theorem with an extra hypothesis. We need a definition.

Definition B.26 *Let $\mathcal{A} \subset B(\mathcal{H})$ be a C^* algebra of operators on a Hilbert space $\mathcal{H}$. A vector $v \in \mathcal{H}$ is said to be* $\mathcal{A}$-cyclic *if $\mathcal{A}(v) = \{Tv : T \in \mathcal{A}\} \subset \mathcal{H}$ is dense in $\mathcal{H}$.*

Theorem B.27 (spectral theorem I) *Let $\mathcal{A} \subset B(\mathcal{H})$ be a commutative C^* algebra of operators in Hilbert space, containing the identity operator, and let $v \in \mathcal{H}$ be an $\mathcal{A}$-cyclic vector. Then there exist a finite measure space (X, μ) and a unitary isometry $U : L^2(X, \mu) \to \mathcal{H}$ such that $U^*TU : L^2(X, \mu) \to L^2(X, \mu)$ is a multiplication operator, for each $T \in \mathcal{A}$.*

Proof. There is no loss of generality in assuming that the cyclic vector v has unit norm. From the Gelfand–Naimark theorem, we know that $\mathcal{A} \cong C(X)$, for some compact metric space X. For each $f \in C(X)$, let us denote by $T_f \in \mathcal{A}$ the corresponding operator in Hilbert space via such $*$-isomorphism. Let $\lambda_v : C(X) \to \mathbb{C}$ be the linear functional given by $\lambda_v(f) = \langle T_f v, v \rangle$. Note that $\lambda_v(1) = \|v\|^2 = 1$. Moreover, if $f \in C(X)$ is ≥ 0, so that $f = g\bar{g}$ for some

$g \in C(X)$, then

$$\begin{aligned}\lambda_v(f) &= \lambda_v(g\overline{g}) = \langle T_{g\overline{g}}(v), v\rangle \\ &= \langle T_g(v), T_g(v)\rangle = \|T_g(v)\|^2 \geq 0\,.\end{aligned}$$

Hence, λ_v is a positive, normalized linear functional on $C(X)$. By the Riesz–Markov theorem, there exists a regular Borel probability measure μ on X such that

$$\lambda_v(f) = \int_X f\,d\mu\,, \quad \text{for all } f \in C(X)\,.$$

Now, consider the Hilbert space $L^2(X, \mu)$, of which $C(X)$ is a *dense* subspace. We define an isometry $U : L^2(X, \mu) \to \mathcal{H}$ as follows. If $f \in C(X)$, let $U(f) = T_f(v)$. Note that U is linear on $C(X)$, and

$$\begin{aligned}\|U(f)\|^2 &= \langle T_f(v), T_f(v)\rangle = \langle T_{f\overline{f}}v, v\rangle = \lambda_v(f\overline{f}) \\ &= \int_X |f|^2\,d\mu = \|f\|^2\,.\end{aligned}$$

Hence, U is norm-preserving on $C(X)$, and therefore it extends uniquely to an isometry $U : L^2(X, \mu) \to \mathcal{H}$. But because $v \in \mathcal{H}$ is cyclic, $U(C(X))$ is dense in $\mathcal{H}$. This shows that U is *onto* $\mathcal{H}$, and hence in fact a *unitary* isometry.

Finally, given $f \in C(X)$, let $M_f : L^2(X, \mu) \to L^2(X, \mu)$ be the multiplication operator $M_f(\varphi) = f\varphi$. Whenever $\varphi \in C(X)$, we have

$$T_f(U\varphi) = T_f T_\varphi(v) = T_{f\varphi}(v) = U(f\varphi) = UM_f(\varphi)\,,$$

and thus $U^*T_fU(\varphi) = M_f(\varphi)$. Because $C(X)$ is dense in $L^2(X, \mu)$, it follows that $U^*T_fU = M_f$, and this finishes the proof. □

In the context of the definition just preceding the above theorem, let us agree to call a closed subspace $W \subset \mathcal{H}$ *cyclic* (or $\mathcal{A}$*-cyclic*) if there exists a vector $w \in W$ such that $\mathcal{A}(w)$ is dense in W.

Lemma B.28 *Let $\mathcal{A} \subset B(\mathcal{H})$ be a unital C^* algebra of operators in Hilbert space. Then there exists a decomposition $H = \bigoplus_{i\in I} H_i$ into mutually orthogonal, $\mathcal{A}$-invariant cyclic subspaces.*

Proof. We apply Zorn's lemma to the family of all direct sums of mutually orthogonal, $\mathcal{A}$-invariant cyclic subspaces of $\mathcal{H}$, partially ordered by inclusion in an obvious way. Let $V = \bigoplus_{i\in I} H_i \subset \mathcal{H}$ be a maximal such direct sum, and suppose $V \neq \mathcal{H}$. Take $v \neq 0$ in the orthogonal complement of V, and let $W = \overline{\mathcal{A}(v)} \subset \mathcal{H}$. Then W is cyclic, and $V' = W \oplus V = W \oplus \bigoplus_{i\in I} H_i$

is a strictly larger direct sum in the family, a contradiction. Therefore $V = \mathcal{H}$. □

With this lemma at hand, we are now in a position to prove a more familiar version of the spectral theorem.

Theorem B.29 (spectral theorem II) *Let $T \in B(\mathcal{H})$ be a self-adjoint operator. Then T is unitarily equivalent to a multiplication operator.*

Proof. Let $\mathcal{A}_T \in B(\mathcal{H})$ be the C^* algebra generated by T and the identity, i.e. the C^* algebra arising as the closure (in the operator norm) of the polynomial algebra generated by T. By Lemma B.28, there exists a decomposition $\mathcal{H} = \bigoplus_{i\in I} H_i$ of the Hilbert space $\mathcal{H}$ into mutually orthogonal, $\mathcal{A}_T$-invariant subspaces. For each $i \in I$, let $T_i = T|_{H_i} : H_i \to H_i$. Then each T_i has a cyclic vector $v_i \in H_i$. By Theorem B.24, there exist a finite measure space (X_i, μ_i) and a unitary isometry $U_i : L^2(X_i, \mu_i) \to H_i$ such that $U_i^* T_i U_i = M_{f_i}$, where $f_i \in L^\infty(X_i, \mu_i)$. Now define

$$\tilde{U} = \bigoplus_{i\in I} U_i \,:\, \bigoplus_{i\in I} L^2(X_i, \mu_i) \to \mathcal{H}\,.$$

This linear operator is a unitary isometry, and it conjugates T to $\bigoplus_{i\in I} M_{f_i}$. To finish the proof, it suffices to show that this last direct sum of multiplication operators is itself (unitarily equivalent to) a multiplication operator. Let (X, μ) be the measure space obtained as the disjoint union of (X_i, μ_i). To wit, a set $E \subset X$ is μ-measurable iff $E \cap X_i$ is μ_i-measurable for each i, and $\mu(E) = \sum_{i\in I} \mu_i(E \cap X_i)$. There is a natural unitary isometry $U' : L^2(X, \mu) \to \bigoplus_{i\in I} L^2(X_i, \mu_i)$. Moreover, if $f : X \to \mathbb{C}$ is defined so that $f|_{X_i} = f_i$ for each i, then $f \in L^\infty(X, \mu)$. Letting $U = \tilde{U} \circ U'$, we see at once that U is unitary and $U^* T U = M_f : L^2(X, \mu) \to L^2(X, \mu)$. This shows that T is unitarily equivalent to a multiplication operator. □

We end this section with a couple of remarks. First, note that the measure space (X, μ) constructed in the above proof will in general be an infinite measure space. Even if $\mathcal{H}$ is separable, the naive construction in the proof will in general yield only a σ-finite measure. This is somewhat unpleasant, but can be circumvented: indeed, one can show that if (X, μ) is σ-finite, then there is a *finite* measure space (Y, ν) such that $L^2(X, \mu)$ is unitarily isometric to $L^2(Y, \nu)$. This is left as an exercise for the reader.

Second, the above version of the spectral theorem can be considerably strengthened if one makes full use of Lemma B.28. The result is an improved version of Theorem B.27 in which the hypothesis that our commutative C^* algebra $\mathcal{A} \subset B(\mathcal{H})$ has a cyclic vector can be dropped. Note that such an improved version contains the case of the C^* algebra generated by any finite collection of commuting operators $T_i : \mathcal{H} \to \mathcal{H}, i = 1, 2, \ldots, n$, i.e. $T_i T_j = T_j T_i$, which are assumed to be either self-adjoint or normal. This form of the spectral theorem is especially suitable for the quantization scheme described in Chapter 2, for a system with a finite number of particles.

B.4 States and GNS representation

The Gelfand–Naimark theorem for commutative C^* algebras provides a concrete realization of such an algebra as a space of continuous functions on some compact space. The proof made essential use of multiplicative functionals, or characters. For non-commutative algebras, this approach will not do, since characters need not exist. For example, the C^* algebra $M_n(\mathbb{C})$ of complex $n \times n$ matrices ($n > 1$) carries no such multiplicative functionals (see Exercise B.2). Thus one needs to consider the next best thing, namely *positive linear functionals*.

Definition B.30 *Let $\mathcal{A}$ be a C^* algebra. A self-adjoint element $a \in \mathcal{A}$ is said to be* positive *if $\sigma(a) \subset [0, \infty)$. The set of all positive elements of $\mathcal{A}$ is denoted by $\mathcal{A}_+$.*

Note that $\mathcal{A}_+$ is a convex cone in $\mathcal{A}$. In particular, positivity induces a partial order in $\mathcal{A}$: given two elements $a, b \in \mathcal{A}$, we say that $a \leq b$ if $b - a$ is positive.

Definition B.31 *A linear functional $\rho : \mathcal{A} \to \mathbb{C}$ is said to be* positive *if $\rho(a) \geq 0$ for all $a \in \mathcal{A}_+$.*

For instance, characters, when they exist, are certainly positive linear functionals.

A positive linear functional is automatically bounded; the proof of this fact is left as an exercise (see Exercise B.5). If $\rho : \mathcal{A} \to \mathbb{C}$ is positive, we can therefore look at its norm $\|\rho\|$.

Definition B.32 *If $\rho : \mathcal{A} \to \mathbb{C}$ is positive and $\|\rho\| = 1$, then we call ρ a* state. *The set of all states of $\mathcal{A}$ is denoted $S(\mathcal{A})$.*

Two of the most important examples of states are *vector states* and *normal states*. These are defined relative to a given *representation* of our C^* algebra in Hilbert space; hence, we first pause for this crucial concept of representation.

Definition B.33 *A* representation *of a C^* algebra $\mathcal{A}$ in a Hilbert space is a $*$-homomorphism $\pi : \mathcal{A} \to B(\mathcal{H})$. If π is injective, then the representation is said to be* faithful.

Example B.34 ***(vector states)*** *Given a representation $\pi : \mathcal{A} \to B(\mathcal{H})$, we say that a state $\varphi : \mathcal{A} \to \mathbb{C}$ is a* vector state *for $(\pi, \mathcal{H})$ if there exists a unit vector $\xi \in \mathcal{H}$ such that $\varphi(a) = \langle \pi(a)\xi, \xi \rangle$ for all $a \in \mathcal{A}$.*

Example B.35 ***(normal states)*** *Again, we consider a representation $\pi : \mathcal{A} \to B(\mathcal{H})$. We say that a state $\varphi : \mathcal{A} \to \mathbb{C}$ is a* normal state *for $(\pi, \mathcal{H})$ if there exists a positive, trace-class operator $\rho : \mathcal{H} \to \mathcal{H}$ such that $\varphi(a) = \mathrm{Tr}(\rho\pi(a))$ for all $a \in \mathcal{A}$. The operator ρ is usually called the density matrix of the state. Note that every vector state is a normal state. Indeed, if $\varphi : \mathcal{A} \to \mathbb{C}$ is a vector state for $(\pi, \mathcal{H})$ with unit vector ξ, take $\rho : \mathcal{H} \to \mathcal{H}$ to be the orthogonal projection onto the one-dimensional subspace generated by ξ. Then ρ is positive and trace-class, and for all $a \in \mathcal{A}$ we have*

$$\begin{aligned} \langle \pi(a)\xi, \xi \rangle &= \langle \pi(a)\xi, \rho^*\xi \rangle \\ &= \langle \rho\pi(a)\xi, \xi \rangle = \mathrm{Tr}(\rho\pi(a)) \,. \end{aligned}$$

Normal states play a key role in the mathematical formulation of quantum statistical mechanics [AJP], as well as in algebraic quantum field theory (see Section B.6 and [H]).

We shall soon see that the set of all states is closed in the weak* topology, and that it is also *convex*. First, we digress a bit to talk about approximate units in a C^* algebra.

Definition B.36 *An* approximate unit *in a C^* algebra $\mathcal{A}$ is a net $\{u_\lambda\}_{\lambda \in \Lambda}$ of positive elements $u_\lambda \in \mathcal{A}_+$ such that*

(i) $\lambda_1 \leq \lambda_2 \Rightarrow u_{\lambda_1} \leq u_{\lambda_2}$;
(ii) for each $a \in \mathcal{A}$ we have $\|a - u_\lambda a\| \to 0$ and $\|a - a u_\lambda\| \to 0$.

An approximate unit in $\mathcal{A}$ always exists: this is not entirely trivial, but is nevertheless left as an exercise (see Exercise B.7).

Example B.37 *Let $\mathcal{H}$ be a separable Hilbert space, and let $\{e_1, e_2, \ldots, e_n, \ldots\} \subset \mathcal{H}$ be an orthonormal basis. For each $n \geq 1$, let $P_n : \mathcal{H} \to \mathcal{H}$ be the orthogonal projection onto the closed linear subspace spanned by the finite set $\{e_1, e_2, \ldots, e_n\}$. Then $(P_n)_{n \geq 1}$ is an approximate unit for $\mathcal{A} = K(\mathcal{H})$, the C^* algebra of compact operators on $\mathcal{H}$, but not for $B(\mathcal{H})$.*

For our purposes in this section, the important fact concerning positive functionals and approximate units is the following.

Lemma B.38 *Let (u_λ) be an approximate unit in $\mathcal{A}$, and let $\rho : \mathcal{A} \to \mathbb{C}$ be a positive linear functional. Then $\rho(u_\lambda) \to \|\rho\|$. In particular, if $\mathcal{A}$ is unital, then $\|\rho\| = \rho(1)$.*

Proof. Because ρ is positive, it is order-preserving, and so $(\rho(u_\lambda))$ is a net. Since we also have $\rho(u_\lambda) \leq \|\rho\| \cdot \|u_\lambda\| \leq \|\rho\|$, and therefore $\alpha = \lim \rho(u_\lambda)$ exists, and $\alpha \leq \|\rho\|$. On the other hand, if $a \in \mathcal{A}$ is such that $\|a\| \leq 1$, then using the multiplicativity of ρ and the self-adjointness of u_λ we have

$$|\rho(au_\lambda)|^2 = |\rho(a^*a)\rho(u_\lambda^2)| \leq \rho(a^*a)\rho(u_\lambda)\,, \qquad \text{(B.2)}$$

where we used also that $u_\lambda^2 \leq u_\lambda$ for all λ. But $\rho(au_\lambda) \to \rho(a)$ and $\rho(u_\lambda) \to \alpha$, and moreover $\rho(a^*a) \leq \|\rho\|$. Hence, from (B.2) we deduce that $|\rho(a)|^2 \leq \alpha\|\rho\|$, for all $a \in \mathcal{A}$ with $\|a\| \leq 1$. Taking the supremum over all such a, we see that $\|\rho\|^2 \leq \alpha\|\rho\|$, i.e. $\alpha \geq \|\rho\|$. This shows that $\|\rho\| = \alpha = \lim \rho(u_\lambda)$ as claimed. The last assertion in the statement is clear. □

Proposition B.39 *The set $S(\mathcal{A})$ of all states of a C^* algebra $\mathcal{A}$ is a weak*-closed, convex subset of $\mathcal{A}^*$.*

Proof. In fact, $S(\mathcal{A})$ is weak*-compact, as follows easily from Alaoglu's theorem. The convexity of $S(\mathcal{A})$ is an easy consequence of Lemma B.38. Indeed, if $\rho_0, \rho_1 : \mathcal{A} \to \mathbb{C}$ are states, then for each $0 \leq t \leq 1$ we have that $\rho_t = (1-t)\rho_0 + t\rho_1$ is a positive linear functional (obvious); if (u_λ) is an approximate unit in $\mathcal{A}$, then on one hand $\rho_t(u_\lambda) \to \|\rho_t\|$, and on the other hand

$$\begin{aligned}\rho_t(u_\lambda) &= (1-t)\rho_0(u_\lambda) + t\rho_1(u_\lambda) \\ &\to (1-t)\|\rho_0\| + t\|\rho_1\| = (1-t) + t = 1\,.\end{aligned}$$

Therefore $\|\rho_t\| = 1$, i.e. ρ_t is a state for all $0 \leq t \leq 1$. □

Another important property of positive linear functionals needed below is the following version of Cauchy–Schwarz's inequality.

Lemma B.40 *If $\rho : \mathcal{A} \to \mathbb{C}$ is a positive linear functional, then for all $a, b \in \mathcal{A}$ we have $|\rho(a^*b)|^2 \leq \rho(a^*a)\rho(b^*b)$.*

The proof is left as an exercise. Here is an immediate consequence of the Cauchy–Schwarz inequality that is crucial in the GNS construction to follow.

Lemma B.41 *If $\rho : \mathcal{A} \to \mathbb{C}$ is a positive linear functional and $a \in \mathcal{A}$, then $\rho(a^*a) = 0$ if and only if $\rho(ba) = 0$ for all $b \in \mathcal{A}$.*

Proof. If $\rho(a^*a) = 0$, then by Lemma B.40 we have $|\rho(ba)|^2 \leq \rho(bb^*) \cdot \rho(a^*a) = 0$, so $\rho(ba) = 0$ for all $b \in \mathcal{A}$. The converse is obvious (take $b = a^*$). □

We are now ready for the so-called GNS construction, which is the basis for the proof of the non-commutative Gelfand–Naimark theorem; here, GNS stands for Gelfand–Naimark–Segal.

Theorem B.42 (GNS) *Let $\mathcal{A}$ be a C^* algebra. Then for each positive linear functional $\rho : \mathcal{A} \to \mathbb{C}$ there exist a Hilbert space $\mathcal{H}_\rho$, a vector $\xi_\rho \in \mathcal{H}_\rho$, and a representation $\pi_\rho : \mathcal{A} \to B(\mathcal{H}_\rho)$ with the following properties:*

(i) *for each $a \in \mathcal{A}$ we have $\rho(a) = \langle \pi_\rho(a)\xi_\rho, \xi_\rho \rangle$, where $\langle , \rangle$ is the inner product in $\mathcal{H}_\rho$;*

(ii) *the orbit $\pi_\rho(\mathcal{A})\xi_\rho$ is dense in $\mathcal{H}_\rho$.*

Proof. Given $\rho : \mathcal{A} \to \mathbb{C}$ as in the statement, let $I_\rho = \{a \in \mathcal{A} : \rho(a^*a) = 0\}$. Then, by Lemma B.41, I_ρ is a left ideal in $\mathcal{A}$: for $\rho(a^*a) = 0$ implies $\rho(a^*b) = 0$ for all $b \in \mathcal{A}$, so in particular, letting $c \in \mathcal{A}$ and taking $b = c^*ca$, we have $\rho((ca)^*ca) = 0$, whence $ca \in I_\rho$ for all $c \in \mathcal{A}$. It follows that the quotient $\tilde{\mathcal{H}}_\rho = \mathcal{A}/I_\rho$ is a complex vector space, and we can endow it with the inner product

$$\langle a + I_\rho, b + I_\rho \rangle \; = \; \rho(a^*b) \, .$$

Let $\mathcal{H}_\rho$ be the completion of $\tilde{\mathcal{H}}_\rho$ with respect to this inner product. For each $a \in \mathcal{A}$, consider the linear map $\mathcal{A} \to \mathcal{A}$ given by left multiplication by a. This map induces a linear map $\mathcal{A}/I_\rho \to \mathcal{A}/I_\rho$ that is clearly bounded and thus induces a bounded linear operator $\pi_\rho(a) : \mathcal{H}_\rho \to \mathcal{H}_\rho$. The map $a \in \pi_\rho(a)$ is the desired representation. Note also that by construction, $\pi_\rho(a)(x + I_\rho) = ax + I_\rho$, for all $x \in \mathcal{A}$.

To finish the proof, we need to find $\xi_\rho \in \mathcal{H}_\rho$ such that (i) and (ii) hold true. If $\mathcal{A}$ has a unit $1 \in \mathcal{A}$, take $\xi_\rho = 1 + I_\rho$; in this case (i) and (ii) are obvious. If $\mathcal{A}$ does not have a unit, let (u_λ) be an approximate unit (one always exists, by exercise), and define $\xi_\rho = \lim_\lambda (u_\lambda + I_\rho) \in \mathcal{H}_\rho$. Then (ii) is a consequence of the fact that $au_\lambda \to a$ for all $a \in \mathcal{A}$. Moreover, for all such a we have

$$\langle \pi_\rho(a)\xi_\rho, \xi_\rho \rangle = \lim_\lambda \rho(au_\lambda \cdot u_\lambda^*) = \rho(a) \, ,$$

so (i) holds true as well. □

The state space $S(\mathcal{A})$ has the following separating property.

Lemma B.43 *If a is a positive element of a C^* algebra $\mathcal{A}$, then there exists $\rho \in S(\mathcal{A})$ such that $\rho(a) = \|a\|$.*

Proof. Let $C^*(a)$ be the C^* algebra generated by a and the unit of $\mathcal{A}$ (if $\mathcal{A}$ is not unital, consider its unitization $\mathcal{A}^+$ instead of $\mathcal{A}$). Because $C^*(a)$ is abelian, the commutative Gelfand–Naimark theorem tells us that $C^*(a) \cong C(\sigma(a))$. Let $\tilde{\rho} : C^*(a) \to \mathbb{C}$ be the linear functional that corresponds, via this $*$-isomorphism, to evaluation at $\|a\| \in \sigma(a)$. Then $\|\tilde{\rho}\| = \tilde{\rho}(1) = 1$, and

$\tilde{\rho}(a) = \|a\|$. By the Hahn–Banach theorem, $\tilde{\rho}$ extends to a linear functional $\rho : \mathcal{A} \to \mathbb{C}$ with the same norm. To prove that $\rho \in S(\mathcal{A})$, it remains to show that ρ is positive. Let $x \in \mathcal{A}_+$ (the positive cone of $\mathcal{A}$), so that x is self-adjoint and $\sigma(x) \subset [0, \infty)$. It suffices to show that $\rho(x)$ is in the convex hull of $\sigma(x)$. If the latter is not true, then we can find a disk $D(z_0, R)$ in the complex plane such that $\sigma(x) \subset D(z_0, R)$ but $\rho(x) \notin D(z_0, R)$. Thus, on one hand we have that the spectral radius $r(x - z_0)$ is $\leq R$, and on the other hand $R < |\rho(x) - z_0| = |\rho(x - z_0)| \leq \|x - z_0\|$ (because $\|\rho\| = 1$). But $x - z_0 = x - z_0 \cdot 1$ is a normal element of $\mathcal{A}$. Hence, applying Lemma B.17, we see that $r(x - z_0) = \|x - z_0\|$. This is a contradiction, and the lemma is proved. □

We are now ready for the central result of this section. The noncommutative Gelfand–Naimark theorem asserts that every C^* algebra can be faithfully represented as an algebra of operators in *some* Hilbert space. Because by Corollary B.23 every injective homomorphism of C^* algebras is an isometry onto its image, the Gelfand–Naimark theorem can be stated as follows.

Theorem B.44 (Gelfand–Naimark) *Every C^* algebra $\mathcal{A}$ is $*$-isomorphic to a $*$-subalgebra of $B(\mathcal{H})$ for some Hilbert space $\mathcal{H}$. If $\mathcal{A}$ is separable, then one can take $\mathcal{H}$ to be separable as well.*

Proof. Let $F \subset S(\mathcal{A})$ be any non-empty family of states with the property that for each $0 \neq a \in \mathcal{A}$ there exists $\rho \in F$ such that $\rho(a) \neq 0$. For example, one can take $F = S(\mathcal{A})$. Define $\mathcal{H} = \bigoplus_{\rho \in F} \mathcal{H}_\rho$, where $\mathcal{H}_\rho$ is given by the GNS construction, alongside the cyclic representation $\pi_\rho : \mathcal{A} \to B(\mathcal{H}_\rho)$. Let $\pi : \mathcal{A} \to B(\mathcal{H})$ be given by the direct sum of representations

$$\pi(a) = \sum_{\rho \in F}^{\oplus} \pi_\rho(a) \; : \; \mathcal{H} \to \mathcal{H} \, .$$

Because $\|\pi_\rho(a)\| \leq \|a\|$ for each $\rho \in F$, it follows that $\|\pi(a)\| \leq \|a\|$ as well, so $\pi(a) \in B(\mathcal{H}$, and so π is a well-defined representation of the C^* algebra $\mathcal{A}$ in $\mathcal{H}$. If $a \neq 0$, then $\pi_\rho(a) \neq 0$ for at least one $\rho \in F$, so $\phi(a)$. This shows that $\pi : \mathcal{A} \to B(\mathcal{H})$ is faithful. This proves the first assertion in the statement. Now suppose that $\mathcal{A}$ is separable. Then each cyclic Hilbert space $\mathcal{H}_\rho$ is separable (by property (ii) of the GNS construction). Let $\{a_n : n \in \mathbb{N}\}$ be a countable dense set in $\{a \in \mathcal{A}_+ : \|a\| = 1\}$, and for each $n \geq 1$ let $\rho_n \in S(\mathcal{A})$ be such that $\rho_n(a_n) = 1$, a state whose existence is guaranteed by Lemma B.43. Then $F = \{\rho_n : n \in \mathbb{N}\} \subset S(\mathcal{A})$ is countable. Therefore, $\mathcal{H} = \bigoplus_{\rho \in F} \mathcal{H}_\rho$ is separable, and the representation $\pi : \mathcal{A} \to B(\mathcal{H})$ as given above is faithful. This finishes the proof. □

B.5 Representations and spectral resolutions

In this section, we present more details on the representations of commutative C^* algebras, and apply the results to the spectral decomposition of a self-adjoint operator on Hilbert space.

By the commutative Gelfand–Naimark theorem, it suffices to consider representations of $C(X)$, where X is compact Hausdorff. It turns out that, for everything we do in the sequel, local compactness suffices. Hence, we shall deal with the C^* algebra $C_0(X)$ of continuous functions vanishing at infinity on a locally compact Hausdorff space X. Given such a space X, let $\mathcal{B}$ be a σ-algebra of Borel subsets of X. The elements of $\mathcal{B}$ are sometimes called *events*.

Definition B.45 *A* projection-valued measure *on $(X, \mathcal{B})$ is a map $P : \mathcal{B} \to B(\mathcal{H})$, where $\mathcal{H}$ is a Hilbert space, with the following properties:*

(i) $P(X) = I_{\mathcal{H}}$, *the identity operator;*
(ii) $P(E_1 \cap E_2) = P(E_1)P(E_2)$ *for all* $E_1, E_2 \in \mathcal{B}$;
(iii) $P(\cup_1^\infty E_i) = \sum_1^\infty P(E_i)$, *whenever the* $E_i \in \mathcal{B}$ *are pairwise disjoint;*
(iv) $P(E)^* = P(E)$ *for all* $E \in \mathcal{B}$.

Note that (ii) implies that $P(E)^2 = P(E)$ for all $E \in \mathcal{B}$. This idempotency, together with the self-adjointness condition (iv), tell us that $P(E)$ is a projection operator on $\mathcal{H}$; hence, the name. Note also that any pair of non-zero vectors $v, w \in \mathcal{H}$ determines a complex measure $\mu_{v,w}$ on $(X, \mathcal{B})$ given by $\mu_{v,w}(E) = \langle P(E)v, w\rangle$. Thus, each projection-valued measure $(P, \mathcal{H})$ on $(X, \mathcal{B})$ gives rise to a family of complex measures on $(X, \mathcal{B})$ called the *spectral measures* of $(P, \mathcal{H})$.

Now, it turns out that continuous functions on X (vanishing at ∞) can be integrated against a projection-valued measure $(P, \mathcal{H})$ and the result is a bounded linear operator on $\mathcal{H}$: this is the content of the following theorem.

Theorem B.46 *Let $P : \mathcal{B} \to B(\mathcal{H})$ be a projection-valued measure over a locally compact Hausdorff space X with Borel σ-algebra $\mathcal{B}$. Then for each $f \in C_0(X)$, there exists an operator $I_P(f) \in B(\mathcal{H})$ such that*

$$\langle I_P(f)v, w\rangle = \int_X f(x)\, d\mu_{v,w}(x)$$

for all $v, w, \in \mathcal{H}$. Moreover, $f \mapsto I_P(f)$ is a representation of $C_0(X)$ in $B(\mathcal{H})$.

Remark B.47 The operator $I_P(f)$ whose existence is asserted by the above theorem is denoted

$$I_P(f) = \int f(x)\, dP(x)$$

and is called the *integral* of f with respect to $(P, \mathcal{H})$. The name and notation are justified by the fact that $I_P(f)$ is the limit in norm of Riemann–Stieltjes sums of the form $\sum_k f(x_k)\, P(E_k)$ (see Exercise B.10).

Proof. Let us consider the sesquilinear form $B_f : \mathcal{H} \times \mathcal{H} \to \mathbb{C}$ given by

$$B_f(v, w) = \int_X f(x)\, d\mu_{v,w}(x)\, .$$

This form is continuous. Indeed, we have $|B_f(v, w)| \leq |\mu_{v,w}| \cdot \|f\|_\infty$, where

$$|\mu_{v,w}| = \sup \sum_i \mu_{v,w}(E_i)\, , \tag{B.3}$$

the supremum being taken over all countable partitions of X into $\mathcal{B}$-measurable sets. Since for each i we have

$$\begin{aligned} |\mu_{v,w}(E_i)| &= |\langle P(E_i)v, w\rangle| = |\langle P(E_i)v, P(E_i)w\rangle| \\ &\leq \|P(E_i)v\| \cdot \|P(E_i)w\|\, , \end{aligned}$$

we get, using the Cauchy–Schwarz inequality in (B.3),

$$\begin{aligned} |\mu_{v,w}| &\leq \sup \left\{ \left(\sum_i \|P(E_i)v\|^2 \right)^{1/2} \cdot \left(\sum_i \|P(E_i)w\|^2 \right)^{1/2} \right\} \\ &= \|v\| \cdot \|w\|\, . \end{aligned}$$

This shows that $|B_f(v, w)| \leq \|f\|_\infty \|v\| \cdot \|w\|$, so B_f is continuous as claimed. Using Riesz representation, we deduce that there exists a unique bounded linear operator $I_P(f) : \mathcal{H} \to \mathcal{H}$ such that $B_f(v, w) = \langle I_P(f)v, w\rangle$ for all $v, w \in \mathcal{H}$. It is straightforward to check that $I_P(\bar{f}) = I_P(f)^*$ for all $f \in C_0(X)$.

It remains to prove that $I_P : C_0(X) \to B(\mathcal{H})$ is multiplicative. First we claim that $\mu_{v,T_f w}$ is absolutely continuous with respect to $\mu_{v,w}$, where $T_f = I_P(f)$. Indeed, for all events $E, F \in \mathcal{B}$ we have

$$\mu_{P(E)v,w}(F) = \langle P(F)P(E)v, w\rangle = \langle P(E \cap F)v, w\rangle = \mu_{v,w}(E \cap F)\, . \tag{B.4}$$

Therefore, we get

$$\begin{aligned} \mu_{v,T_f w}(E) &= \langle P(E)v, T_f w\rangle \\ &= \langle T_{\bar{f}} P(E)v, w\rangle \\ &= \int_X \bar{f}\, d\mu_{P(E)v,w} \\ &= \int_E \bar{f}\, d\mu_{v,w}\, , \end{aligned}$$

where in the last step we used (B.4). This shows that $\mu_{v,T_f w} \ll \mu_{v,w}$, with Radon–Nikodym derivative a.e. equal to $\bar{f}$. But then we see that, given $f, g \in C_0(X)$,

$$\begin{aligned} \langle T_{fg} v, w \rangle &= \int_X fg \, d\mu_{v,w} \\ &= \int_X g \, d\mu_{v,T_{\bar{f}} w} \\ &= \langle T_g v, T_{\bar{f}} w \rangle \\ &= \langle T_f T_g v, w \rangle \ . \end{aligned}$$

Because this holds for all $v, w \in \mathcal{H}$, it follows that $T_{fg} = T_f T_g$, i.e. $I_P(fg) = I_P(f) I_P(g)$. This shows that I_P is a representation as claimed. □

We have just proved that every projection-valued measure on $(X, \mathcal{B})$ gives rise to a representation of the commutative C^* algebra $C_0(X)$ in Hilbert space. Let us now prove a converse to this result. We shall say that a projection-valued measure $(P, \mathcal{H})$ is *regular* if the variations of the measures $\mu_{v,w}$ are regular Borel measures on $(X, \mathcal{B})$ for all $v, w \in \mathcal{H}$.

Theorem B.48 *Let X be locally compact and Hausdorff, and let $\pi : C_0(X) \to B(\mathcal{H})$ be a non-degenerate*[3] *representation. Then there exists a unique regular projection-valued measure P on the Borel sets of X with values in $B(\mathcal{H})$ such that*

$$\pi(f) = \int f(x) \, dP(x)$$

for all $f \in C_0(X)$.

Proof. We know from Proposition B.20 that π is norm-contracting, in other words $\|\pi(f)\| \le \|f\|_\infty$. Thus, for each $v, w \in \mathcal{H}$, the correspondence $f \mapsto \langle \pi(f)v, w \rangle$ defines a bounded linear functional on $C_0(X)$, with norm $\le \|v\| \cdot \|w\|$. Applying the Riesz–Markov theorem, we get a unique, regular complex Borel measure $\mu_{v,w}$ on X such that

$$\langle \pi(f)v, w \rangle = \int_X f \, d\mu_{v,w}$$

and $|\mu_{v,w}| \le \|v\| \cdot \|w\|$. Note that the correspondence is sesquilinear as a map from $\mathcal{H} \times \mathcal{H}$ into the space of complex Borel measures on X. If $E \in \mathcal{B}$ is a Borel set, the map $(v, w) \mapsto \mu_{v,w}(E)$ is also sesquilinear (into $\mathbb{C}$), and therefore the

[3] Here *non-degenerate* means that $\pi(f)v = 0$ for all $f \in C_0(X) \Rightarrow v = 0$.

Riesz representation theorem yields $\mu_{v,w}(E) = \langle P(E)v, w\rangle$ for some operator $P(E) \in B(\mathcal{H})$ with $\|P(E)\| \le 1$.

To prove that $P : \mathcal{B} \to B(\mathcal{H})$ as defined above is indeed a projection-valued measure, we need to verify conditions (i)–(iv) of Definition B.45. Let us first verify condition (ii). We note that if $\varphi \in C_0(X)$ then

$$d\mu_{v,\pi(\bar{\varphi})w}(x) = \varphi(x)\, d\mu_{v,w}(x) \tag{B.5}$$

for all $v, w \in \mathcal{H}$. This happens because, for all $\psi \in C_0(X)$, we have

$$\begin{aligned}\int_X \psi(x)\, d\mu_{v,\pi(\bar{\varphi})w}(x) &= \langle \pi(\psi)v, \pi(\bar{\varphi})w\rangle \\ &= \langle \pi(\varphi)\pi(\psi)v, w\rangle \\ &= \langle \pi(\varphi\psi)v, w\rangle \\ &= \int_X \psi(x)\varphi(x)\, d\mu_{v,w}(x)\,.\end{aligned}$$

Let us now suppose that $E \in \mathcal{B}$. We claim that

$$d\mu_{P(E)v,w}(x) = \mathbf{1}_E(x)\, d\mu_{v,w}(x)\,. \tag{B.6}$$

Indeed, for all $\varphi \in C_0(X)$ we have

$$\begin{aligned}\int_X \varphi(x)\, d\mu_{P(E)v,w}(x) &= \langle \pi(\varphi)P(E)v, w\rangle = \langle P(E)v, \pi(\bar{\varphi})w\rangle \\ &= \mu_{v,\pi(\bar{\varphi})w}(E) = \int_X \mathbf{1}_E(x)\, d\mu_{v,\pi(\bar{\varphi})w}(x) \\ &= \int_X \varphi(x)\mathbf{1}_E(x)\, d\mu_{v,w}(x)\,,\end{aligned}$$

where we have used (B.5), and so (B.6) follows. Hence, if $F \in \mathcal{B}$ is any other Borel set, we have

$$\mu_{P(E)v,w}(F) = \int_F \mathbf{1}_E(x)\, d\mu_{v,w}(x) = \mu_{v,w}(E \cap F)\,.$$

This shows that $\langle P(F)P(E)v, w\rangle = \langle P(E \cap F)v, w\rangle$ for all $v, w \in \mathcal{H}$, and therefore $P(E \cap F) = P(E)P(F)$.

Next, we verify that condition (i) holds. Note that $\mu_{v,P(X)w} = \mu_{v,w}$ for all $v, w \in \mathcal{H}$, because for each $E \in \mathcal{B}$ we have

$$\begin{aligned}\mu_{v,P(X)w}(E) &= \langle P(E)v, P(X)w\rangle \\ &= \langle P(X)P(E)v, w\rangle \\ &= \langle P(E)v, w\rangle \\ &= \mu_{v,w}(E)\,,\end{aligned}$$

where we have used condition (ii), already established, and the self-adjointness of $P(X)$, which will be proved below. Hence, for all $f \in C_0(X)$, we have

$$\begin{aligned}\langle \pi(f)v, P(X)w\rangle &= \int_X f(x)\, d\mu_{v,P(X)w}(x) \\ &= \int_X f(x)\, d\mu_{v,w}(x) = \langle \pi(f)v, w\rangle \ .\end{aligned}$$

Thus, we see that $\langle P(X)\pi(f)v, w\rangle = \langle \pi(f)v, w\rangle$, for all $v, w \in \mathcal{H}$. This shows that $P(X)\pi(f)v = \pi(f)v$ for all $f \in C_0(X)$ and all $v \in \mathcal{H}$. But because by hypothesis π is non-degenerate, $\{\pi(f)v : f \in C_0(X),\ v \in \mathcal{H}\}$ is dense in $\mathcal{H}$ (check!). Therefore $P(X) = I_{\mathcal{H}}$, and condition (i) holds true, provided condition (iv) is true.

But the self-adjointness condition (iv) is easy to check. Because π is a $*$-morphism, we see that $\langle \pi(f)w, v\rangle = \overline{\langle \pi(f)v, w\rangle}$ for all $f \in C_0(X)$ and all $v, w \in \mathcal{H}$; that is

$$\begin{aligned}\int_X \bar{f}\, d\mu_{w,v} &= \overline{\int_X f\, d\mu_{v,w}} \\ &= \int_X \bar{f}\, d\bar{\mu}_{v,w} \ .\end{aligned}$$

From this, it follows that $\bar{\mu}_{v,w} = \mu_{w,v}$, and in particular

$$\begin{aligned}\langle P(E)v, w\rangle &= \mu_{v,w}(E) = \overline{\mu_{w,v}(E)} \\ &= \overline{\langle P(E)w, v\rangle} = \langle v, P(E)w\rangle\end{aligned}$$

for all $v, w \in \mathcal{H}$, and therefore $P(E)^* = P(E)$, as required.

Finally, the σ-additivity condition (iii) is also easy. If $\{E_i\}$ is a countable family of pairwise disjoint elements of $\mathcal{B}$, then for each pair $(v, w) \in \mathcal{H} \times \mathcal{H}$ we have[4]

$$\begin{aligned}\langle P(\cup_i E_i)v, w\rangle &= \mu_{v,w}(\cup_i E_i) = \sum_i \mu_{v,w}(E_i) \\ &= \sum_i \langle P(E_i)v, w\rangle = \left\langle \sum_i P(E_i)v, w \right\rangle .\end{aligned}$$

This shows that $P(\cup_i E_i) = \sum_i P(E_i)$. This finishes the proof of our theorem. □

[4] The reader should check that all series appearing here are convergent.

The results presented above allow us to establish the following spectral decomposition theorem for bounded self-adjoint operators in Hilbert space.

Theorem B.49 (spectral theorem III) *Let $T \in B(\mathcal{H})$ be a self-adjoint operator, and let $\mathcal{B}_T$ be the Borel σ-algebra of $\sigma(T)$. Then there exists a unique projection-valued measure $P : \mathcal{B}_T \to B(\mathcal{H})$ such that*

$$T = \int_{\sigma(T)} \lambda \, dP(\lambda) \, .$$

Moreover, T is in the norm closure of the set of all orthogonal projections that commute with all bounded operators commuting with T.

Proof. The idea is to combine the commutative Gelfand–Naimark theorem with Theorem B.48 above. Let $\mathcal{A}_T = C^*(I, T) \subset B(\mathcal{H})$ be the C^* algebra generated by $I = I_{\mathcal{H}}$ and T. Since this algebra is abelian, the commutative GN theorem yields a $*$-isomorphism $\mathcal{A}_T \cong C(\sigma(\mathcal{A}_T))$, where $\sigma(\mathcal{A}_T)$ is the Gelfand spectrum of $\mathcal{A}_T$. We claim that $\sigma(\mathcal{A}_T)$ is homeomorphic to $\sigma(T) \subset \mathbb{R}$. To see why, let $\Phi : \sigma(\mathcal{A}_T) \to \sigma(T)$ be given by $\Phi(\varphi) = \varphi(T) \in \mathbb{R}$. This is clearly continuous. It is injective, because $\Phi(\varphi_1) = \Phi(\varphi_2)$ implies $\varphi_1(T) = \varphi_2(T)$, which in turn yields $\varphi_1(p(T)) = \varphi_2(p(T))$ for every polynomial $p \in \mathbb{C}[X]$. Because $\{p(T) : \ p \in \mathbb{C}[X]\}$ is *dense* in $\mathcal{A}_T$, it follows that $\varphi_1(A) = \varphi_2(A)$ for all $A \in \mathcal{A}_T$, and so $\varphi_1 = \varphi_2$. The map Φ is also sujective: if $\lambda \in \sigma(T)$, then $T - \lambda I$ is non-invertible, so $\mathcal{J} = (T - \lambda I)\mathcal{A}_T$ is a proper ideal in $\mathcal{A}_T$. Hence $\mathcal{J}$ is contained in a maximal ideal; this is equivalent to saying that there exists a multiplicative linear functional $\varphi : \mathcal{A}_T \to \mathbb{C}$ with $\mathcal{J} \subset \ker \varphi$, i.e. $\varphi(T - \lambda I) = 0$, whence $\varphi(T) = \lambda$. Thus, being a continuous, bijective map between $\sigma(\mathcal{A}_T)$ and $\sigma(T)$, which are both Hausdorff spaces, Φ is a homeomorphism. This homeomorphism in turn yields a C^* isomorphism $C(\sigma(\mathcal{A}_T)) \cong C(\sigma(T))$: given $f \in C(\sigma(T))$, let $\hat{\Phi}(f) = f \circ \Phi \in C(\sigma(\mathcal{A}_T))$.

Now, there is an obvious representation of $\mathcal{A}_T$ in $B(\mathcal{H})$, namely the identity representation $\tilde{\pi} : \mathcal{A}_T \to B(\mathcal{H})$ given by $\tilde{\pi}(A) = A$. We know also that there is a $*$-isomorphism $\Psi : \mathcal{A}_T \cong C(\sigma(T))$ given by $\Psi(A) = \hat{A} \circ \Phi^{-1}$, where $\hat{A}(\varphi) = \varphi(A)$ is the Gelfand transform of A. Hence we define $\pi : C(\sigma(T)) \to B(\mathcal{H})$ by $\pi = \tilde{\pi} \circ \Psi^{-1}$, which is easily seen to be a non-degenerate representation. Everything has been set up so that $\pi(p(\lambda)) = \tilde{\pi}(p(T)) = p(T)$ for each polynomial $p = p(\lambda \in C(\sigma(T))$. On the other hand, by Theorem B.48, we have

$$\pi(f) = \int_{\sigma(T)} f(\lambda) \, dP(\lambda) \, ,$$

where P is the projection-valued measure associated to π. In particular, taking $f = \mathrm{id}_{\sigma(T)}$ (the identity polynomial), we see that

$$\pi(\mathrm{id}_{\sigma(T)}) = T = \int_{\sigma(T)} \lambda \, dP(\lambda) \,. \tag{B.7}$$

This establishes the first assertion of the theorem, except for the uniqueness of P, which is left as an exercise (see Exercise B.11).

Let us now suppose that $S \in B(\mathcal{H})$ commutes with T. Hence it commutes with every $A \in \mathcal{A}_T$. Using the isomorphism Ψ defined above, let us write $f_A = \Psi(A)$ for each $A \in \mathcal{A}_T$. We know that

$$\langle Av, w\rangle = \langle \pi(A)v, w\rangle = \int_{\sigma(T)} f_A(\lambda)\, d\mu_{v,w}(\lambda) \,.$$

Now, from $SA = AS$ we have $\langle Av, S^*w\rangle = \langle SAv, w\rangle = \langle ASv, w\rangle$, and therefore

$$\int_{\sigma(T)} f_A(\lambda)\, d\mu_{v,S^*w}(\lambda) = \int_{\sigma(T)} f_A(\lambda)\, d\mu_{Sv,w}(\lambda)$$

for all $A \in \mathcal{A}_T$. But $\sigma(T)$ is compact and second countable, i.e. a Polish space, and in such a space any finite Borel measure is regular, hence, determined by the values of its integral against continuous functions. Therefore, $\mu_{v,S^*w} = \mu_{Sv,w}$, for all $v, w \in \mathcal{H}$. In particular, $\mu_{v,S^*w}(E) = \mu_{Sv,w}(E)$ for all Borel sets $E \subset \sigma(T)$, that is

$$\big\langle P(E)v, S^*w\big\rangle = \langle P(E)Sv, w\rangle \ ,$$

or even

$$\langle SP(E)v, w\rangle = \langle P(E)Sv, w\rangle \ .$$

Because this holds for all $v, w \in \mathcal{H}$, we see that $P(E)S = SP(E)$, for all Borel sets $E \subset \sigma(T)$. In other words, each spectral projection $P(E)$ commutes with each operator that commutes with T. This proves the second assertion in the statement of the theorem, and we are done. □

B.6 Algebraic quantum field theory

In this section, we present a *very* brief description of the algebraic approach to QFT. This approach has been developed by Haag, Kastler, Araki, Borchers, Buchholz, Fredenhagen, Doplicher, Roberts, among many others (see [H and the references therein]). Now that we have the basic language of C^* algebras at hand, the description is not too difficult to present.

B.6.1 The algebraic approach

The algebraic approach makes free use of the theory of operator algebras, both concrete C^* algebras and von Neumann algebras. Let us start with some basic observations, which will eventually lead to a (tentative) definition of an algebraic QFT.

(i) In the Wightman formulation of QFT (see Chapter 6), the fields are operator-valued distributions. From a physical standpoint, such fields are supposed to represent *local* operations.

(ii) Locality suggests that we consider for each open neighborhood $\mathcal{O} \subset \mathbb{R}^4$ in Minkowski space the (topological vector) space[5] $\mathcal{D}(\mathcal{O})$ of C^∞ test functions $f : \mathbb{R}^4 \to \mathbb{C}$ with $\mathrm{supp}(f) \subset \mathcal{O}$. The corresponding *smeared fields*

$$\Phi(f) = \int_{\mathbb{R}^4} \Phi(x) f(x)\, d^4x \ ,$$

which, we recall, are *unbounded* operators in some Hilbert space $\mathcal{H}$, generate a polynomial algebra over $\mathbb{C}$ with monomials of the form $\Phi(f_1)\Phi(f_2)\cdots\Phi_n(f)$. Let us call this algebra $\mathcal{P}(\mathcal{O})$.

(iii) The algebra $\mathcal{P}(\mathcal{O})$, and *a fortiori* the C^* algebra it generates, can be quite wild. One can look instead at the spectral projections of the operators $\Phi(f)$, or bounded functions of them, via functional calculus. These projections generate a C^* algebra $\mathcal{A}(\mathcal{O})$ that is (one hopes) more amenable to analysis.

(iv) We are thus led to consider a *net* of algebras $\mathcal{A} = \{\mathcal{A}(\mathcal{O}) : \mathcal{O} \subset \mathbb{R}^4\}$ of bounded linear operators in Hilbert space. Here $\mathcal{O}$ varies over all open subsets of Minkowski spacetime.

An analysis of Wightman's axioms for QFT suggests that the net of algebras $\mathcal{O} \mapsto \mathcal{A}(\mathcal{O})$ should satisfy the following properties:

(i) *Isotony:* $\mathcal{O}_1 \subset \mathcal{O}_2 \Rightarrow \mathcal{A}(\mathcal{O}_1) \subset \mathcal{A}(\mathcal{O}_2)$.

(ii) *Additivity:* $\mathcal{A}(\mathcal{O}_1 \cup \mathcal{O}_2) = \mathcal{A}(\mathcal{O}_1) \vee \mathcal{A}(\mathcal{O}_2)$.

(iii) *Hermiticity:* Each $\mathcal{A}(\mathcal{O})$ is involutive, i.e. a $*$-algebra.

(iv) *Poincaré covariance:* There is a representation $(a, \Lambda) \mapsto U(a, \Lambda)$ of the Poincaré group $\mathcal{P}$ into the direct limit $\mathcal{A} = \varinjlim \mathcal{A}(\mathcal{O})$ such that, for each $(a, \Lambda) \in \mathcal{P}$ and each open set $\mathcal{O} \subset \mathbb{R}^4$,

$$U(a,\Lambda)\mathcal{A}(\mathcal{O})U(a,\Lambda)^{-1} = \mathcal{A}(\Lambda\mathcal{O} + a)\ .$$

[5] In fact, in the Wightman formulation, one considers the Schwartz space $\mathcal{S}(\mathcal{O})$ corresponding to *tempered* distributions, but here we stick to $\mathcal{D}(\mathcal{O})$ as it seems more suitable to encode locality.

(v) If $\mathcal{O}_1$ and $\mathcal{O}_1$ are spacelike separated in $\mathbb{R}^4$, then for each $A_1 \in \mathcal{A}(\mathcal{O}_1)$ and each $A_2 \in \mathcal{A}(\mathcal{O}_2)$, we have $[A_1, A_2] = 0$.

(vi) Given $\mathcal{O} \subset \mathbb{R}^4$, let $\widehat{\mathcal{O}} \subset \mathbb{R}^4$ be its causal completion (see Section B.6.2 below). Then $\mathcal{A}(\widehat{\mathcal{O}}) = \mathcal{A}(\mathcal{O})$.

This formulation should in fact be slightly changed, to accommodate general symmetries. We discuss this point further below, in Section B.6.3. One should also take into account that any reasonable theory is supposed to accommodate both bosons and fermions, and these have different symmetries. Thus, a rotation of 2π around an axis in spacetime leaves a bosonic field unchanged, but changes the sign of a fermionic field. Therefore it is more appropriate to consider, instead of the Poincaré group $\mathcal{P}$, its universal covering group $\widetilde{\mathcal{P}}$.

B.6.2 Causality structure

As we have seen before in this book, for a quantum theory to be compatible with relativity it must address the issue of causality in an appropriate manner. It must face the fact that the propagation of any signal is limited by the speed of light.

Let us briefly discuss the causality relation in Minkowski space $\mathbb{M} = \mathbb{R}^4$ (see Chapter 3). We denote by

$$\langle x, y\rangle_{\mathbb{M}} = x^0y^0 - x^1y^1 - x^2y^2 - x^3y^3$$

the usual Minkowski inner product of $\mathbb{M}$. Given an open set $\mathcal{O} \subset \mathbb{M}$, we define its causal complement to be

$$\mathcal{O}' = \{x \in \mathbb{M} : \langle x - y, x - y\rangle_{\mathbb{M}} < 0 \; \forall \, y \in \mathcal{O}\} .$$

In other words, $\mathcal{O}'$ is the set of all points in Minkowski space that are *spacelike* with all points of $\mathcal{O}$. A set $\mathcal{O} \subset \mathbb{M}$ is said to be *causally complete* if $\mathcal{O}'' = \mathcal{O}$. Note that if $\mathcal{O}$ is causally complete, then so is $\mathcal{O}'$. Let us denote by $\mathbb{K}$ the set of all causally complete subsets of $\mathbb{M}$. Given $\mathcal{O}_1, \mathcal{O}_2 \in \mathbb{K}$, we define their *wedge* as

$$\mathcal{O}_1 \wedge \mathcal{O}_2 = \mathcal{O}_1 \cap \mathcal{O}_2$$

and their *join* as

$$\mathcal{O}_1 \vee \mathcal{O}_2 = (\mathcal{O}_1 \cup \mathcal{O}_2)'' = (\mathcal{O}_1' \wedge \mathcal{O}_2')' .$$

With these lattice operations, we have the following simple fact.

Lemma B.50 *The set $\mathbb{K}$ of all causally complete subsets of Minkowski space has the structure of an orthocomplemented lattice.*

Proof. Exercise. □

The simplest (and smallest) causally complete regions are *double cones*, also called *diamonds*. Given $x \in \mathbb{M}$, let $C^+(x) \subset \mathbb{M}$ be the *forward light-cone* of x, i.e. the set of all $y \in \mathbb{M}$ that lie timelike with respect to x, in the sense that $y^0 - x^0 > |\mathbf{y} - \mathbf{x}|$. Let the *backward light-cone* $C^-(x)$ be similarly defined. If $x, y \in \mathbb{M}$ are two distinct points with $y \in C^+(x)$, we define the double cone with vertices x, y to be $K_{x,y} = C^+(x) \cap C^-(y)$. These and their causal complements determine a sub-lattice of $\mathbb{K}$, and this sub-lattice is oftentimes quite sufficient for the deployment of the causal structure in algebraic QFT.

B.6.3 Von Neumann algebras in QFT

A $*$-subalgebra $\mathcal{V} \subset B(\mathcal{H})$ is called a *von Neumann algebra* if it is closed in the weak topology of $B(\mathcal{H})$. Because every weakly closed subset of $B(\mathcal{H})$ is also closed in any stronger topology, it follows that every von Neumann algebra is a C^* algebra, but not conversely. The theory of von Neumann algebras seems especially suitable to help incorporate the causality relation of Minkowski space into QFT in a natural way.

Given any subset $\mathcal{S} \subset B(\mathcal{H})$, we define its *commutant* to be

$$\mathcal{S}' = \{T \in B(\mathcal{H}) : TS = ST \text{ for all } S \in \mathcal{S}\} .$$

The following result due to von Neumann is fundamental. See [AS] for a proof.

Theorem B.51 (von Neumann) *Let $\mathcal{S} \subset B(\mathcal{H})$ be a non-empty subset. Then*

(i) $\mathcal{S}'$ is a von Neumann algebra;
(ii) $\mathcal{S}'' \supset \mathcal{S}$, and in fact $\mathcal{S}''$ is the smallest von Neumann algebra containing $\mathcal{S}$;
(iii) $\mathcal{S}''' = \mathcal{S}'$.

This theorem allows us, in particular, to regard the set $\mathcal{V}^\sharp$ of von Neumann subalgebras of a given von Neumann algebra $\mathcal{V}$ as an *orthocomplemented lattice*. The lattice operations are defined as follows. The prime operation is the commutant. Given $\mathcal{V}_1, \mathcal{V}_2 \subset \mathcal{V}$, let

$$\mathcal{V}_1 \wedge \mathcal{V}_2 = \mathcal{V}_1 \cap \mathcal{V}_2$$

and

$$\mathcal{V}_1 \vee \mathcal{V}_2 = (\mathcal{V}_1 \cup \mathcal{V}_2)''$$

One easily checks (exercise) that $(\mathcal{V}^\sharp, \wedge, \vee, ')$ is a lattice. The attentive reader will not fail to notice that this lattice structure is akin to the lattice structure of causally complete subsets of Minkowski space. This is one of the main reasons that the theory of von Neumann algebras is especially suitable for a proper formulation of algebraic QFT.

In summary an algebraic quantum field theory should consist of a net of von Neumann algebras $\mathfrak{U} = \{\mathfrak{U}(\mathcal{O})\}$ satisfying the postulates formulated above. We are being rather sketchy here, but let us add a bit more information in the following question/answer format.

(i) *Why is the weak topology the relevant topology to be used?* Here is a (partial) justification. Given a state ω over $\mathcal{V} = \mathfrak{U}(\mathcal{O})$, we know that $\omega(A^*A) \geq 0$ for every $A \in \mathcal{V}$. Let us agree to call A an *operation* if A is norm non-increasing as an element of $B(\mathcal{H})$. Here $\mathcal{H}$ is the Hilbert space into which the algebra $\mathcal{V}$ is represented. Thus, if A is an operation, then $\omega(A^*A) \leq 1$. Hence we may interpret $\omega(A^*A)$ as a probability, namely the probability that a transition from the state ω to the state $A\omega$ (given by $(A\omega)(B) = \omega(A^*BA)$) occurs. Now, such a transition probability can never be measured with absolute precision. Instead, it is only determined up to an error. Thus, if $A_1, \ldots, A_n$ are operations and $p_1, \ldots, p_n$ are the corresponding *measured* transition probabilities, all we can say is that $|\omega(A_i^*A_i) - p_i| < \epsilon_i$, where ϵ_i is the error in the ith measurement. But this is tantamount to saying that ω lies in some weak neighborhood, namely

$$\{\varphi \in \mathcal{V}' : |\varphi(A_i^*A_i) - p_i| < \epsilon_i,\ i = 1, \ldots, n\} .$$

In other words, the very physical limitations of measurement dictate that the "right" topology in the space of states is the weak topology.

(ii) *How is the causality structure formalized in algebraic QFT?* First, we are given a net of von Neumann algebras $\mathfrak{U} = \{\mathfrak{U}(\mathcal{O}) : \mathcal{O} \in \mathbb{K}\}$. These von Neumann algebras are supposed to represent *observables*. In QFT, not all fields can be observed (e.g. the strength field of an electron), so each von Neumann algebra is to be regarded as a subalgebra of a C^* algebra $\mathfrak{F}(\mathcal{O})$ of local fields in $\mathcal{O}$. Causality is incorporated into the theory by requiring that the natural map $\mathbb{K} \to \mathfrak{U}$ given by $\mathcal{O} \mapsto \mathfrak{U}(\mathcal{O})$ be a lattice homomorphism.

(iii) *How does one take account of symmetries in this theory?* Let us say a few words about that. First, let us agree to call a map $\tau : \mathfrak{U} \to \mathfrak{U}$ a *net automorphism* if it is on-to-one and onto and respects the net structure, and if, for each $\mathcal{O} \in \mathbb{K}$, there is an isomorphism $\tau_{\mathcal{O}} : \mathfrak{U}(\mathcal{O}) \to \tau(\mathfrak{U}(\mathcal{O}))$. The automorphisms of $\mathfrak{U}$ form a group, denoted $\mathrm{Aut}(\mathfrak{U})$. Given a group G, which we want to impose as a group of symmetries of our theory, we simply require that there be a representation $\alpha : G \to \mathrm{Aut}(\mathfrak{U})$. This approach should also be natural from the point of view of *gauge theory*. As we know, the gauge group of a QFT is supposed to represent the *internal* symmetries of the theory (recall for instance the example of electromagnetism). Here, given a symmetry group G, and the corresponding representation α, we may consider the subgroup $G_{\text{int}} \subset G$ consisting of those g's such that the automorphism α_g maps each local algebra $\mathfrak{U}(\mathcal{O})$ to itself. This is the gauge group of the theory.

Exercises

B.1 *Unitization.* Let $\mathcal{A}$ be a C^* algebra without unit. Then there exists a C^* algebra $\mathcal{A}^+$ with unit containing $\mathcal{A}$ as a closed ideal such that $\mathcal{A}^+/\mathcal{A} \cong \mathbb{C}$. Prove this statement by working through the following steps.
(a) Let $\pi : \mathcal{A} \to B(\mathcal{A})$ be the map given by $\pi(x)y = xy$ for all $x, y \in \mathcal{A}$. Show that π is a homomorphism with $\|\pi(x)\| = \|x\|$ for all x.
(b) Let $\mathcal{B} \subset B(\mathcal{A})$ be the algebra of operators of the form $\pi(x) + \lambda I$, where $I : \mathcal{A} \to \mathcal{A}$ is the identity operator, for all $x \in \mathcal{A}$ and all $\lambda \in \mathbb{C}$. Show that $\mathcal{B}$ is a C^* algebra.
(c) Show that $\mathcal{B}/\pi(\mathcal{A}) \cong \mathbb{C}$.

B.2 Show that $M_n(\mathbb{C})$ has no proper ideals. Deduce that there are no non-trivial characters $\varphi : M_n(\mathbb{C}) \to \mathbb{C}$.

B.3 Let $\rho : M_n(\mathbb{C}) \to \mathbb{C}$ be a positive linear functional. Show that there exists $B \in M_n(\mathbb{C})_+$ such that $\rho(X) = \text{tr}(BX)$ for all $X \in M_n(\mathbb{C})$. [B is called the *density matrix* of ρ.]

B.4 Let $\rho : \mathcal{A} \to \mathbb{C}$ be a positive linear functional on a C^* algebra $\mathcal{A}$, and let $a \in \mathcal{A}$. Show that the linear functional $x \mapsto \rho(a^*xa)$ is positive.

B.5 Prove that every positive linear functional on a $C*$ algebra is bounded and self-adjoint.

B.6 Give an example of two bounded, self-adjoint operators $T_1, T_2 : \mathcal{H} \to \mathcal{H}$ on a Hilbert space $\mathcal{H}$ such that $\sigma(T_1) = \sigma(T_2)$ and yet T_1 and T_2 are *not* unitarily equivalent.

B.7 *Approximate units.* Let $\mathcal{A}$ be a C^* algebra and let $\mathcal{A}_+$ be its positive cone.
(a) Show that $\mathcal{A}_+$ is a directed set.
(b) Show that $\Lambda = \mathcal{A}_+ \cap \{a \in \mathcal{A} : \|a\| < 1\}$ is order isomorphic to $\mathcal{A}_+$.
(c) Deduce that Λ is an approximate unit in $\mathcal{A}$.

B.8 If $a \in \mathcal{A}$ is normal, show that $r(a) = \|a\|$.

B.9 Fill in the details of Example B.34, supplying the proofs.

B.10 Let $(P, \mathcal{H})$ be a projection-valued measure over a locally compact measurable space $(X, \mathcal{B})$. Let $f \in C_0(X)$. Given $n \geq 1$, let $s_n : X \to \mathbb{C}$ be a simple function of the form

$$s_n(x) = \sum_{k=1}^{m} f(x_k)\mathbf{1}_{E_k}(x)$$

(where $\{E_1, E_2, \ldots, E_m\}$ is a measurable partition of X and $x_k \in E_k$ for all k) with the property that $|f(x) - s_n(x)| \leq n^{-1}$ for all $x \in X$. Prove that

$$\left\| \int f(x)\, dP(x) - \sum_{k=1}^{m} f(x_k) P(E_k) \right\| \leq \frac{1}{n}.$$

This shows that the integral of f relative to $(P, \mathcal{H})$ is the norm limit of Riemann–Stieltjes sums.

B.11 Show that the projection-valued measure whose existence we proved in Theorem B.49 is unique, by working through the following steps.

(a) Let $(Q, \mathcal{H})$ be another projection-valued measure such that $T = \int_{\sigma(T)} \lambda\, dQ(\lambda)$. Show that $p(T) = \int_{\sigma(T)} p(\lambda)\, dQ(\lambda)$ for every polynomial $p(\lambda)$.

(b) Using the Stone–Weiertrass theorem, deduce from (a) that

$$\int_{\sigma(T)} f(\lambda)\, dP(\lambda) = \int_{\sigma(T)} f(\lambda)\, dQ(\lambda)$$

for all $f \in C_0(X)$.

(c) Denoting by $\mu^P_{v,w}$ and $\mu^Q_{v,w}$ the spectral measures for P and Q respectively, show that $\mu^P_{v,w}(G) = \mu^P_{v,w}(G)$ for each closed set $G \subset \sigma(T)$. [Hint: Let $f \in C_0(X)$ be such that $0 \leq f \leq 1$ everywhere, with $f(\lambda) = 1$ iff $\lambda \in G$. Look at $f_n(\lambda) = f(\lambda)^n$ an integrate.]

(d) Deduce from (c) that the measures $\mu^P_{v,w}$ and $\mu^Q_{v,w}$ agree on all Borel sets, for each pair $(v, w) \in \mathcal{H} \times \mathcal{H}$, and therefore that $P(E) = Q(E)$ for all $E \in \mathcal{B}$.

B.12 Let $K(\mathcal{H})$ be the space of compact operators on a Hilbert space $\mathcal{H}$. Show that $K(\mathcal{H})$ is a $*$-ideal in $B(\mathcal{H})$.

Bibliography

[AJP] S. Attal, A. Joye, C.-A. Pillet. *Open Quantum Systems I: The Hamiltonian Approach*. Lecture Notes in Mathematics, **1880**. Springer-Verlag, 2006.

[AJPS] S. Albeverio, J. Jost, S. Paycha, S. Scarlatti. *A Mathematical Introduction to String Theory*. LMS Lecture Notes Series, **225**. Cambridge University Press, 1997.

[Al] O. Alvarez. Lectures in quantum mechanics and the index theorem. In *Geometry and Quantum Field Theory*, D. Freed and K. Uhlenbeck, eds. American Mathematical Society, 273–322.

[AS] E. Alfsen, F. Shultz. *State Spaces of Operator Algebras*. Birkhäuser, 2001.

[B] T. Banks. *Modern Quantum Field Theory: A Concise Introduction*. Cambridge University Press, 2008.

[BFLS] F. Bayen, M. Flato, C. Fronsdal, A. Lichnerowicz, D. Sternheimer. Deformation theory and quantization I, II. *Ann. Phys.*, **11** (1978), 61–110, 111–51.

[BL] D. Bailin, A. Love. *Introduction to Gauge Field Theory*. Graduate Student Series in Physics (gen. ed. Douglas F. Brewer). Taylor & Francis Group, 1993.

[Bl] D. Bleecker. *Gauge Theory and Variational Principles*. Dover, 2005.

[BP] N. Bogoliubov, O. Parasiuk. *Acta Math.*, **97** (1957) 227.

[BV] I.A. Batalin, G.A.Vilkovisky. Gauge algebra and quantization. *Phys. Lett. B* **102** (1981), 27–31.

[Ca] E. Cartan. *The Theory of Spinors*. Dover, 1981.

[CCM] A.H. Chamseddine, A. Connes, M. Marcolli. Gravity and standard model with neutrino mixing. arXiv:hep-th/0610241v1.

[CG] W. Cottingham, D. Greenwood. *An Introduction to the Standard Model of Particle Physics*. 2nd ed. Cambridge University Press, 2007.

[Ch] A. Chamseddine. A brief introduction to particle interactions. arXiv:hep-th/0511073 v1, 2005.

[CK1] A. Connes, D. Kreimer. Renormalization in quantum field theory and the Riemann–Hilbert problem I. The Hopf algebra structure of graphs and the main theorem. *Comm. Math. Phys.*, **210** (2000), 249–73.

[CK2] A. Connes, D. Kreimer. Renormalization in quantum field theory and the Riemann–Hilbert problem II. The β function, diffeomorphisms and the renormalization group. *Comm. Math. Phys.*, **216** (2001), 215–41.

[Dy] F. Dyson. The S-matrix in quantum electrodynamics. *Phys. Rev. (2)*, **75** (1949), 1736–55.

[Dy1] F. Dyson. *Selected Papers of Freeman Dyson with Commentary*. American Mathematical Society, 1996.

[Ehr] L. Ehrenpreis. On the theory of the kernels of Schwartz. *Proc. Amer. Math. Soc.*, **7** (1956), 713–18.

[Fa] L. Fadeev. Elementary introduction to quantum field theory. In *Quantum Fields and Strings: A Course for Mathematicians*, P. Deligne, ed., 513–50.

[FHM] A. Fuster, M. Henneaux, A. Maas. BRST-antifield quantization: A short review. arXiv:hep-th/050698 v2 08/15/2005.

[FJ] G. Friedlander, M. Joshi. *Introduction to the Theory of Distributions*. 2nd ed., Cambridge University Press, 1998.

[FKT] J. Feldman, H. Knörrer, E. Trubowitz. *Fermionic Functional Integrals and the Renormalization Group*. American Mathematical Society, 2002.

[FP] L. Fadeev, V. Popov. Feynman diagrams for Yang–Mills fields. *Phys. Lett. B*, **25** (1967), 29–30.

[Fr] T. Frankel. *The Geometry of Physics*. Cambridge University Press, 1997.

[FS] L. Fadeev, A. Slavnov. *Gauge Fields: An Introduction to Quantum Theory*. Frontiers in Physics, **83**, 2nd ed., Addison–Wesley, 1991.

[FY] L. Fadeev, O. Yakubovskii. *Lectures on Quantum Mechanics for Mathematics Students*. Student Mathematical Library, **47**. American Mathematical Society, 2009.

[GJ] J. Glimm, A. Jaffe. *Quantum Physics. A Functional Integral Point of View*. 2nd ed. Springer-Verlag, 1987.

[GMS] I. Gelfand, R. Minlos, Z. Shapiro. *Representations of the Rotation and Lorentz Groups and Their Applications*. Pergamon, 1963.

[GO] B.R. Greene, H. Ooguri. Geometry and quantum field theory: A brief introduction. *AMS/IP Stud. Adv. Math.*, **1** (1997), 3–27.

[Gr] D. Gross. Renormalization groups. In *Quantum Fields and Strings: A Course for Mathematicians*, P. Deligne, ed., 551–93.

[Gu] S. Gudder. *Stochastic Methods in Quantum Mechanics*. Elsevier North Holland, 1979 [Dover edition 2005].

[GuS] S.J. Gustafson, I.M. Sigal. *Mathematical Concepts of Quantum Mechanics*. Universitext, Springer-Verlag, 2003.

[GW] D.J. Gross, F. Wilczek. Ultraviolet behavior of non-abelian gauge theories, *Phys. Rev. Lett.*, **30** (1973), 1343–6.

[H] R. Haag. Local *Quantum Physics: Fields, Particles, Algebras*. 2nd ed., Springer-Verlag, 1996.

[He] K. Hepp. Proof of the Bogoliubov–Parasiuk theorem on renormalization. *Comm. Math. Phys.*, **2** (1966), 301–26.

[HT] M. Henneaux, C. Teitelboim. *Quantization of Gauge Systems*. Princeton University Press, 1992.

[IZ] C. Itzykson, J.-B. Zuber. *Quantum Field Theory*. Dover, 2005.

[Ja1] A. Jaffe. Quantum theory and relativity. Group representations, ergodic theory, and mathematical physics: A tribute to George W. Mackey. *Contemp. Math.*, **449** (2008), 209–45.

[Ja2] A. Jaffe. Constructive quantum field theory. In *Mathematical Physics 2000*. Imperial College Press, 2000, 111–27.

[JL] G. Johnson, M. Lapidus. *The Feynman Integral and Feynman's Operational Calculus*. Oxford Science Publications, 2000.

[JaW] A. Jaffe, E. Witten. Quantum Yang–Mills theory. *The Millennium Prize Problems*. Clay Math. Inst., 2006, 129–52.

[K] C. Kopper. Renormalization theory based on flow equations. arXiv:hep-th/0508143.

[Kh] D. Kazhdan. Introduction to QFT. In *Quantum Fields and Strings: A Course for Mathematicians*, P. Deligne, ed., 377–418.

[KK] G. Keller, C. Kopper. Renormalizability proof for QED based on flow equations. *Comm. Math. Phys.*, **176** (1996), 193–226.

[KM] C. Kopper, V.F. Muller. Renormalization proof for spontaneous broken Yang–Mills theory with flow equations. *Comm. Math. Phys.*, **209** (2000), 477–515.

[KM1] C. Kopper, V.F. Muller. Renormalization of spontaneously broken SU(2) Yang–Mills theory with flow equations. *Rev. Natg. Ohys.*, **21** (2009), 781–820 (arXiv:0902.2486).

[Kon] M. Kontsevich. Deformation quantization of Poisson manifolds. *Lett. Math. Phys.*, **66** (2003) 157–216.

[L] S. Lang. *Real and Functional Analysis*. Graduate Texts in Mathematics, **142**. 3rd ed., Springer-Verlag, 1993.

[LM] B. Lawson, M. L. Michelson. *Spin Geometry*. Princeton University Press.

[M] G. Mackey. *Mathematical Foundations of Quantum Mechanics*. Benjamim, 1963.

[Ma] M. Maggiore. *A Modern Introduction to Quantum Field Theory*. Oxford University Press, 2005.

[Mul] V.F. Muller. Perturbative renormalization by flow equations. arXiv:hep-th/0208211v5 2009.

[Na1] G. Naber. *The Geometry of Minkowski Spacetime*. Dover, 2003.

[Na2] G. Naber. *Topology, Geometry and Gauge Fields: Foundations*. Springer-Verlag, 1997.

[Na3] G. Naber. *Topology, Geometry and Gauge Fields: Interactions*. Springer-Verlag, 2000.

[Nak] M. Nakahara. *Geometry, Topology and Physics*. IOP Publishing, 1990.

[OS] K. Osterwalder, R. Schrader. Axioms for Euclidean Green's functions. *Comm. Math. Phys*. **31** (1973), 83–112.

[P] H.D.Politzer. Reliable perturbative results for strong interactions. *Phys. Rev. Lett.*, **30** (1973), 1346–9.

[Pen] R. Penrose. *The Road to Reality: A Complete Guide to the Laws of the Universe*. Alfred A. Knopf, 2006.

[Po] M. Polyak. Feynmann graphs for pedestrians and mathematicians. arXiv:math.GT/0406251 v1 06/12/2004.

[Pol] J. Polchinski. Renormalization and effective Lagrangians. *Nucl. Phys. B*, **231** (1984), 269–95.

[Polc] J. Polchinski. *String Theory I, II*. Cambridge University Press, 1998.

[PS] M. Peskin, D. Schroeder. *An Introduction to Quantum Field Theory*. Addison–Wesley, 1995.

[Ra] J. Rabin. Introduction to quantum field theory for mathematicians. In *Geometry and Quantum Field Theory*, D. Freed and K. Uhlenbeck, eds. American Mathematical Society, 185–269.

[Ri] V. Rivasseau. *From Perturbative to Constructive Renormalization*. Princeton Series in Physics. Princeton University Press, 1991.

[Roe] J. Roe. *Elliptic Operators, Topology and Asymptotic Methods*. Pitman Research Notes in Mathematics Series, **395**. Longman, 1998.

[Ros] W. Rossmann. *Lie Groups: An Introduction through Linear Groups*. Oxford Graduate Texts in Mathematics, **5**. Oxford University Press, 2002.

[RS1] M. Reed, B. Simon. *Methods of Mathematical Physics I: Functional Analsysis*. Academic Press, 1980.

[RS2] M. Reed, B. Simon. *Methods of Mathematical Physics II: Fourier Analysis, Self-Adjointness*. Academic Press, 1975.

[S] A. Schwarz. *Quantum Field Theory and Topology*. Springer-Verlag, 1993.

[Sal] M. Salmhofer. *Renormalization: An Introduction*. Springer-Verlag, 1999.

[Sch] M. Schottenloher. *A Mathematical Introduction to Conformal Field Theory*. Springer-Verlag, 1997.

[Scw] J. Schwinger. On the Euclidean structure of relativistic field theory. *Proc. Nat. Acad. Sci. U.S.A.*, **44** (1958), 956–65.

[Sh] D.V. Shirkov. Evolution of the Bogoluibov renormalization group. arXiv:hep-th/9909024.

[St] S. Sternberg. *Group Theory and Physics*. Cambridge University Press, 1994.

[Sth] J. Stasheff. Deformation theory and the Batalin–Vilkovsky master equation. In *Deformation Theory and Symplectic Geometry (Ascona, 1996). Math. Phys. Studies*, **20** (1997), 271–84. Kluwer Academic, 1997.

[Str] F. Strocchi. *An Introduction to the Mathematical Structure of Quantum Mechanics – A Short Course for Mathematicians*. Advanced Series in Mathematical Physics, **27**. World Scientific, 2005.

[SW] R. Streater, A. Wightman. *PCT, Spin and Statistics, and All That*. Princeton University Press, 1964.

[Th] J. Thayer. *Operadores Auto-adjuntos e Equações Diferenciais Parciais*. Projeto Euclides, IMPA, 2007.

[tH1] G. t'Hooft. Gauge theory and renormalization. arXiv:hep-th/9410038 v2 03/03/1997.

[tH2] G. t'Hooft, ed. *50 Years of Yang–Mills*. World Scientific, 2005.

[tHV] G. t'Hooft, M. Veltman. Regularization and renormalization of gauge fields. *Nucl. Phys. B*, **44** (1972), 189–213.

[Ti] R. Ticciati. *Quantum Field Theory for Mathematicians*. Cambridge University Press, 1999.

[V] M. Veltman, *Facts and Mysteries in Elementary Particle Physics*. World Scientific, 2003.

[vB] P. van Baal. *A Course in Field Theory*. Notes, Instituut Lorentz for Theoretical Physics, University of Leiden, 1998 [corrected version 2001].

[We] S. Weinberg. *The Quantum Theory of Fields, I, II, III*. Cambridge University Press, 1995.

[Wein] S. Weinberg. High-energy behavior in quantum field theory. *Phys. Rev.*, **118** (1960), 838–49.

[Wi] E. Witten. Perturbative quantum field theory. In *Quantum Fields and Strings: A Course for Mathematicians*, P. Deligne, ed., 419–511.

[Wi1] E. Witten. Physical law and the quest for mathematical understanding. In *Mathematical Challenges of the 21st Century (Los Angeles, CA, 2000). Bull. Amer. Math. Soc. (N.S.)*, **40** (2003), 21–9 [electronic].

[Wil] K. Wilson, J.B. Kogut. The renormalization group and the ϵ-expansion. *Phys. Rep. C*, **12** (1974), 75–199.

[WL] M. de Wilde, P.B.A. Leconte. Existence of *-products and of formal deformations of Poisson Lie algebra of arbitrary symplectic manifols. *Lett. Math. Phys.*, **7** (1983) 487–96.

[Zi] W. Zimmermann. Convergence of Bololiubov's method of renormalization in momentum space. *Comm. Math. Phys.*, **15** (1969), 2008–2034.

[Zim] W. Zimmermann. The fower counting theorem for Minkovski metric. *Comm. Math. Phys.*, **11** (1968), 1–8.

Index

Printed in the United States
By Bookmasters